PU M [...] CS

3 C4

David
Paul V

First published 2004 by
Elmwood Press
80 Attimore Road
Welwyn Garden City
Herts AL8 6LP
Tel. 01707 333232

Reprinted 2005, 2009, 2010, 2011

British Library Cataloguing in publication Data

Elmwood Press
The moral rights of the author have been asserted.
Database right Elmwood press (maker)

ISBN 978 1 902 214 467

Typeset and illustrated by Tech-Set Ltd, Gateshead, Tyne and Wear
Printed and bound by Bookwell

CONTENTS

Part 10 Integration

Part 11 Vectors

PREFACE

This book together with **Pure Mathematics C1 C2** is for candidates working towards A2 in Mathematics. It can be used both in the classroom and by students working on their own. There are explanations, worked examples and numerous exercises which, it is hoped, will help students to build up confidence. The authors believe that people learn mathematics by doing mathematics. The questions are graded in difficulty throughout the exercises.

The work is collected into sections on Algebra, Coordinate Geometry, Sequences and Series, Calculus, Trigonometry, Exponentials and Logarithms, Numerical methods and Vectors. Mathematics at this level can be very satisfying for students and an effort has been made to provide enough 'easy' questions for weaker students for whom mathematics is not a specialist subject. It is hoped that such students can maintain their motivation by experiencing regular success. More able students will find plenty to stimulate them in the later questions in exercises as well as the past examination questions. The revision exercises provide further practice and teachers may find them useful for setting homework or for general revision.

In response to the requests of many teachers a section of past examination questions has been included. Teachers will note that the specifications and questions set by all the main boards are very similar. This book can be used with confidence when preparing for the A2 examinations set by all the main boards.

Thanks are due to AQA, EDEXCEL, MEI and OCR for kindly allowing the use of questions from their past mathematics papers. The answers are solely the work of the authors and are not ratified by the examining groups.

The authors would especially like to thank George Lane for his extensive contribution to the text.

D. Rayner
P. Williams

Part 1

Algebra and functions 1

1.1 Simplifying rational expressions

A rational expression is a fraction. We simplify algebraic fractions using the same rules that we use when simplifying numerical fractions.

Example 1

Simplify

a $\dfrac{12}{15}$ **b** $\dfrac{7a}{5a^2}$ **c** $\dfrac{3x+9}{6x}$

d $\dfrac{x^2+2x}{x+2}$ **e** $\dfrac{x^2+2x-3}{x^2+5x+6}$

a $\dfrac{12}{15}$. Divide numerator and denominator by 3, we have $\dfrac{12}{15} = \dfrac{4}{5}$.

b $\dfrac{7a}{5a^2}$. Dividing numerator and denominator by a, we have $\dfrac{7a}{5a^2} = \dfrac{7}{5a}$

c $\dfrac{3x+9}{6x}$. A common mistake is often made with fractions like this
Do **not** cancel the xs.

Instead we write $\dfrac{3x+9}{6x} = \dfrac{3(x+3)}{6x} = \dfrac{(x+3)}{2x}$.

d Factorising the numerator, $\dfrac{x^2+2x}{x+2} = \dfrac{x(x+2)}{x+2} = x$

e Factorising $\dfrac{x^2+2x-3}{x^2+5x+6} = \dfrac{(x-1)(x+3)}{(x+2)(x+3)} = \dfrac{x-1}{x+2}$

Exercise 1

Simplify as far as possible.

1 $\dfrac{24}{30}$ **2** $\dfrac{4a^2}{a}$ **3** $\dfrac{x}{3x}$ **4** $\dfrac{8x^2}{2x^2}$

5 $\dfrac{2a}{4b}$ **6** $\dfrac{6m}{2m}$ **7** $\dfrac{5ab}{10b}$ **8** $\dfrac{8ab^2}{4ab}$

9 $\dfrac{15y}{20y^2}$ **10** $\dfrac{11xy}{12x^2}$ **11** $\dfrac{8ya^2}{12a}$ **12** $\dfrac{12m^2n^2}{3mn^3}$

13 Sort these into four pairs of equivalent expressions.

A $\dfrac{x^2}{4x}$ **B** $\dfrac{x(x+1)}{x^2}$ **C** $\dfrac{x^2+x}{x^2-x}$ **D** $\dfrac{3x+6}{3x}$

E $\dfrac{x+1}{x}$ **F** $\dfrac{x+2}{x}$ **G** $\dfrac{x}{4}$ **H** $\dfrac{x+1}{x-1}$

In questions **14** to **29** simplify as far as possible.

14 $\dfrac{7a^2b}{35ab^2}$ **15** $\dfrac{(2a)^2}{4a}$ **16** $\dfrac{7yx}{8xy}$ **17** $\dfrac{3x}{4x-x^2}$

18 $\dfrac{5x+2x^2}{3x}$ **19** $\dfrac{9x+3}{3x}$ **20** $\dfrac{4a+5a^2}{5a}$ **21** $\dfrac{5ab}{15a+10a^2}$

22 $\dfrac{3x-x^2}{2x}$ **23** $\dfrac{12x+6}{6y}$ **24** $\dfrac{3(2x^2+5x)}{6x}$ **25** $\dfrac{xy+x^2y}{x}$

26 $\dfrac{5x^3+4x^2}{x(3x^2-2x)}$ **27** $\dfrac{a+4a^2}{ab+ab^2}$ **28** $\dfrac{4+8x+8x^2}{4x}$ **29** $\dfrac{54mn^2-27m^2n}{18(mn)^2}$

30 Look at each of the following and decide whether or not the statement is 'true' or 'false'.

a $\dfrac{4x^2+8}{2x}=\dfrac{4x+8}{2}$ **b** $\dfrac{x^2y+xy^2}{xy}=x+y$

c $\dfrac{4ab^2c^3}{a^3b^2c}=\dfrac{4c^2}{a^2}$ **d** $\dfrac{5x+6}{6y}=\dfrac{5x+1}{y}$

31 a Factorise the expression x^2-x-6.

b Hence simplify the expression $\dfrac{x^2-x-6}{(x+2)}$

32 Simplify as far as possible.

a $\dfrac{x^2-3x-1}{x+1}$ **b** $\dfrac{x^2+x-6}{x^2+2x-3}$ **c** $\dfrac{x^2+3x-10}{x^2-4}$

d $\dfrac{x^2-3x}{x^2-2x-3}$ **e** $\dfrac{x^2+4x}{2x^2-10x}$ **f** $\dfrac{x^2+6x+5}{x^2-x-2}$

g $\dfrac{2x^2+x}{4x^2-1}$ **h** $\dfrac{2x^2-5x-3}{2x^2-3x-9}$ **i** $\dfrac{3x^2+x-4}{9x^2-16}$

33 Write the expression $\dfrac{x+\frac{1}{2}}{x+\frac{1}{3}}$ in a more simple form without fractions on the numerator and denominator.
[Hint: Multiply numerator and denominator by 6].

34 Write the following in a more simple form without fractions.

a $\dfrac{x+\dfrac{1}{x}}{x}$ **b** $\dfrac{2x-\dfrac{1}{x}}{\dfrac{1}{x}}$ **c** $\dfrac{x-\dfrac{1}{2}}{\dfrac{1}{2}}$

d $\dfrac{3x + \dfrac{1}{4}}{\dfrac{1}{4}}$ **e** $\dfrac{5x - \dfrac{1}{3}}{\dfrac{1}{6}}$ **f** $\dfrac{\dfrac{1}{4} - x}{\dfrac{1}{2}}$

g $\dfrac{3x + \dfrac{1}{x}}{x + \dfrac{2}{x}}$ **h** $\dfrac{x - \dfrac{4}{x}}{x - 2}$ **i** $\dfrac{2x - 3}{4x - \dfrac{9}{x}}$

j $\dfrac{1 - \dfrac{1}{x^2}}{\dfrac{1}{x^2}}$ **k** $\dfrac{1 - \dfrac{4}{x^2}}{1 - \dfrac{2}{x}}$ **l** $\dfrac{1 + \dfrac{2}{x} - \dfrac{3}{x^2}}{1 + \dfrac{3}{x} - \dfrac{4}{x^2}}$

Addition and subtraction of algebraic fractions

Example 2

a Write as a single fraction $\dfrac{2}{x} + \dfrac{3}{y}$

The L.C.M. of x and y is xy.

$$\therefore \frac{2}{x} + \frac{3}{y} = \frac{2y}{xy} + \frac{3x}{xy} = \frac{2y + 3x}{xy}$$

b Write as a single fraction $\dfrac{4}{x} + \dfrac{5}{x - 1}$

The L.C.M. of x and $(x - 1)$ is $x(x - 1)$

$$\therefore \frac{4}{x} + \frac{5}{x - 1} = \frac{4(x - 1) + 5x}{x(x - 1)}$$
$$= \frac{9x - 4}{x(x - 1)}$$

EXERCISE 2

1 Write as a single fraction.

a $\dfrac{2x}{5} + \dfrac{x}{5}$ **b** $\dfrac{2}{x} + \dfrac{1}{x}$ **c** $\dfrac{x}{7} + \dfrac{3x}{7}$ **d** $\dfrac{1}{7x} + \dfrac{3}{7x}$

e $\dfrac{5x}{8} + \dfrac{x}{4}$ **f** $\dfrac{5}{8x} + \dfrac{1}{4x}$ **g** $\dfrac{2x}{3} + \dfrac{x}{6}$ **h** $\dfrac{2}{3x} + \dfrac{1}{6x}$

2 Sort into four pairs of equivalent fractions.

A $\dfrac{x}{2} + \dfrac{x}{4}$ **B** $\dfrac{2}{x} + \dfrac{2}{x}$ **C** $\dfrac{9x}{8} - \dfrac{x}{4}$ **D** $\dfrac{7x}{8}$

E $\dfrac{5x}{x^2} - \dfrac{2}{x}$ **F** $\dfrac{3x}{4}$ **G** $\dfrac{3}{x}$ **H** $\dfrac{4}{x}$

3 Simplify.

a $\dfrac{3x}{4} + \dfrac{2x}{5}$ **b** $\dfrac{3}{4x} + \dfrac{2}{5x}$ **c** $\dfrac{3x}{4} - \dfrac{2x}{3}$

d $\dfrac{3}{4x} - \dfrac{2}{3x}$ **e** $\dfrac{x}{2} + \dfrac{x + 1}{3}$ **f** $\dfrac{x - 1}{3} + \dfrac{x + 2}{4}$

4 Work out these subtractions.

a $\frac{x+1}{3} - \frac{(2x+1)}{4}$ **b** $\frac{x-3}{3} - \frac{(x-2)}{5}$ **c** $\frac{(x+2)}{2} - \frac{(2x+1)}{7}$

5 Write as a single fraction.

a $\frac{2x-1}{5} + \frac{x+3}{2}$ **b** $\frac{1}{x} + \frac{2}{x+1}$ **c** $\frac{3}{x-2} + \frac{4}{x}$

d $\frac{5}{x-2} + \frac{3}{x+3}$ **e** $\frac{7}{x+1} - \frac{3}{x+2}$ **f** $\frac{x}{2} + \frac{x+3}{x+2}$

Multiplication and division

The method is similar to that used when multiplying and dividing ordinary numerical fractions.

Example 3

a $x^2 \div \frac{x}{2} = x^2 \times \frac{2}{x} = 2x$

b $\frac{2a}{3x} \times \frac{2x^2}{5a} = \frac{4\cancel{a}x^{\cancel{2}}}{15\cancel{a}\cancel{x}} = \frac{4x}{15}$

c $\frac{(x-1)}{(x+2)} \div \frac{2(x-1)}{(x+3)} = \frac{\cancel{(x-1)}}{(x+2)} \times \frac{(x+3)}{2\cancel{(x-1)}}$

$= \frac{(x+3)}{2(x+2)}$

Exercise 3

Simplify the following:

1 $\frac{3x}{2} \times \frac{2a}{3x}$ **2** $\frac{5mn}{3} \times \frac{2}{n}$ **3** $\frac{3y^2}{x^2} \times \frac{2x}{9y}$ **4** $\frac{5ab^2}{2} \times \frac{3}{2a^2b}$

5 $\frac{2}{a} \div \frac{a}{2}$ **6** $\frac{4x}{3} \div \frac{x}{2}$ **7** $\frac{x-1}{2} \div \frac{x+2}{2}$ **8** $\frac{x}{5} \times \frac{y^2}{x^2}$

9 $\frac{x-1}{4} \times \frac{x}{2x-2}$ **10** $\frac{x^2-4}{3} \times \frac{9}{x-2}$

Questions **11** onwards involve addition, subtraction, multiplication and division.

11 Copy and complete.

a $\frac{x^2}{3} \times \frac{\square}{x} = 2x$ **b** $\frac{8}{x} \div \frac{2}{x} = \square$ **c** $\frac{a}{3} + \frac{a}{3} = \frac{\square}{3}$

12 Sort these into four pairs of equivalent expressions.

A $\frac{x^2}{3x}$ **B** $\frac{x}{2} \times \frac{x}{2}$ **C** $\frac{12x+6}{6}$ **D** $\frac{x}{5} - \frac{2}{5}$

E $\frac{2x^2+x}{x}$ **F** $\frac{x(x+1)}{3x+3}$ **G** $\frac{x-2}{5}$ **H** $\frac{ax^2}{4a}$

13 Write as a single fraction in its simplest form.

a $\frac{3x}{2} \times \frac{2a}{3x}$ **b** $\frac{5mn}{3} \times \frac{2}{n}$ **c** $\frac{3y^2}{3} \times \frac{2x}{9y}$

d $\frac{2}{q} \div \frac{a}{2}$ **e** $\frac{4x}{3} \div \frac{x}{2}$ **f** $\frac{x}{5} \times \frac{y^2}{x^2}$

g $\frac{a^2}{5} \div \frac{a}{10}$ **h** $\frac{x^2}{x^2 + 2x} \div \frac{x}{x + 2}$

14 Simplify the following.

a $\frac{x}{y} \times \frac{xy}{z} \times \frac{z}{x^2}$ **b** $\frac{ab}{x} \times \frac{xb^2}{a^2} \times \frac{a^2}{x}$

c $\frac{x^2 + 7x}{x^2 - 1} \times \frac{x + 1}{x + 7}$ **d** $\frac{\left(\frac{x}{y}\right)}{z} \times \frac{z^3}{x}$

15 Copy and complete.

a $\frac{x}{5} - \frac{\square}{\square} = \frac{x}{10}$ **b** $\frac{\square}{2} + \frac{x}{4} = \frac{7x}{4}$ **c** $\frac{3}{2x} - \frac{1}{8x} = \frac{\square}{8x}$

16 The perimeter of the rectangle shown is 24 units.
Form an equation and solve it to find x.

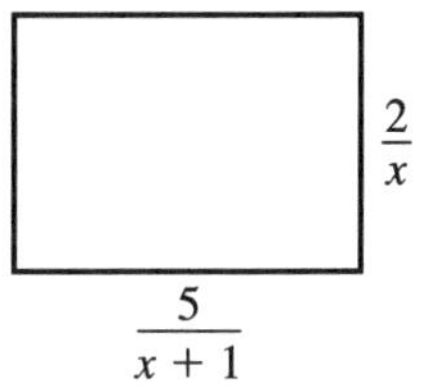

17 A rectangle measuring $\frac{3}{x}$ by $\frac{6}{x + 1}$ has an area of 75 square units.
Find x.

1.2 Functions and mappings

If A and B are non-empty sets then a *mapping* from A to B is a rule which associates an element of B with every element of A.

Here are two types of mappings.

Type 1

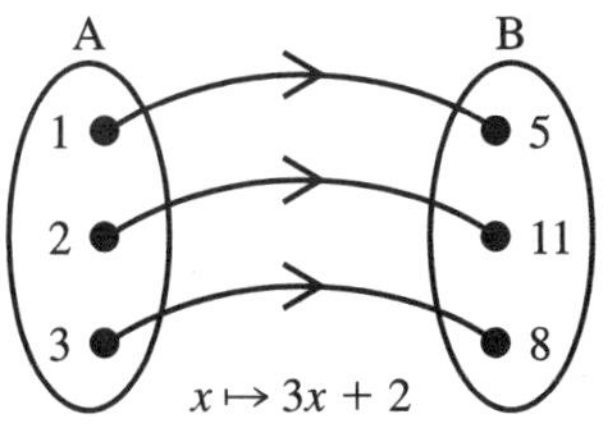

This is called a **one-to-one mapping**.

Type 2

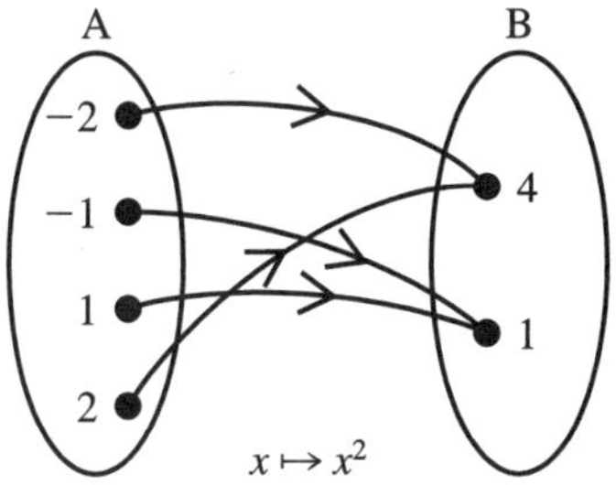

This is called a **two-to-one mapping** or a **many-to-one mapping**.

A *function* is a mapping which associates one and only one element of B with every element of A.
So a function is one-to-one mapping or a many-to-one mapping.

The mapping below is *not* a function.

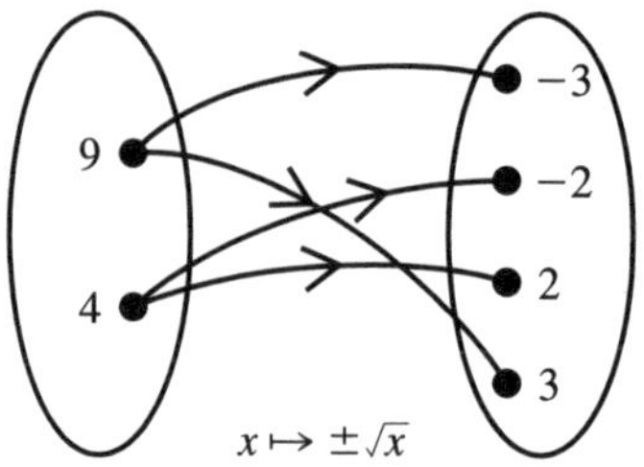

This is a one-to-many mapping.

Our main interest in this section is with *functions*.

The notations used for functions are either **a** $f(x) = x^2$
or **b** $f: x \mapsto x^2$

We use 'an arrow with a tail' $\mapsto$ to avoid confusion with a simple arrow such as '$x \to 0$' which means '*x tends towards* 0'.

The graph of $y = 2x + 1$ is a one-to-one mapping

The graph of $y = \dfrac{1}{x}$ is a one-to-one mapping

The graph of $y = \sin x$ is a many-to-one mapping.

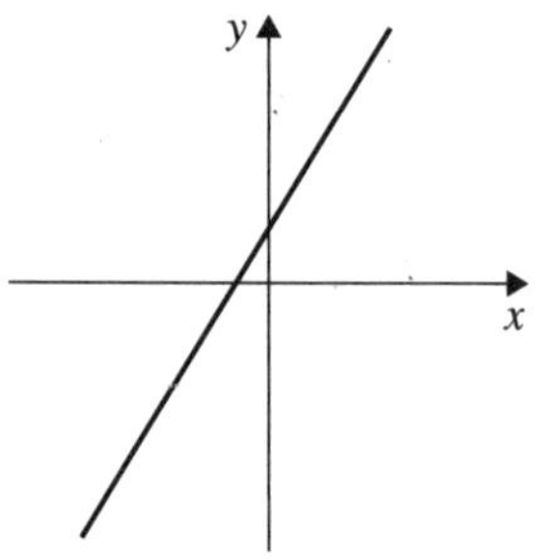

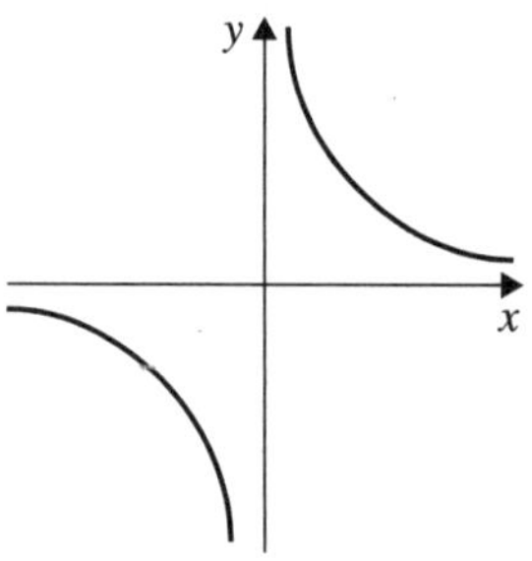

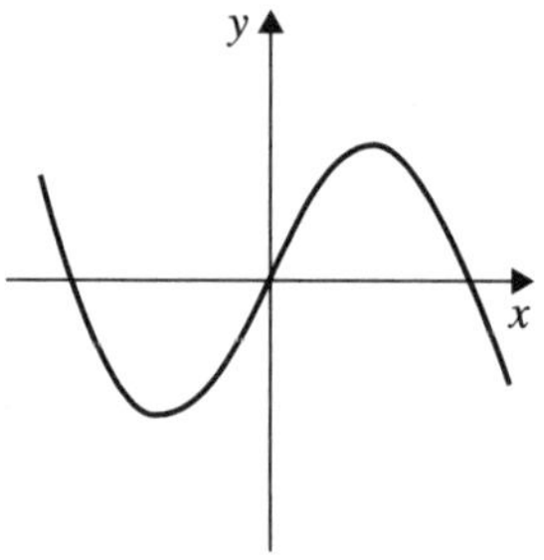

The **Domain** of a function is the set of all the points on which the function acts.

The **Range** of a function is the set of all the points to which the function maps.

In general the best way to find the range of a function is to sketch a graph of the function.

Example 1

Find the range of each function below.

a $f(x) = x^2$, for $x \in \mathbb{R}$

The domain $[x \in \mathbb{R}]$ is the set of all real numbers.
The graph of $f(x) = x^2$ is shown.
The range of the function is $f(x) \geqslant 0$.

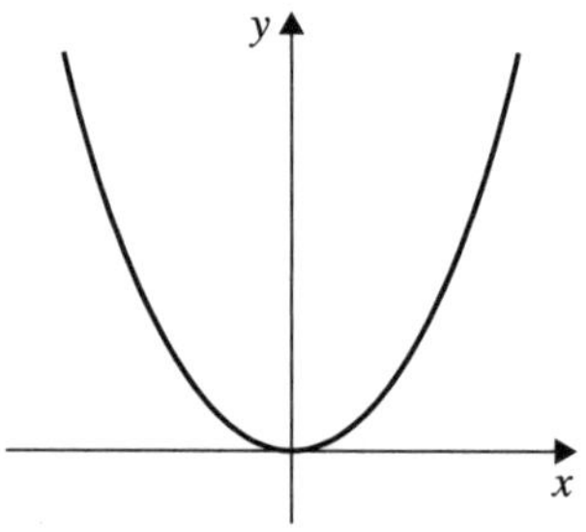

b $f(x) = 2x + 1$, for $x \in \mathbb{R}, x > 0$

The domain is all values of x greater than 0.
From the graph of $f(x) = 2x + 1$, we see that the range is $f(x) > 1$.
Notice that since $x = 0$ is not in the domain, the range is $f(x) > 1$ and *not* $f(x) \geqslant 1$.

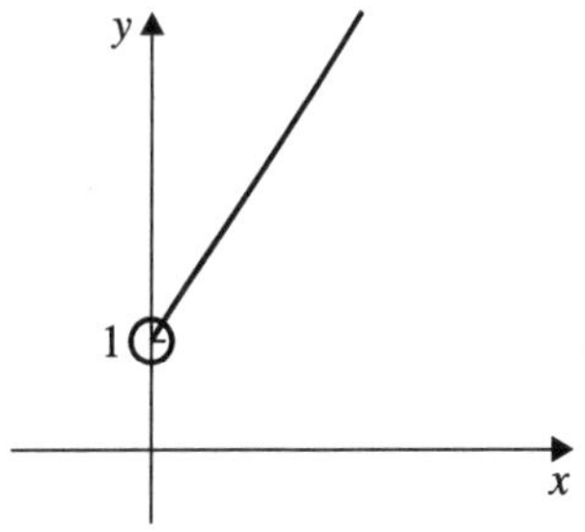

c $f(x) = \sin x$, for $0° \leqslant x \leqslant 360°$

The range of the function is $-1 \leqslant f(x) \leqslant 1$.

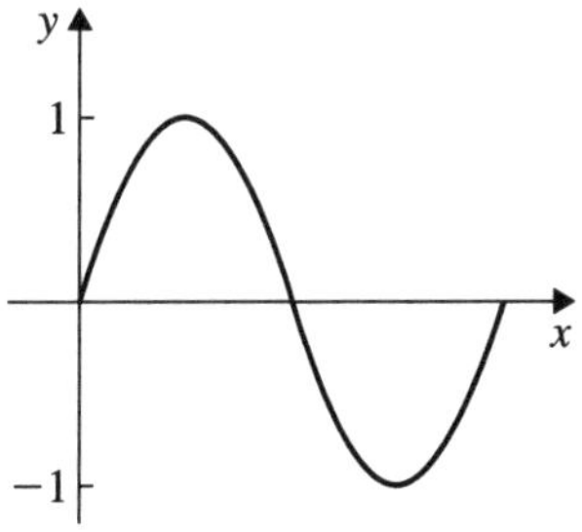

EXERCISE 4

1 Decide which of these functions are one-to-one and which are many-to-one.

a $f: x \mapsto 2x - 5, \quad x \in \mathbb{R}$
b $f: x \mapsto x^2 + 2, \quad x \in \mathbb{R}$
c $f: x \mapsto \frac{1}{x}, \quad x \in \mathbb{R}, x \neq 0$
d $f: x \mapsto x^3, \quad x \in \mathbb{R}$
e $f: x \mapsto x(x - 1), \quad x \in \mathbb{R}$
f $f: x \mapsto \cos x, \quad 0° \leqslant x \leqslant 360°$
g $f: x \mapsto (x - 1)^2 + 2, \quad x \in \mathbb{R}$
h $f: x \mapsto x(x - 1)(x - 2), \quad x \in \mathbb{R}$
i $f: x \mapsto x^3 + 10, \quad x \in \mathbb{R}$
j $f: x \mapsto \sqrt{x}, \quad x \in \mathbb{R}, x \geqslant 0$

2 Find the range of each function.

a $f(x) = 2x + 1, \quad x \in \mathbb{R}, x \geqslant 1$
b $f(x) = x^2 + 2, \quad x \in \mathbb{R}$
c $f(x) = x + 1, \quad x \in \mathbb{R}, x \geqslant 0$
d $f(x) = x^3, \quad x \in \mathbb{R}, x \geqslant 0$
e $f(x) = \sin x, \quad 0° \leqslant x \leqslant 180°$
f $f(x) = x(x - 3), \quad 0 \leqslant x \leqslant 3$
g $f(x) = 5x - x^2, \quad 0 \leqslant x \leqslant 5$
h $f(x) = \sqrt{x}, \quad 0 \leqslant x \leqslant 25$
i $f(x) = \frac{1}{x}, \quad x \geqslant 1$
j $f(x) = \frac{1}{1 + x^2}, \quad x \in \mathbb{R}$

3 **a** By completing the square, write $f(x) = x^2 + 6x + 4$ in the form $(x + a)^2 + b$.
b Sketch the graph of $y = f(x)$ and state its range for $x \in \mathbb{R}$.

4 Find the range for each of these functions for $x \in \mathbb{R}$.

a $x^2 - 6x + 12$
b $x^2 + 20x + 50$
c $\frac{1}{x^2 + 2}$
d $\frac{1}{x^2 + 4x + 5}$

Composite functions

The function $f : x \mapsto 3x + 2$ is itself a composite function, consisting of two simpler functions: 'multiply by 3' and 'add 2'.

If $f : x \mapsto 3x + 2$ and $g : x \mapsto x^2$ then fg is a composite function where g is performed first and then f is performed on the result of g.

The function fg may be found using a flow diagram.

$$x \rightarrow \boxed{\text{square}} \xrightarrow{x^2} \boxed{\text{multiply by 3}} \xrightarrow{3x^2} \boxed{\text{add 2}} \xrightarrow{3x^2 + 2}$$
$$\text{'g'} \qquad\qquad \text{'f'}$$

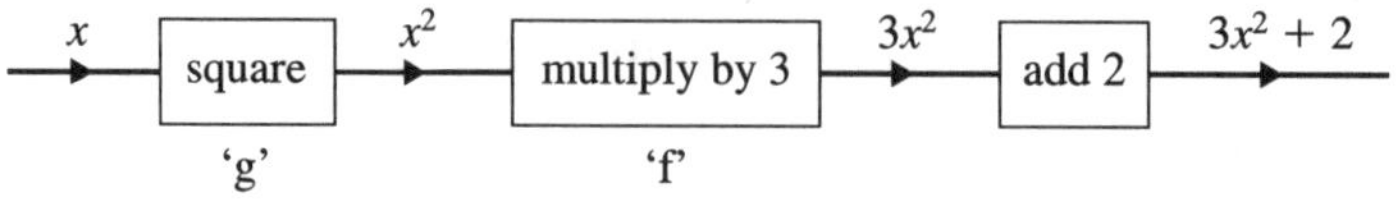

Thus $fg : x \mapsto 3x^2 + 2$ (or we can write $fg(x) = 3x^2 + 2$)

It is helpful to remember that fg means 'do g first and then f'.

Example 2

Given $f(x) = x^2$, $g(x) = x - 3$, $h(x) = 3x + 2$.

Find **a** $fg(x)$

     b $gf(x)$

     c $hf(x)$

     d $gfh(x)$

a $fg(x)$ 'do g first, and then f'
    So $fg(x) = (x - 3)^2$

b $gf(x) = x^2 - 3$

c $hf(x) = 3x^2 + 2$

d $gfh(x) = (3x + 2)^2 - 3$

Inverse functions

If $f(x)$ is a **one-to-one function** then it has an **inverse function** denoted by $f^{-1}(x)$ such that $f^{-1}[f(x)] = x$ for all values of x [and also $f[f^{-1}(x)] = x$].

The inverse of a given function can be found by two different methods:

Method A using a flow diagram or
Method B by letting $y = f(x)$ and rearranging to make x the subject.

The two methods are illustrated below.

Example 3   (Method A)

Find the inverse of f where $f(x) = \dfrac{5x - 2}{3}$

a Draw a flow diagram for f

$$x \rightarrow \boxed{\text{multiply by 5}} \xrightarrow{5x} \boxed{\text{subtract 2}} \xrightarrow{5x - 2} \boxed{\text{divide by 3}} \xrightarrow{\frac{5x - 2}{3}}$$

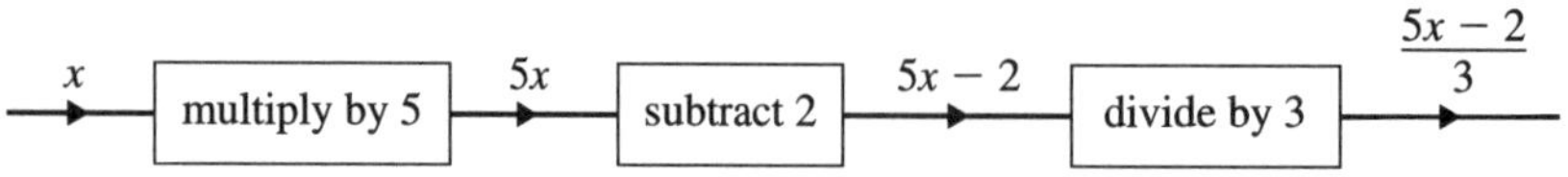

b Draw a new flow diagram with each operation replaced by its inverse. Start with x on the right.

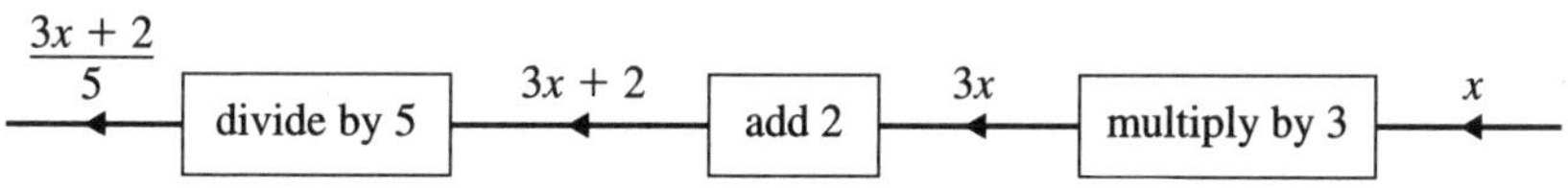

Thus the inverse of f is given by

$$f^{-1}(x) = \frac{3x + 2}{5}$$

Example 4 (Method B)

Find the inverse of the function $f(x) = \dfrac{5x - 2}{3}$.

Let $y = \dfrac{5x - 2}{3}$.

Rearrange to make x the subject.

$$5x - 2 = 3y$$
$$x = \frac{3y + 2}{5}$$

So the inverse function is $f^{-1}(x) = \dfrac{3x + 2}{5}$

Why does this method work?

In the example above given x we can find $y = f(x) = \dfrac{5x - 2}{3}$.

Now if we are given y, the value of x can be found using $x = \dfrac{3y + 2}{5}$.

So the inverse function is given by $f^{-1}(x) = \dfrac{3x + 2}{5}$.

Example 5

The function f is defined by $f(x) = \dfrac{x + 1}{x}$ for $x \in \mathbb{R}, x \neq 0$.
Find an expression for f^{-1}.

$$f(x) = \frac{x + 1}{x}$$

Let $y = \dfrac{x + 1}{x}$ and rearrange to make x the subject.

$$\begin{aligned} xy &= x + 1 \\ xy - x &= 1 \\ x(y - 1) &= 1 \\ x &= \frac{1}{y - 1} \end{aligned}$$

So $f^{-1}(x) = \dfrac{1}{x - 1}$

We can check the answer by substituting any value for x, say 3.

$$f(3) = \frac{3+1}{3} = \frac{4}{3}$$

$$f^{-1}\left(\frac{4}{3}\right) = \frac{1}{\frac{4}{3} - 1} = \frac{1}{\frac{1}{3}} = 3 \text{ (as expected)}$$

EXERCISE 5

In this exercise the domain for each function is the set of real numbers unless otherwise stated.

For questions **1** and **2**, the functions f, g and h are as follows:

$f: x \mapsto 4x \qquad g: x \mapsto x + 5 \qquad h: x \mapsto x^2$

1 Find the following in the form '$x \mapsto \ldots$'

a fg **b** gf **c** hf **d** fh

e gh **f** fgh **g** hfg

2 Find

a x if $hg(x) = h(x)$ **b** x if $fh(x) = gh(x)$

For questions **3**, **4** and **5**, the functions f, g and h are as follows:

$f: x \mapsto 2x \qquad g: x \mapsto x - 3 \qquad h: x \mapsto x^2$

3 Find the following in the form '$x \mapsto \ldots$'

a fg **b** gf **c** gh

d hf **e** ghf **f** hgf

4 Evaluate

a fg(4) **b** gf(7) **c** gh(−3)

d fgf(2) **e** ggg(10) **f** hfh(−2)

5 Find

a x if $f(x) = g(x)$ **b** x if $hg(x) = gh(x)$

c x if $gf(x) = 0$ **d** x if $fg(x) = 4$

For questions **6**, **7** and **8**, the functions l, m and n are as follows:

$l: x \mapsto 2x + 1 \qquad m: x \mapsto 3x - 1 \qquad n: x \mapsto x^2$

6 Find the following in the form '$x \mapsto \ldots$'

a lm **b** ml **c** ln

d nm **e** lnm **f** mln

7 Find

a lm(2) **b** nl(1) **c** mn(−2)

d mm(2) **e** nln(2) **f** llm(0)

8 Find

a x if $\mathrm{l}(x) = \mathrm{m}(x)$

b two values of x if $\mathrm{nl}(x) = \mathrm{nm}(x)$

c x if $\mathrm{ln}(x) = \mathrm{mn}(x)$

In questions **9** to **20**, find the inverse of each function in the form '$x \mapsto \ldots$'

9 $\mathrm{f}: x \mapsto 5x - 2$

10 $\mathrm{f}: x \mapsto 5(x - 2)$

11 $\mathrm{f}: x \mapsto 3(2x + 4)$

12 $\mathrm{g}: x \mapsto \dfrac{2x + 1}{3}$

13 $\mathrm{f}: x \mapsto \dfrac{3(x - 1)}{4}$

14 $\mathrm{g}: x \mapsto 2(3x + 4) - 6$

15 $\mathrm{h}: x \mapsto \frac{1}{2}(4 + 5x) + 10$

16 $\mathrm{k}: x \mapsto -7x + 3$

17 $\mathrm{j}: x \mapsto \dfrac{12 - 5x}{3}$

18 $\mathrm{l}: x \mapsto \dfrac{4 - x}{3} + 2$

19 $\mathrm{m}: x \mapsto \dfrac{\left[\dfrac{(2x - 1)}{4} - 3\right]}{5}$

20 $\mathrm{n}: x \mapsto \dfrac{12}{x}, x \neq 0$

For questions **21** to **24**, the functions f, g and h are defined as follows:

$\mathrm{f}: x \mapsto 3x \qquad \mathrm{g}: x \mapsto x - 5 \qquad \mathrm{h}: x \mapsto 2x + 1$

21 Find the form '$x \mapsto \ldots$'

a f^{-1} **b** g^{-1} **c** h^{-1}

d fg **e** $(\mathrm{fg})^{-1}$ **f** $\mathrm{g}^{-1}\mathrm{f}^{-1}$

22 Find the form '$x \mapsto \ldots$'

a hf **b** hf^{-1} **c** $\mathrm{f}^{-1}\mathrm{h}^{-1}$

d hg **e** $(\mathrm{hg})^{-1}$ **f** $\mathrm{g}^{-1}\mathrm{h}^{-1}$

23 Find

a $\mathrm{g}^{-1}(2)$ **b** $\mathrm{fg}^{-1}(2)$ **c** $(\mathrm{gf})^{-1}(10)$

d $\mathrm{f}^{-1}\mathrm{g}^{-1}(10)$ **e** $\mathrm{f}^{-1}\mathrm{f}^{-1}\mathrm{ff}(2)$

24 Find

a x if $\mathrm{h}(x) = \mathrm{f}(x)$

b the set of values of x for which

i $\mathrm{f}(x) > \mathrm{g}(x)$

ii $\mathrm{fg}(x) > 0$

25 The function f is defined by $\mathrm{f}: x = \dfrac{2x + 5}{3}$

a Find f^{-1} in the form '$x \mapsto \ldots$'

b Find $\mathrm{f}^{-1}(3)$

c Show that $\mathrm{f}^{-1}(3)$ is the solution of the equation $\mathrm{f}(x) = 3$.

Inverse functions and their graphs

Consider $f(x) = x + 2$ and its inverse $f^{-1}(x) = x - 2$ and $g(x) = 2x + 1$ and its inverse $g^{-1}(x) = \frac{x-1}{2}$

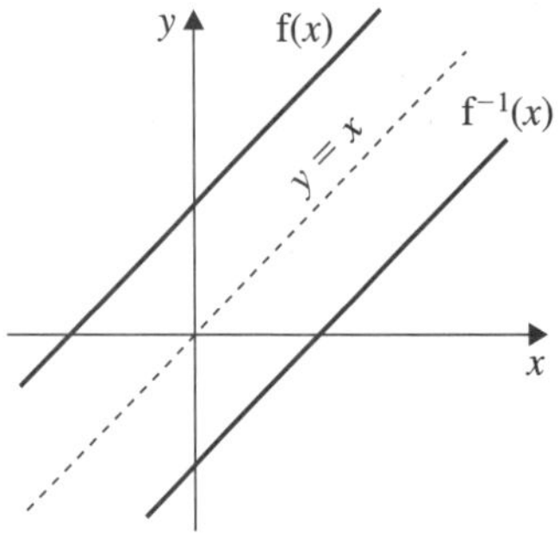

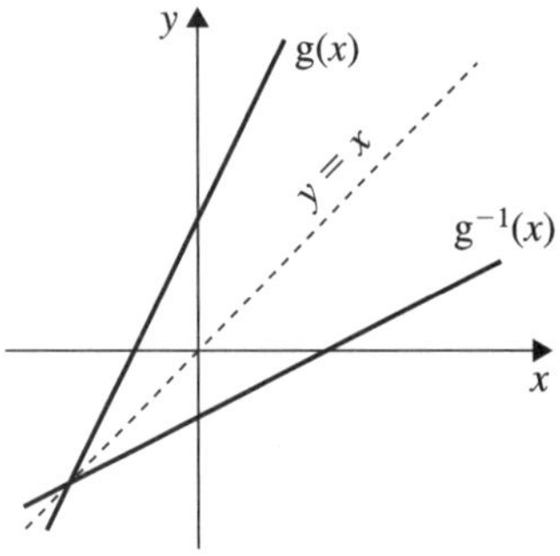

You see that:

$f^{-1}(x)$ is the reflection of $f(x)$ in the line $y = x$
$g^{-1}(x)$ is the reflection of $g(x)$ in the line $y = x$.

This is a general result for any function and its inverse.

For every point (x, y) on the graph of the function if there is a point (y, x) on the graph of f^{-1}. Also the range of f is the domain of f^{-1} and the domain of f^{-1} is the range of f.

Example 6

Find the inverse of the function $f(x) = x^2, x \geqslant 0$

Note that $f(x)$ is a one-to-one mapping as the domain is $x \geqslant 0$.

To find the inverse, let $y = x^2$

Rearranging $x = \pm\sqrt{y}$
We require the positive square root, $x = \sqrt{y}$
So $f^{-1}(x) = \sqrt{x}$.

Here are the graphs of $f(x)$ and $f^{-1}(x)$. Notice that $f^{-1}(x)$ is the reflection of $f(x)$ in the line $y = x$.

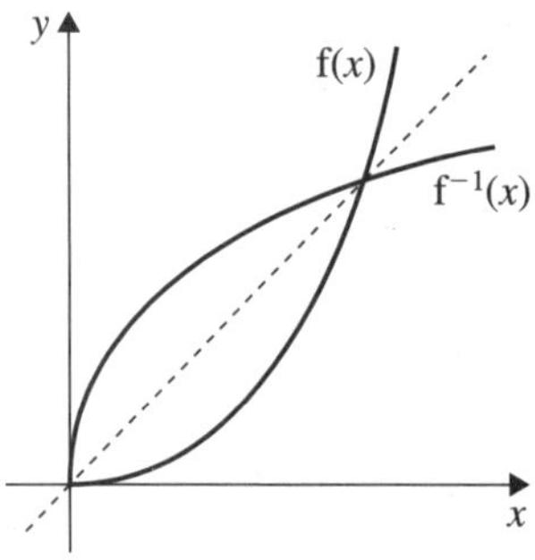

Example 7

Given $f(x) = (x - 1)^2 + 3, x \geqslant p$ is a one-to-one function,

a Find the minimum value of p.

b Find $f^{-1}(x)$ and state its domain.

a Sketch the curve $y = (x - 1)^2 + 3$

We see that f(x) is a one-to-one function for $x \geqslant 1$

$\therefore \quad p = 1$

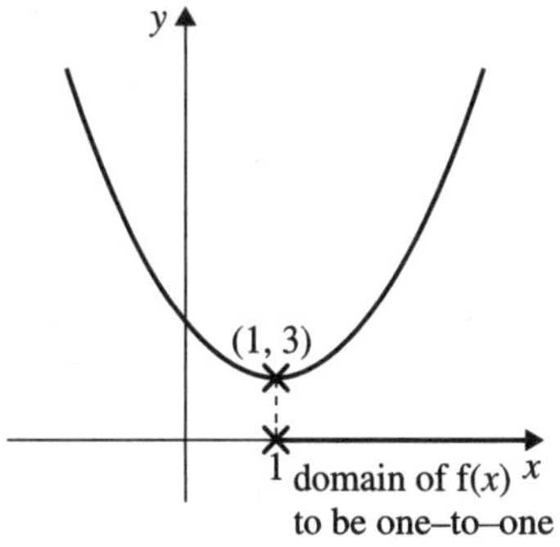

b $y = (x - 1)^2 + 3$

Rearrange to make x the subject,

$$(x - 1)^2 = y - 3$$
$$x - 1 = \pm\sqrt{y - 3}$$
$$x = 1 \pm \sqrt{y - 3}$$

We require the positive square root, $x = 1 + \sqrt{y - 3}$

$\therefore \quad f^{-1}(x) = 1 + \sqrt{x - 3}$

The domain of $f^{-1}(x)$ is $x \geqslant 3$

Notice that the *range* of f(x) is $f(x) \geqslant 3$.

EXERCISE 6

1 Find the inverse of the following functions

a $f(x) = 2 - x, \quad x \in \mathbb{R}$

b $f(x) = \dfrac{13}{2x}, \quad x \in \mathbb{R}, x \neq 0$

c $f(x) = 3 - 5x, \quad x \in \mathbb{R}$

d $f(x) = \dfrac{2x + 5}{x - 7}, \quad x \in \mathbb{R}, x \neq 7$

e $f(x) = \dfrac{3x - 5}{2x + 7}, \quad x \in \mathbb{R}, x \neq -\dfrac{7}{2}$

2 The function f(x) is defined by $f(x) = x^2 + 1, x \geqslant 0$. On the same diagram draw a sketch of $y = f(x)$ and $y = f^{-1}(x)$.

3 **a** If $f(x) = 3x - 1$ and $g(x) = x + 7$, calculate $f^{-1}g(x)$ and $gf^{-1}(x)$.

b Find the values of $f^{-1}g(4)$ and $gf^{-1}(2)$.

4 A function is defined by $f(x) = 3x - 1$ for all x.

a Find ff(x) and $f^{-1}(x)$.

b Show that $(ff)^{-1}(x)$ is identically equal to $f^{-1}f^{-1}(x)$.

5 A function is defined by $f(x) = \dfrac{x + 10}{x - 8}, x \neq 8$.

a Find $f^{-1}(x)$.

b Find $f^{-1}(5)$.

c Find a positive integer p such that $f(p) = p$.

d Find also $f^{-1}(p)$ for the positive integer found in part **c**.

6 **a** Sketch the curve of $y = x^2 - 2x - 3$ showing where the curve cuts the x-axis.

b You are given the function $f(x) = x^2 - 2x - 3, x \geqslant 1$.

i Find the range of f.

ii State the range and domain of f^{-1}.

iii Sketch the graph of f^{-1}, showing where the graph meets the coordinate axes.

7 A function f is defined by $f : x \mapsto 2 - \frac{1}{x}, x \neq 0$.

a Find f^{-1} and state the value of x for which f^{-1} is not defined.

b Find the value of x for which $f(x) = f^{-1}(x)$.

8 The function f is defined by $f : x = (x - 2)^2, x \geqslant 2$.

i State the domain and range of the inverse function f^{-1}.

ii Find an expression for $f^{-1}(x)$.

iii Find a solution to the equation $f(x) = f^{-1}(x)$.

9 The one-to-one functions f and g are defined by $f(x) = \sqrt{x - 2}$ for $x \geqslant a$ and $g(x) = \frac{1}{x^2}$ for $x > b$.

a Find the smallest possible values of a and b.

b Find $f^{-1}(x)$ and $g^{-1}(x)$ and state the restrictions, if any, on the domains of these inverse functions.

10 The functions f and g are defined by $f(x) = 2x + 1$ for all x and $g(x) = \frac{5}{x - 3}$, $x \neq 3$. Find $f^{-1}(x)$ and $g^{-1}(x)$ and state the restrictions, if any, on the domains of these inverse functions.

11 Copy and complete the domains for the following functions so that the domains are as large as possible for the functions to be one-to-one:

a $f(x) = 3x - 1$ $\quad \ldots \leqslant x \leqslant \ldots$

b $f(x) = 2x^2$ $\quad x \geqslant \ldots$

c $f(x) = \sin x$ $\quad -90° < x \leqslant \ldots$

d $f(x) = (x - 1)^2 + 3$ $\quad x \geqslant \ldots$

e $f(x) = x^2 + 2x - 3$ $\quad x \leqslant \ldots$

f $f(x) = x^2 + 8x + 13$ $\quad x \geqslant \ldots$

1.3 The modulus function

The **modulus** of x is written $|x|$ and means 'the positive value of x'.
For example, $|-2| = 2$, $|7| = 7$

Example 1

Sketch the graph of $y = |x - 2|$

a Sketch $y = x - 2$

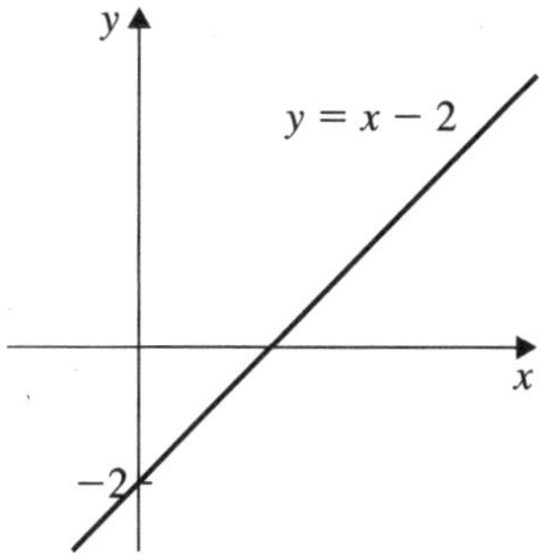

b Where y is negative, draw the reflection of $y = x - 2$ in the x-axis

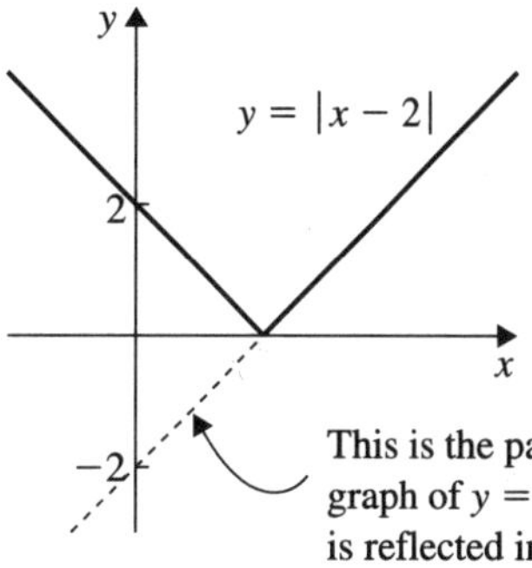

Example 2

Sketch the graph of $y = |\sin x|$ for $-360° \leq x \leq 360°$

a Sketch $y = \sin x$

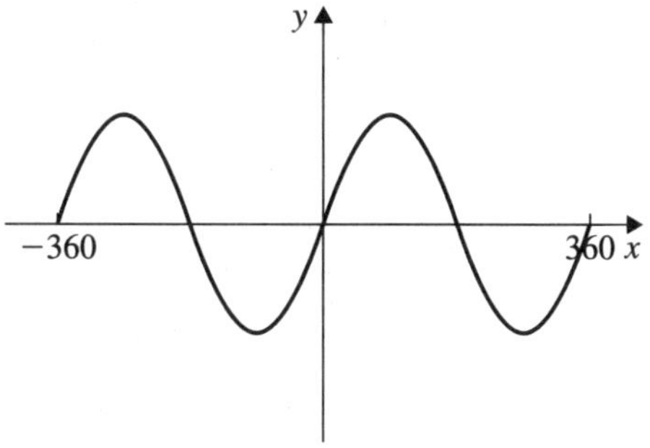

b Where y is negative, reflect the curve in the x-axis

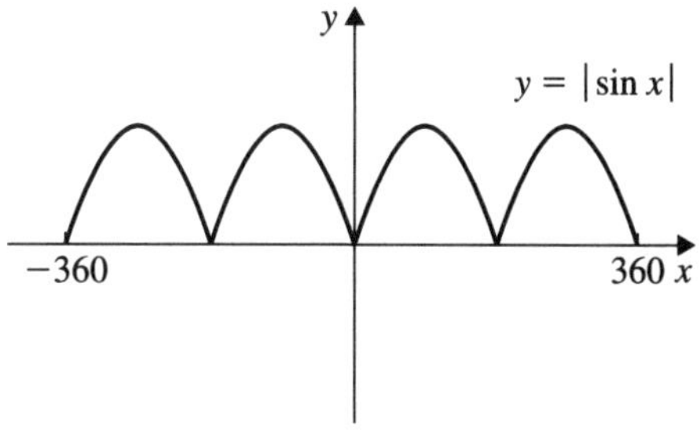

Example 3

Solve the equation $|x - 2| = 3$

Method A

We have either $x - 2 = 3 \Rightarrow x = 5$

or $-(x - 2) = 3$

$-x + 2 = 3 \Rightarrow x = -1$

Method B

Sketch the graphs of $y = |x - 2|$ and $y = 3$

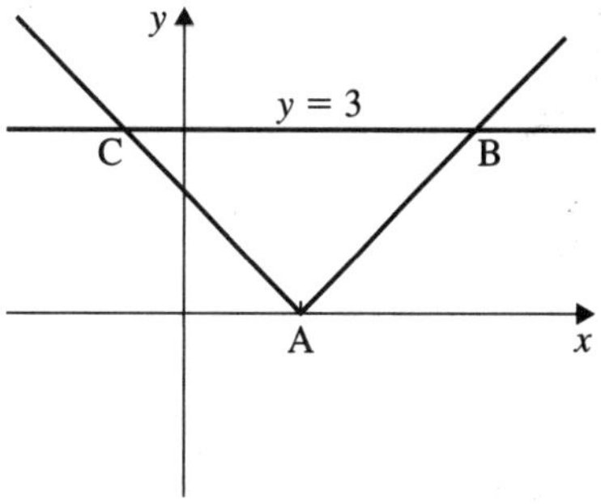

Line AB has equation $y = x - 2$

At B, $x - 2 = 3$

$x = 5$

Line AC has equation $y = -(x - 2)$

At C, $-(x - 2) = 3$

$x = -1$

You can use whichever method you prefer.

Example 4

Solve the inequality $|2x - 1| < 5$

Sketch the graph of $y = |2x - 1|$

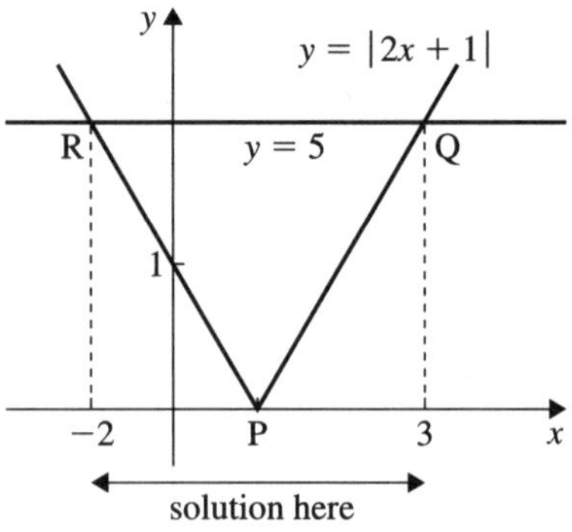

The inequality is satisfied where the graph of $y = |2x - 1|$ is below $y = 5$.

At Q, $2x - 1 = 5 \Rightarrow x = 3$
At R, $-(2x - 1) = 5$
$-2x + 1 = 5 \Rightarrow x = -2$

The solution is $-2 < x < 3$

Example 5

Solve the inequality $|x + 2| > |2x - 1|$

This question could be done like Example 4 by sketching the graphs of $y = |x + 2|$ and $y = |2x - 1|$.
Alternatively we can adopt an algebraic approach.

Since both sides of the inequality are positive, we can square both sides

$$|x+2|^2 > |2x-1|^2$$
$$(x+2)^2 > (2x-1)^2$$
$$x^2 + 4x + 4 > 4x^2 - 4x + 1$$
$$3x^2 - 8x - 3 < 0$$
$$(3x+1)(x-3) < 0$$

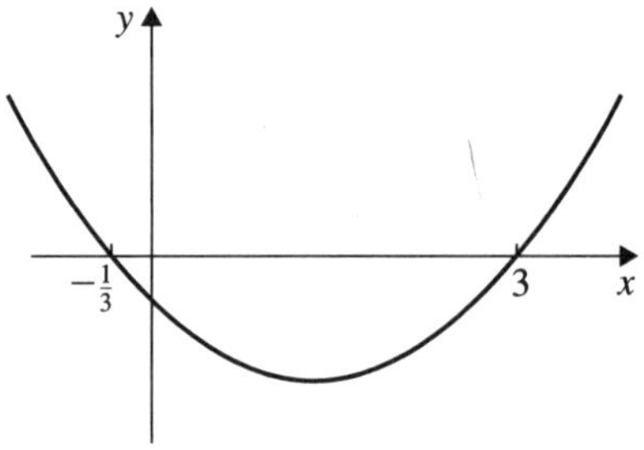

Sketch the curve $y = (3x + 1)(x - 3)$.

The solution is $-\frac{1}{3} < x < 3$.

The graph of $y = f(|x|)$

We sketch the graph of $y = f(x)$ for $x \geqslant 0$ and then reflect this graph in the y axis.

Example 6

Sketch the graph of $y = |x| - 1$.

a Sketch $y = x - 1$ for $x \geqslant 0$

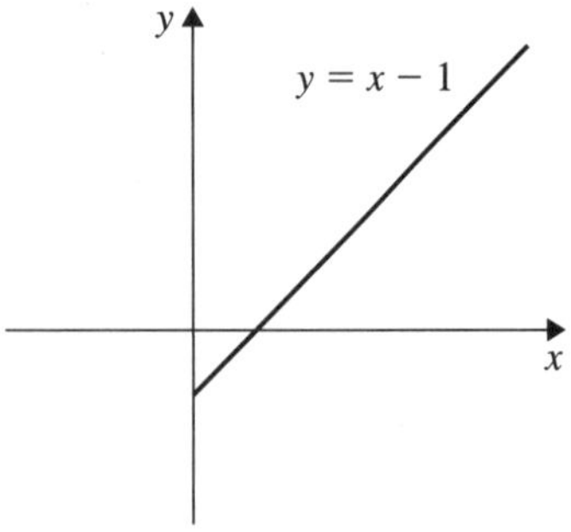

b Reflect in the y axis

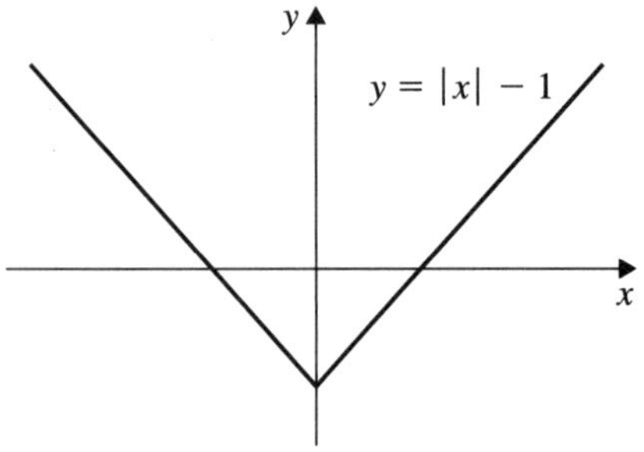

Example 7

Sketch the graph of $y = |x|^2 - |x|$.

a Sketch $y = x^2 - x$ for $x \geqslant 0$

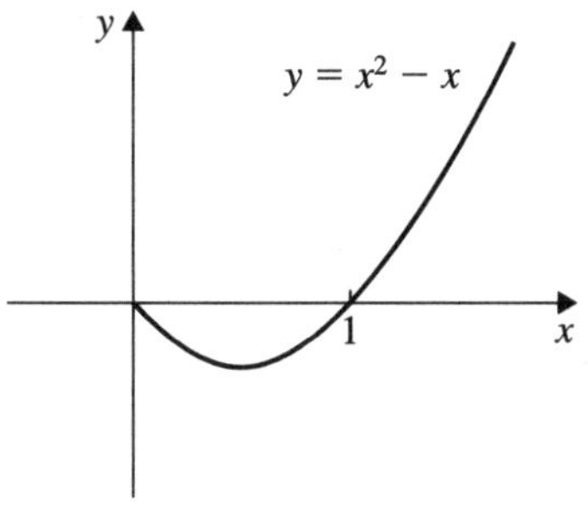

b Reflect in the y axis

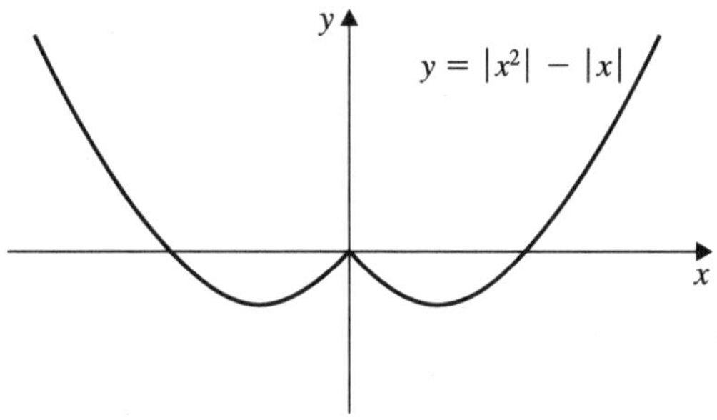

EXERCISE 7

1 Sketch the graphs of the following, marking where the graphs cross the axes.

a $y = |x - 4|$ **b** $y = |x + 2|$ **c** $y = |3x - 1|$

d $y = |x^2 - 2x|$ **e** $y = |x^2 - 9|$ **f** $y = |x^2 - 5x + 4|$

g $y = \left|\frac{1}{x}\right|$ **h** $y = |x^3 - x^2 - 6x|$

2 Solve the following equations.

a $|x - 2| = 5$ **b** $|2x - 3| = 3$ **c** $|1 - 2x| = 5$

d $|x^2 - 5| = 4$ **e** $|x^2 - 37| = 12$ **f** $2|x + 3| = 8$

g $3|2x - 1| = 9$ **h** $\left|\frac{1}{x}\right| = 2$

3 Solve the following inequalities.

a $|x - 4| > 5$ **b** $|x + 1| < 3$ **c** $|2x + 3| > 7$

d $|3x - 1| > -1$ **e** $|x + 2| < \frac{1}{2}x + 5$ **f** $|3x - 2| < 6 - x$

g $|x| > |2x - 1|$ **h** $|x + 4| > |x - 2|$ **i** $2|x - 1| < |x - 2|$

4 Sketch the graphs of $y = |3x - 1|$ and $y = |x^2 - 5x + 4|$ and hence solve the equation $|3x - 1| = |x^2 - 5x + 4|$.

5 Sketch the graphs of the following, marking where the graph crosses the axes.

a $y = |x| - 2$ **b** $y = |x| + 3$ **c** $y = 2|x| - 1$

d $y = |x|^2 - 2|x|$ **e** $y = \cos|x|$ **f** $y = \frac{1}{|x|}$

6 The graphs of $y = f(x)$ and $y = g(x)$ are shown below.

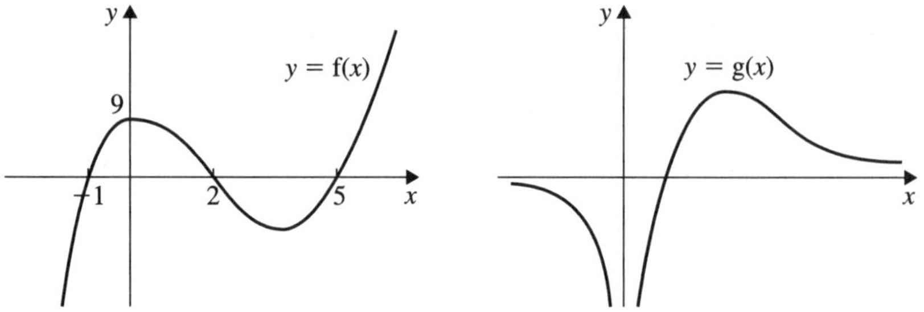

Sketch the graphs of

a $y = |f(x)|$ **b** $y = f(|x|)$ **c** $y = |g(x)|$ **d** $y = g(|x|)$

7 **a** Sketch the graph of $y = \sin |x|$.

b How many solutions are there to the equation

$$\sin |x| = \tfrac{1}{2}, \text{ for } -180° \leqslant x \leqslant 180°?$$

8 **a** Sketch the graphs of $f(x) = |x| - 3$ and $g(x) = \frac{1}{3}x$.

b Hence solve the inequality $|x| - 3 > \frac{1}{3}x$.

9 **a** On the same diagram, sketch the graphs of

$$y = \frac{1}{x - 1} \text{ and } y = 9\,|x - 1|.$$

b Hence find the set of values of x for which

$$\frac{1}{x - 1} < 9\,|x - 1|.$$

10 **a** Sketch the graphs of the following, marking where the graphs cut the axes.

a **i** $y = 2 - |x + 1|$

ii $y = 5 - |x - 1|$

b State the range of each function.

11 Given that $f(x) = x^2 - 3x$, $x \in \mathbb{R}$, sketch in separate diagrams the graphs of

a $y = f(x)$ **b** $y = |f(x)|$ **c** $y = f(|x|)$

1.4 Transformations of the graph of a function

In 'Pure Mathematics C1 C2' we discussed the important single transformations of the graph of a function.
Here is a summary:

$y = f(x) + a$	Translation by a units parallel to the y-axis.
$y = f(x - a)$	Translation by a units parallel to the x-axis (note the negative sign).
$y = af(x)$	Stretch parallel to the y-axis by a scale factor a.

$y = f(ax)$	Stretch parallel to the x-axis by a scale factor $\frac{1}{a}$. (note the reciprocal of a).
$y = -f(x)$	Reflect $f(x)$ in the x-axis.
$y = f(-x)$	Reflect $f(x)$ in the y-axis.

Example 1

The diagram shows a graph of $y = f(x)$.
Sketch the graphs of

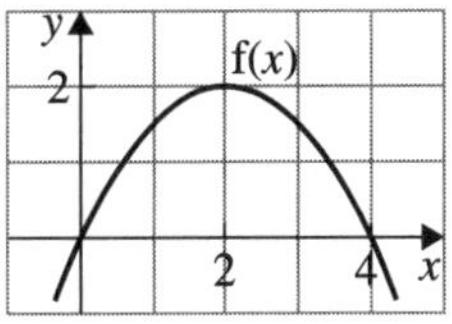

a $y = f(x) + 1$ **b** $y = f(x + 1)$ **c** $y = 2f(x)$
d $y = f(2x)$ **e** $y = -f(x)$ **f** $y = f(-x)$

a Translation 1 unit ↑

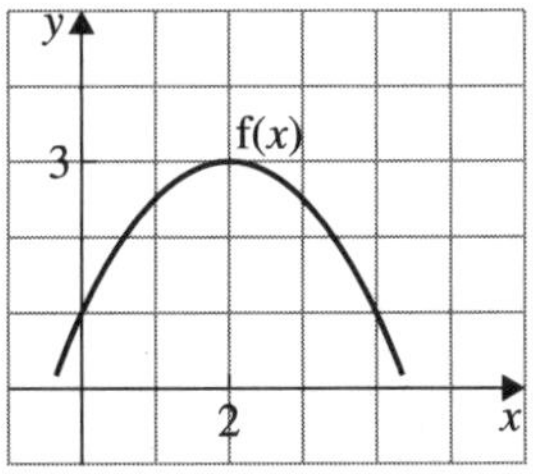

b Translation 1 unit ←

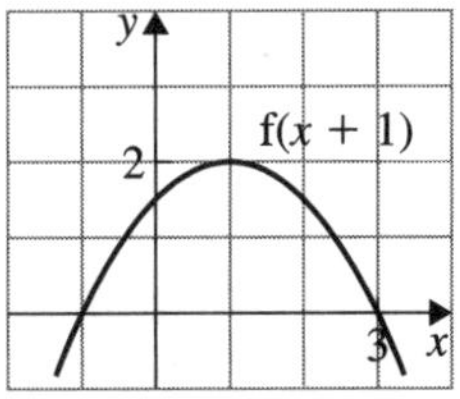

c Stretch ↕, scale factor 2

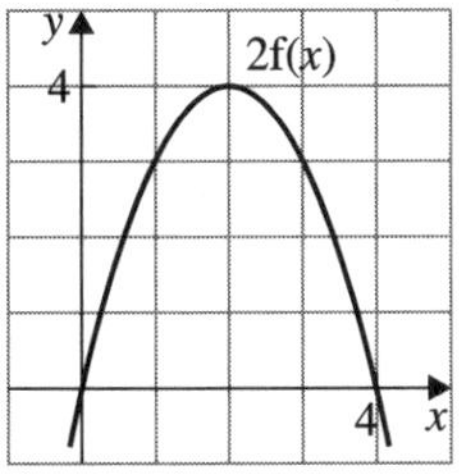

d Stretch ↔, scale factor $\frac{1}{2}$

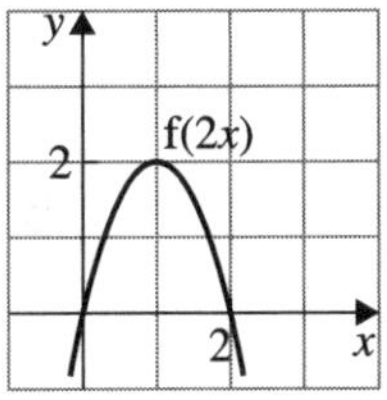

e Reflection in x-axis

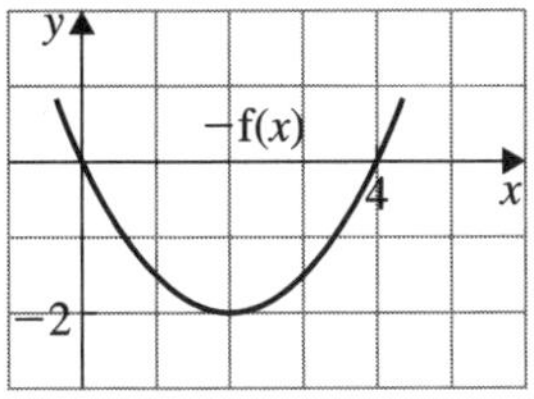

f Reflection in y-axis

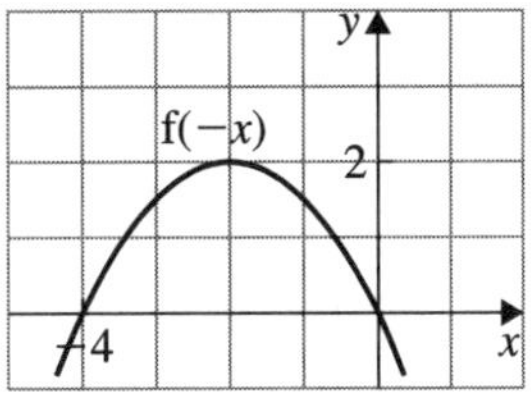

Combination of transformations

We now consider a combination of transformations.

a Suppose we begin with $y = x^2$ (i.e. $f(x) = x^2$)

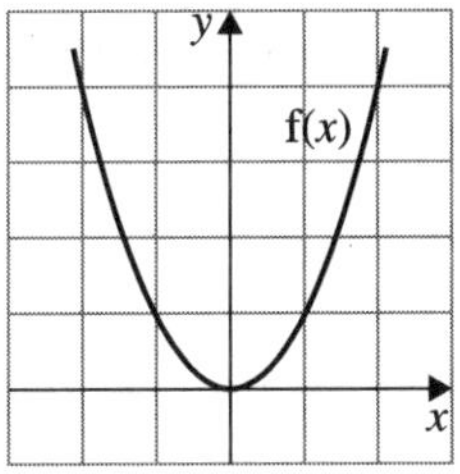

b We now translate the curve by 3 units in the positive x-direction.
This graph has equation $y = (x - 3)^2$

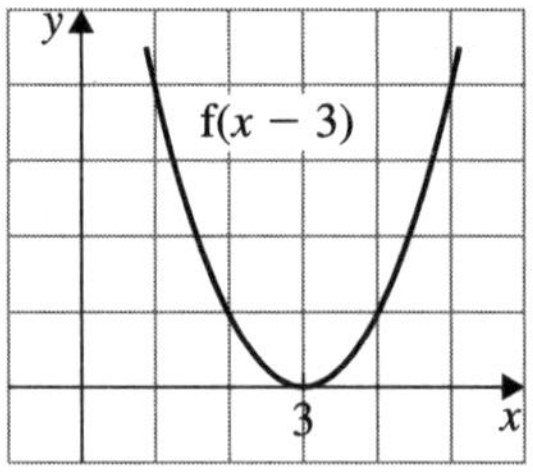

c We now translate this new graph by 2 units in the positive y-direction.
This graph has equation $y = (x - 3)^2 + 2$

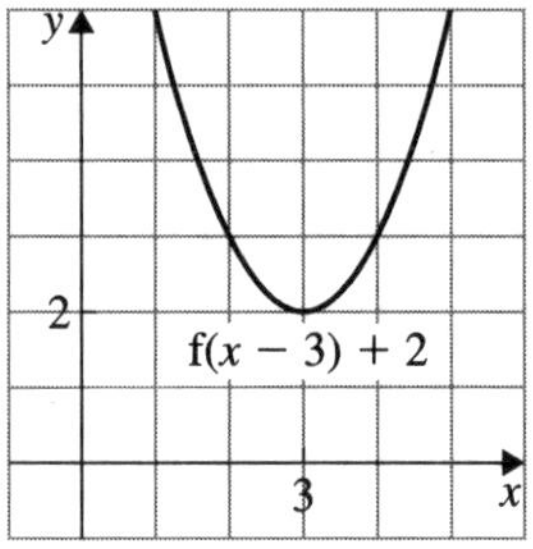

Example 2

A sketch of the curve $y = \sin x$ is shown.
Sketch the graphs of

a $y = \sin \frac{1}{2}x$

b $y = \sin \frac{1}{2}x + 1$

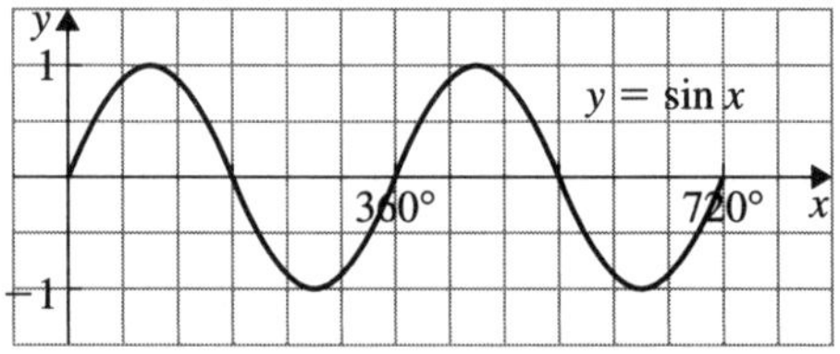

a $y = \sin \frac{1}{2}x$: Stretch ↔, scale factor 2

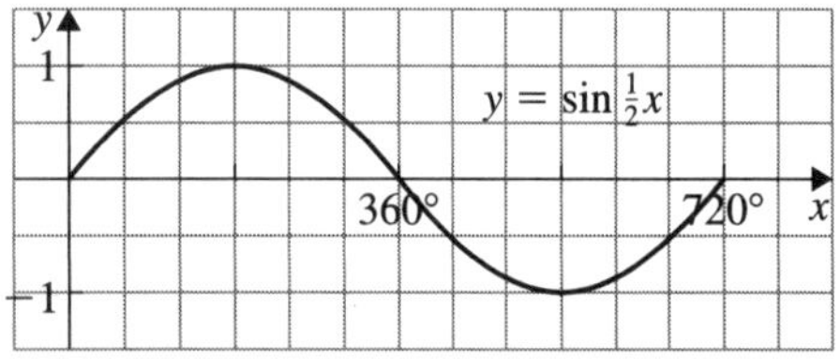

b $y = \sin \frac{1}{2}x + 1$ Translation 1 unit ↑

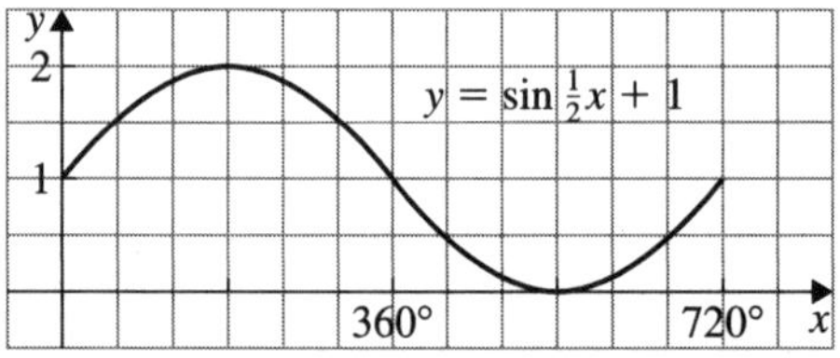

Example 3

This is a sketch of the curve $y = f(x)$.
The only vertex of the curve is at $A(2, -4)$.
The curve $y = x^2$ has been translated to give the curve $y = f(x)$.
Find $f(x)$ in terms of x.

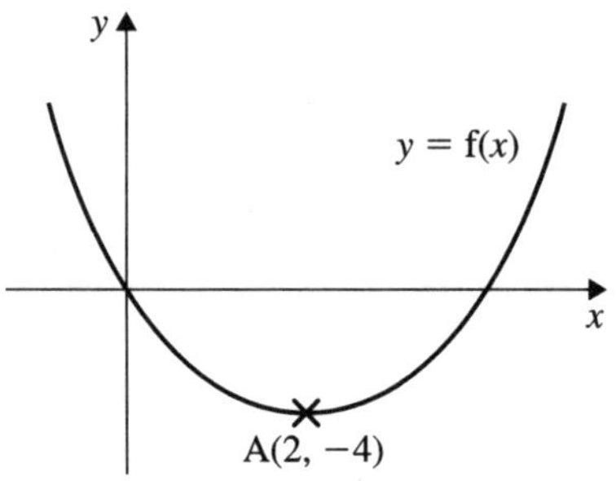

Answer:
The curve $y = x^2$ has been translated 2 units in the positive x direction and 4 units in the negative y direction.

$$\therefore \quad f(x) = (x - 2)^2 - 4$$
$$= x^2 - 4x + 4 - 4$$
$$f(x) = x^2 - 4x.$$

EXERCISE 8

1 The graph of $y = f(x)$ is shown.
Sketch the graphs of,

a $y = f(x + 2)$

b $y = -f(x)$

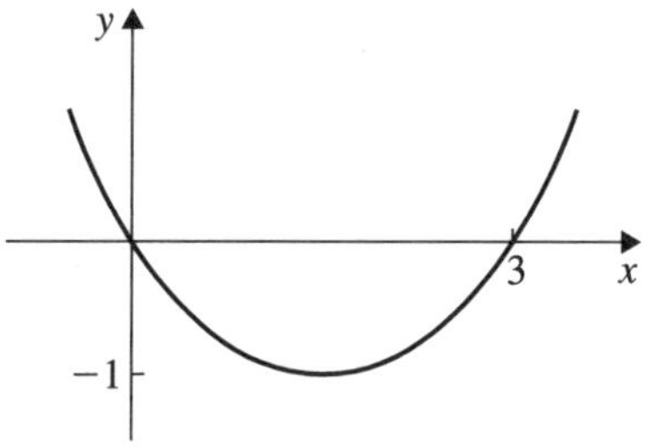

2 This is the sketch graph of $y = f(x)$.

a Sketch the graph of $y = f(x) + 2$

b Sketch the graph of $y = f(x - 1)$

c Sketch the graph of $y = -f(x)$

d Sketch the graph of $y = f(-x)$.

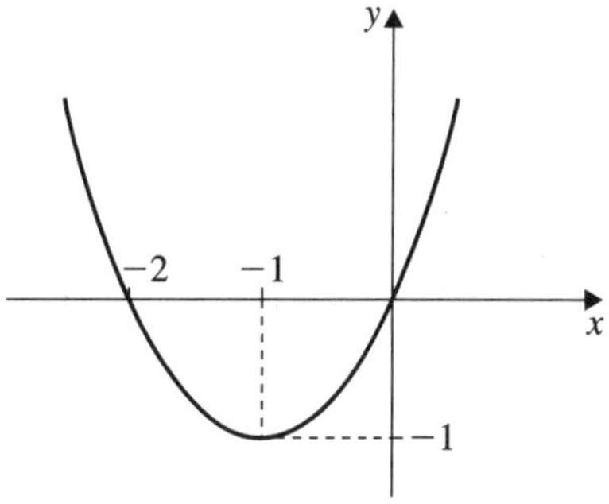

3 This is the sketch of $y = f(x)$ which passes through A, B, C.
Sketch the following curves, giving the new coordinates of A, B, C in each case.

a $y = 2f(x)$

b $y = f(x - 2)$

c $y = f(\frac{1}{2}x)$

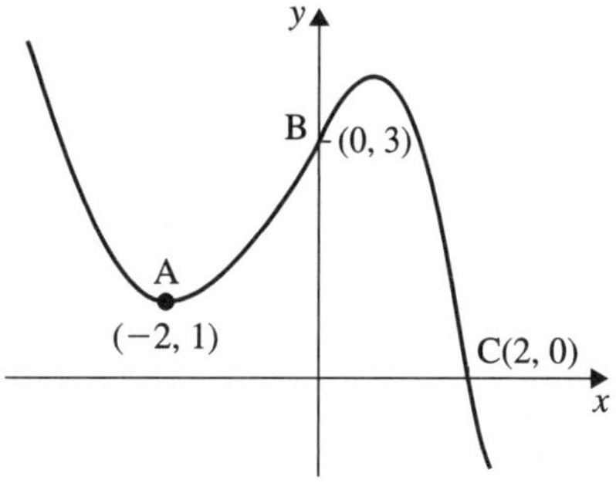

In questions 4 to 9 each graph shows a different function $f(x)$. On squared paper draw a sketch to show the given transformation. The scales are 1 square = 1 unit on both axes.

4

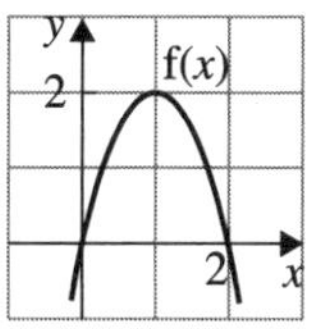

Sketch $f(x + 2)$

5

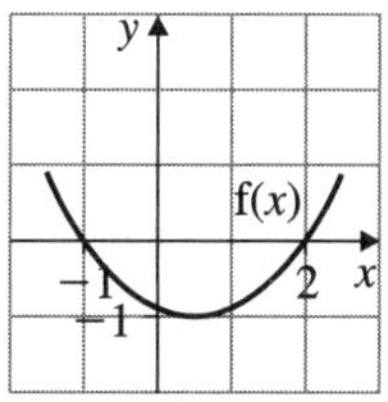

Sketch $f(-x)$

6

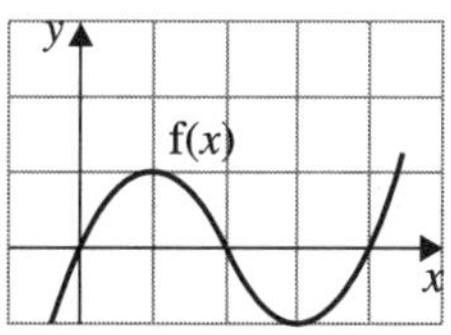

Sketch $f(\frac{1}{2}x)$

7

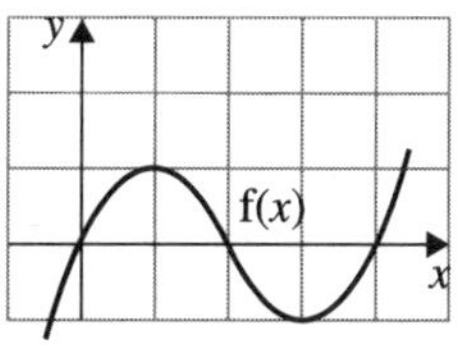

Sketch $2f(x)$

8

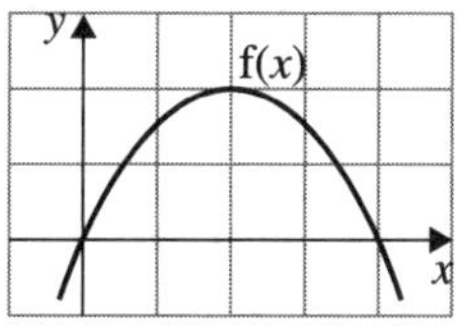

Sketch $f(2x)$

9

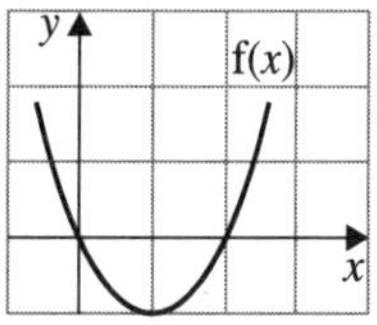

Sketch $-f(x)$

10 This is the sketch graph of $y = f(x)$.

a Sketch $y = f(x - 3)$

b Sketch $y = f(x - 3) + 4$

Give the new coordinates of the point A on the two sketches.

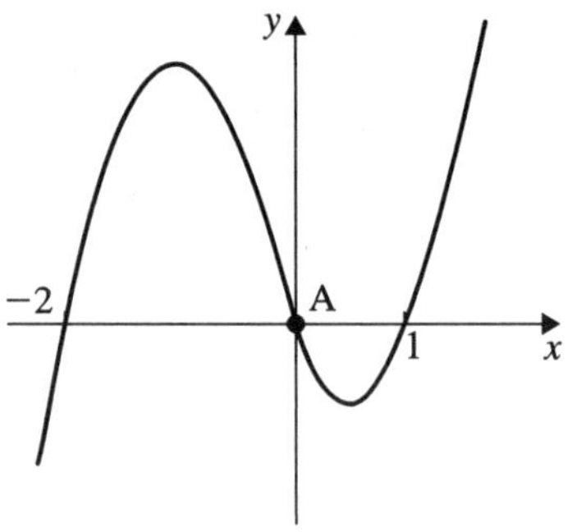

11 On the same axes, sketch and label:

a $y = x^2$

b $y = (x - 2)^2$

c $y = (x - 2)^2 + 4$

12 a Sketch and label $y = f(x)$, where $f(x) = x(x - 4)$.

b On the same axes, sketch and label:

i $y = f(2x)$ **ii** $y = -f(2x)$

13 This is a sketch of $y = f(x)$.

a Sketch $y = f(x + 1) + 5$

b Sketch $y = f(x - 3) - 4$

c Sketch $y = 1 + f(2x)$.

Show the new coordinates of the point A on each sketch.

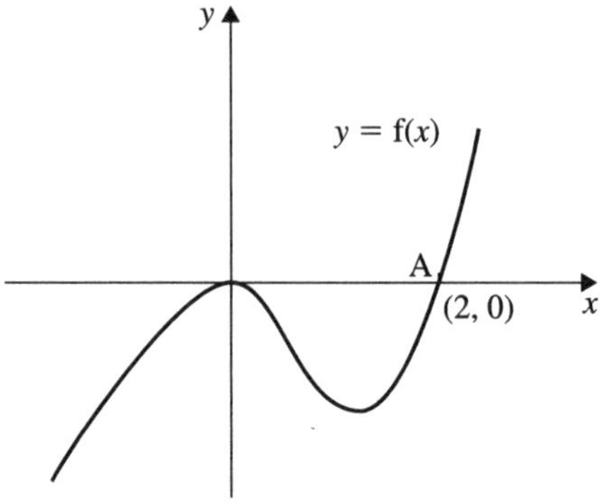

14 a Find the equation of the curve when $y = x^3$ is translated 5 units in the direction ↑.

b Find the equation of the curve when $y = x^3$ is translated 2 units in the direction →.

c Find the equation of the curve when $y = x^3$ is translated by the vector $\begin{pmatrix} 2 \\ 5 \end{pmatrix}$.

15 a Sketch the graph of $y = \sin x$, for $0 \leqslant x \leqslant 360°$.

b Sketch the graph of $y = \sin \frac{1}{2}x$, for $0 \leqslant x \leqslant 360°$.

16 a Sketch the graph of $y = \tan x$, for $0 \leqslant x \leqslant 2\pi$.

b Sketch the graph of $y = \tan\left(x - \frac{\pi}{2}\right)$, for $0 \leqslant x \leqslant 2\pi$.

17 You are given $f(x) = 2x + 1$ and $g(x) = 6x - 1$.
Find a pair of successive transformations which, applied to $f(x)$, will give $g(x)$.

18 You are given $f(x) = \frac{1}{2}x - 2$ and $g(x) = 2x + 1$.
Find a pair of successive transformations which, applied to $f(x)$, will give $g(x)$.

19 $f(x) = x^2$ and $g(x) = x^2 - 4x + 7$.

a If $g(x) = f(x - a) + b$, find the values of a and b.

b Hence sketch the graphs of $y = f(x)$ and $y = g(x)$ showing the transformations from f to g.

20 $f(x) = x^2$ and $g(x) = x^2 + 8x + 17$.

a If $g(x) = f(x + a) + b$, find the values of a and b.

b Hence sketch the graphs of $y = f(x)$ and $y = g(x)$ showing the transformation from f to g.

Questions **21** and **22** involve the graphs of $y = e^x$ and $y = \log_e x$ which are discussed in Part 3 of this book.

21 This is the graph of $y = e^x$.

Sketch $y = e^{2x} - 1$.

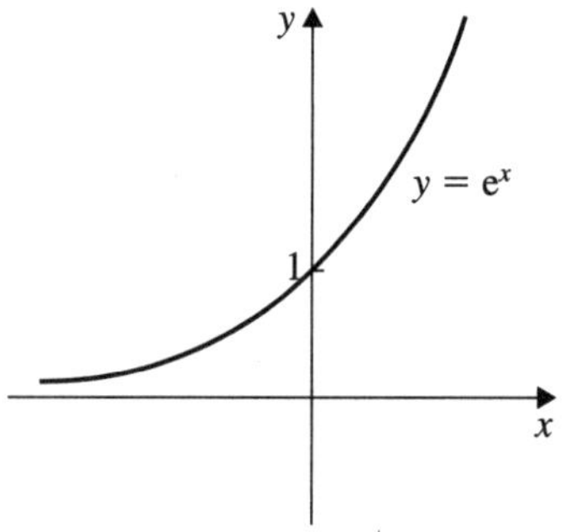

22 This is the graph of $y = \log_e x$

a Sketch $y = \log_e(x - 2)$

b Sketch $y = 2\log_e(x - 2)$.

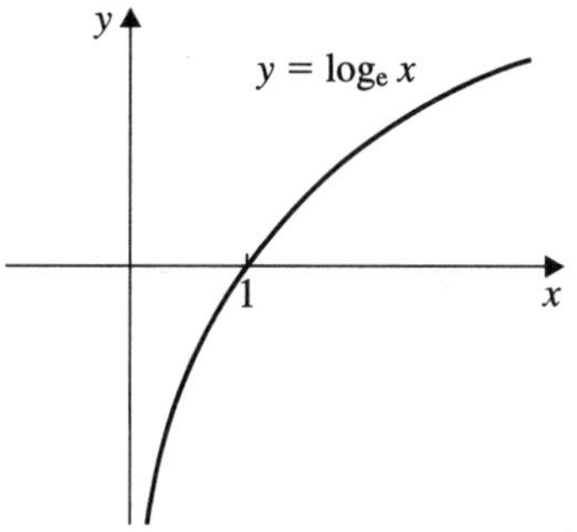

Review Exercise 1

1 Simplify as far as possible.

a $\dfrac{5x}{3x - x^2}$

b $\dfrac{5x^2y}{10y}$

c $\dfrac{9x + 6}{3x}$

d $\dfrac{1 - x^2}{1 - x}$

e $\dfrac{x + 4x^2}{xy + xy^2}$

f $\dfrac{(3x)^2}{6x}$

2 Factorise completely $4x^2 - 4$ and hence simplify $\dfrac{x^2 - 5x + 4}{4x^2 - 4}$.

3 Simplify as far as possible.

a $\dfrac{x^2+x-2}{x^2-x}$ **b** $\dfrac{x^2-5x+6}{x^2-x-2}$ **c** $\dfrac{2x^2+7x-4}{x^2-16}$

d $\dfrac{2x+\dfrac{1}{x}}{x}$ **e** $\dfrac{3x-\dfrac{1}{x}}{\dfrac{1}{x}}$ **f** $\dfrac{4-\dfrac{1}{x^2}}{\dfrac{1}{x}}$

4 Write as a single fraction.

a $\dfrac{x+1}{3}+\dfrac{2x+1}{2}$ **b** $\dfrac{1}{x}+\dfrac{3}{x+1}$ **c** $\dfrac{2}{x+1}+\dfrac{x}{x-1}$

d $\dfrac{x+1}{3}-\dfrac{x}{4}$ **e** $\dfrac{x}{5}-\dfrac{x+1}{2}$ **f** $\dfrac{x-1}{3}\div\dfrac{x+3}{6}$

5 You are given $f(x) = 6x^2 - x - 1$ and $g(x) = 4x^3 - x$.

a By finding $f(\frac{1}{2})$ and $g(\frac{1}{2})$, or otherwise, show that $f(x)$ and $g(x)$ have a common linear factor.

b Hence write $\dfrac{f(x)}{g(x)}$ in its simplest form.

6 The function f is defined by $f: x \mapsto 1 - x^2, x \in \mathbb{R}$.

a State the range of f.

b Find the values of x for which $ff(x) = 0$.

7 The function f is defined by $f: x \mapsto \dfrac{2x+1}{x}, x \in \mathbb{R}, x \neq 0$.

Find, in a similar form, the functions

a ff **b** f^{-1}

8 A function f is defined by

$$f: x = 4 - \frac{3}{x}, x \neq 0$$

a Find f^{-1} and state the value of x for which f^{-1} is not defined.

b Find the values of x for which $f(x) = f^{-1}(x)$.

9 The function f is defined by $f: x \mapsto \sqrt{\left(1 - \dfrac{9}{x^2}\right)}$ with domain $x \in \mathbb{R}$ and $x \geqslant 3$.

a Find the range of f.

b Find an expression for $f^{-1}(x)$ and state the range and domain of f^{-1}.

10 The functions f and g are defined with their respective domains by

$f: x \mapsto \dfrac{3}{2x-1}, x \in \mathbb{R}, x \neq \frac{1}{2}$.

$g: x \mapsto x^2 + 3, x \in \mathbb{R}$.

a State the range of g.

b Find $fg(x)$, giving your answer in its simplest form.

c Find an expression for $f^{-1}(x)$.

d Solve the equation $f(x) = f^{-1}(x)$.

11 The functions f and g are defined by

$$f(x) = \frac{2}{x+1}, x > 0 \qquad g(x) = 1 - x^2, x \in \mathbb{R}.$$

a Find $fg(x)$, giving your answer in its simplest form.

b Find $f^{-1}(x)$.

c Solve the equation $f(x) = f^{-1}(x)$.

12 The function f is defined by $f(x) = x^2 - 4x - 5, x \in \mathbb{R}, x \geqslant 2$.

a Find the range of f.

b Write down the domain and range of f^{-1}.

c Sketch the graph of f^{-1}, showing any points where the graph intersects the coordinate axes.

13 Sketch the graphs of the following, marking where the graphs cross the axes.

a $y = |x - 3|$ **b** $y = |3x + 1|$

c $y = |x^2 - 1|$ **d** $y = |\sin x|$ for $0 \leqslant x \leqslant 2\pi$.

14 The graphs of $y = f(x)$ and $y = g(x)$ are shown.

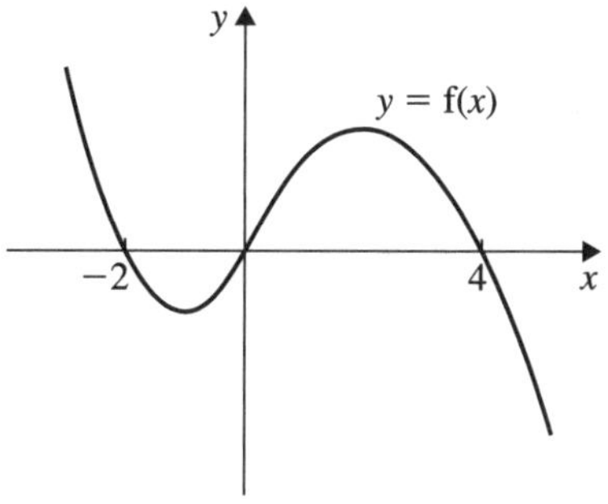

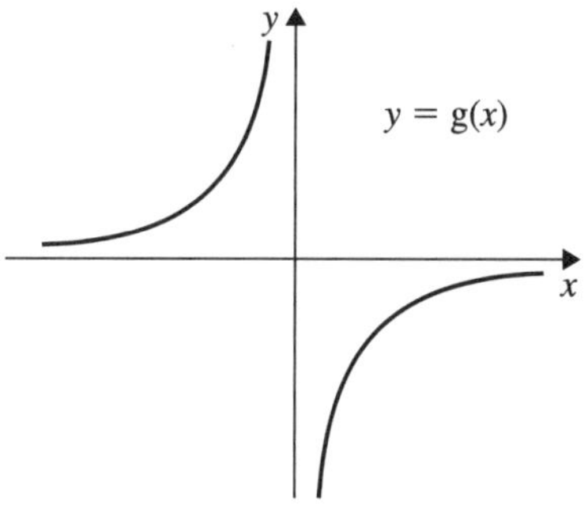

Sketch the graphs of

a $y = |f(x)|$ **b** $y = |g(x)|$ **c** $y = f(|x|)$ **d** $y = g(|x|)$.

15 Solve the inequalities.

a $|x - 1| < |x + 2|$

b) $|2x + 1| > |x - 1|$

16 Solve the inequalities.

a $|x - 1| < 5$

b) $|2x - 3| > 1|$

17 **a** Sketch the graph of $y = |x^2 - 4|$.

b Solve the inequality $|x^2 - 4| > 5$.

18 **a** On the same axes, sketch the graphs of $y = |2x + 1|$ and $y = \frac{2}{x}$.

b How many solutions are there to the equation $|2x + 1| = \frac{2}{x}$?

19 Sketch the graphs of

a $y = |x| - 2$

b $y = 2|x| - 3$

c $y = |x|^2 - |x|$

20 **a** Sketch the graphs of $f(x) = |x| - 5$ and $g(x) = \frac{1}{2}x$.

b Hence solve the inequality $|x| - 5 > \frac{1}{2}x$.

21 The function f is defined for all real values of x by

$$f(x) = |x - 3| + 2$$

a Sketch the graph of $y = f(x)$ and state the range of f.

b Solve the equation $f(x) = 5$.

c Evaluate ff(1).

22 **a** Sketch the graph of $y = 4 - |x + 1|$, showing where the graph crosses the axes.

b State the range of the function $f(x) = 4 - |x + 1|$.

23 **a** Sketch the graph of $y = 5 - |2x + 1|$.

b How many roots are there to the equation $5 - |2x + 1| = 3$?

c State the range of the function $g(x) = 5 - |2x + 1|$.

24 The diagram shows part of the graph of $y = f(x)$.

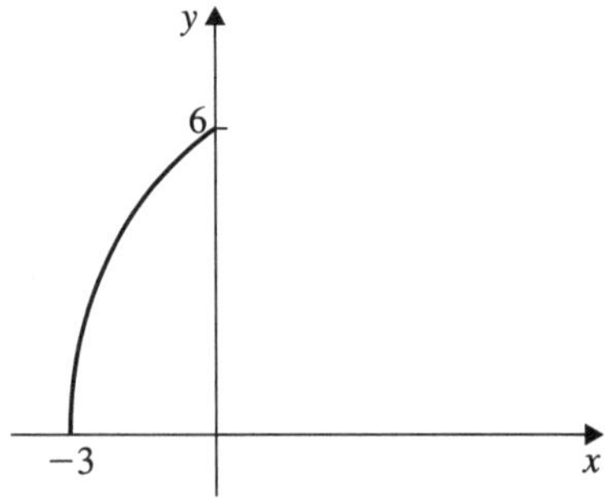

On separate diagrams, sketch the curve with equation

a $y = \frac{1}{2}f(x)$

b $y = f(2x)$

c $y = f(x - 5)$

d $y = f(-x)$

Indicate clearly the new positions of the points $(-3, 0)$ and $(0, 6)$ for each function.

25 The sketch shows the curve with equation $y = f(x)$. It passes through the origin O.

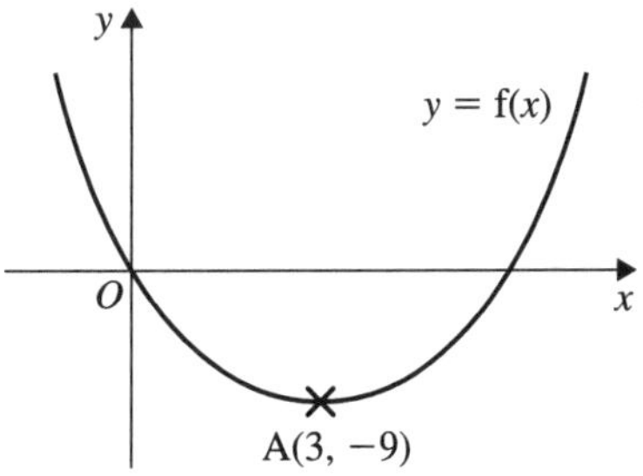

The only vertex of the curve is at A(3, −9).

a Write down the coordinates of the vertex of the curve with equation

i $y = f(x + 1)$

ii $y = f(x) + 5$

iii $y = -f(x)$

iv $y = f(3x)$

v $y = f(-x)$

The curve with equation $y = x^2$ has been translated to give the curve $y = f(x)$ shown in the diagram.

b Find $f(x)$ in terms of x.

26 The sketch shows the curve with equation $y = f(x)$.

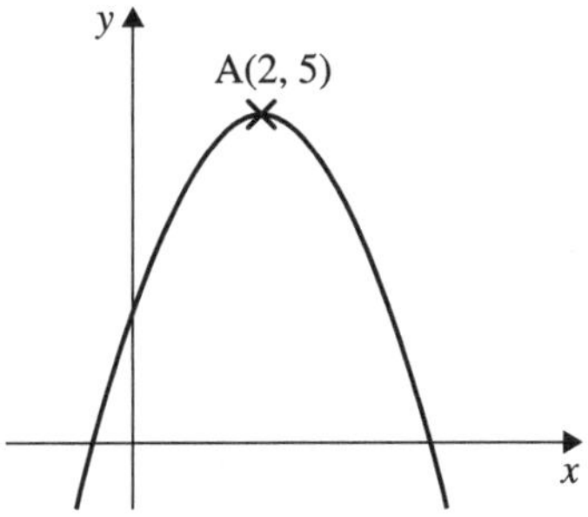

The only maximum point of the curve $y = f(x)$ is A(2, 5).

a Write down the coordinates of the maximum point for the curves with each of the following equations.

i $y = f(x + 3)$

ii $y = f(x) - 6$

iii $y = f(-x)$

The curve with equation $y = -x^2$ has been translated to give the curve $y = f(x)$.

b Find $f(x)$ in terms of x.

27 Describe a series of geometrical transformations that maps the graph of $y = x^2$ onto the graph of $y = 2(x - 3)^2 - 4$.

28 Describe a series of geometrical transformations that maps the graph of $y = \cos x$ onto the graph of $y = -1 + \cos 2x$.

29 The functions $f(x)$ and $g(x)$ are such that

$$g(x) = f(\tfrac{1}{2}x) + 3$$

Describe a sequence of two transformations that maps the graph of $y = f(x)$ onto the graph of $y = g(x)$.

30

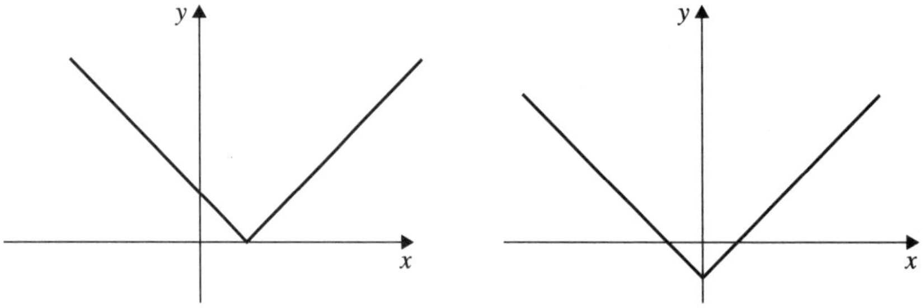

The diagrams show the graphs of $y = f(x)$ and $y = g(x)$, where

$f(x) = |x - 4|$ and $g(x) = |x| - 3, x \in \mathbb{R}$

Describe the geometrical transformations by which each of the above graphs can be obtained from the graph of $y = |x|$.

EXAMINATION EXERCISE 1

1 Simplify $\dfrac{x^3 - 3x^2}{x^3 - 9x}$. [OCR]

2 Express $\dfrac{x}{(x+1)(x+3)} + \dfrac{x+12}{x^2-9}$ as a single fraction in its simplest form. [EDEXCEL]

3 Express $\dfrac{3}{x^2+2x} + \dfrac{x-4}{x^2-4}$ as a single fraction in its simplest form. [EDEXCEL]

4 The function f is defined by

$$f: x \mapsto \frac{1}{\sqrt{x}} + 2, \quad x > 0.$$

i State the range of f.

ii Find an expression for $f^{-1}(x)$. [OCR]

5 The functions f and g are defined with their respective domains by

$$f: x \mapsto \frac{4}{3+x}, \quad x > 0$$

$$g: x \mapsto 9 - 2x^2, \quad x \in \mathbb{R}$$

a Find $fg(x)$, giving your answer in its simplest form.

b i Sketch the graph of $y = g(x)$.

ii Find the range of g.

c i Solve the equation $g(x) = 1$.

ii Explain why the function g does not have an inverse.

d The inverse of f is f^{-1}.

i Find $f^{-1}(x)$.

ii Solve the equation $f^{-1}(x) = f(x)$. [AQA]

6 $$f(x) = \frac{2}{x-1} - \frac{6}{(x-1)(2x+1)}, x > 1.$$

a Prove that $f(x) = \frac{4}{2x+1}$.

b Find the range of f.

c Find $f^{-1}(x)$.

d Find the range of $f^{-1}(x)$. [EDEXCEL]

7 The function f is given by

$$f: x \mapsto \frac{x}{x^2-1} - \frac{1}{x+1}, x > 1.$$

a Show that $f(x) = \frac{1}{(x-1)(x+1)}$.

b Find the range of f.

The function g is given by

$$g: x \mapsto \frac{2}{x}, x > 0.$$

c Solve $gf(x) = 70$ [EDEXCEL]

8 The functions f and g are defined for all real values of x by

$$f(x) = 3x - 10,$$
$$g(x) = |x - 2|.$$

i State the range of each function.

ii Show that the value of $fg(-4)$ is 8.

iii Determine the value of $f^{-1}(7)$.

iv Solve the equation $gf(x) = 1$. [OCR]

9 The functions $f(x)$ and $g(x)$ are defined by

$$f(x) = \ln x, g(x) = x^2.$$

i Write down expressions for $fg(x)$ and $gf(x)$.

ii Show that $fg(x) = kf(x)$, for some constant k, and find k. [MEI]

10 a Sketch, on the same diagram, the graphs of

$$y = |2x + 3| \text{ and } y = 2x^2 - 9,$$

stating the coordinates of any points where the graphs meet the axes.

b Deduce the number of roots of the equation

$$|2x + 3| = 2x^2 - 9.$$

Determine the value of each of these roots. [AQA]

11

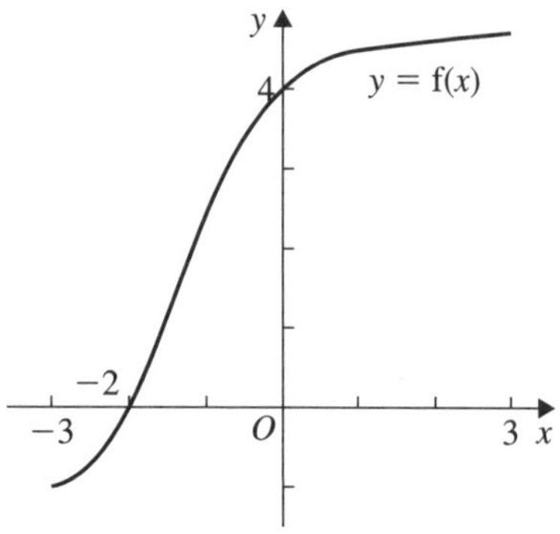

Fig. 1

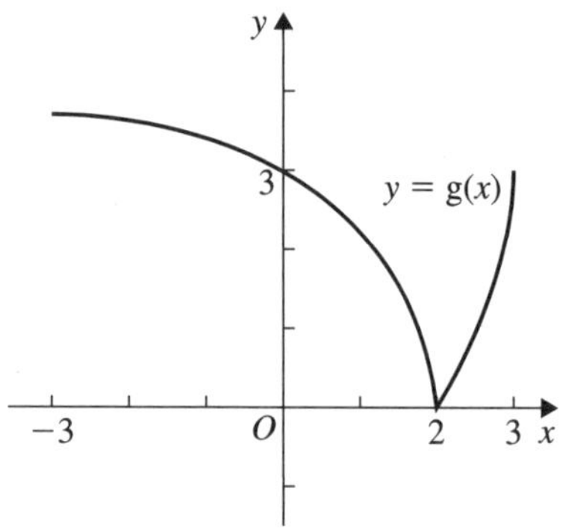

Fig. 2

Each of the functions f and g has domain $-3 \leqslant x \leqslant 3$. Fig. 1 shows the curve $y = \mathrm{f}(x)$, which meets the x-axis at $(-2, 0)$ and the y-axis at $(0, 4)$. Fig. 2 shows the curve $y = \mathrm{g}(x)$, which meets the x-axis at $(2, 0)$ and the y-axis at $(0, 3)$.

i Determine the value of $\mathrm{gf}(-2)$.

ii State which one of the functions f and g is one–one and sketch the graph of its inverse. You should indicate on your sketch the coordinates of intersections with the axes.

iii Given that $\mathrm{g}(x) = |\mathrm{h}(x)|$, where $\mathrm{g}(x) \neq \mathrm{h}(x)$, sketch the graph of a possible curve $y = \mathrm{h}(x)$. [OCR]

12 The functions f and g are defined by

$$\mathrm{f}: x \mapsto x^2 - 2x + 3,\ x \in \mathbb{R},\ 0 \leqslant x \leqslant 4,$$
$$\mathrm{g}: x \mapsto \lambda x^2 + 1, \text{ where } \lambda \text{ is a constant, } x \in \mathbb{R}.$$

a Find the range of f.

b Given that $\mathrm{gf}(2) = 16$, find the value of λ. [EDEXCEL]

13 The function f has domain $x \geqslant 4$ and is defined by $\mathrm{f}(x) = (x - 3)^2 + 1$.

a i Find the value of f(4) and sketch the graph of $y = \mathrm{f}(x)$.

ii Hence find the range of f.

b Explain why the equation $\mathrm{f}(x) = 1$ has no solution.

c The inverse function of f is f^{-1}. Find $\mathrm{f}^{-1}(x)$. [AQA]

14 Figure 3 shows a sketch of the curve with equation $y = \mathrm{f}(x)$, $-1 \leqslant x \leqslant 3$. The curve touches the x-axis at the origin O, crosses the x-axis at the point $A(2, 0)$ and has a maximum at the point $B(\frac{4}{3}, 1)$.

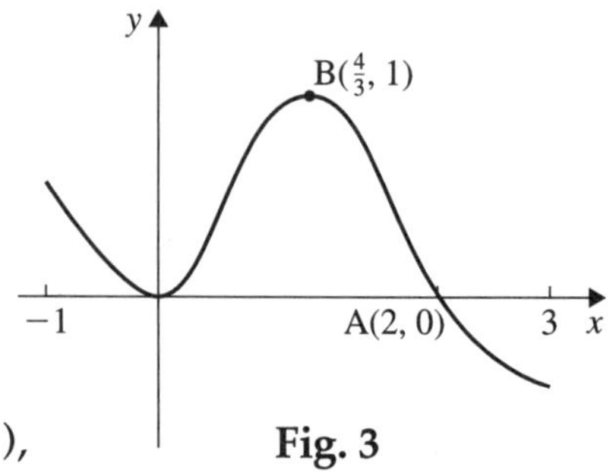

Fig. 3

In separate diagrams, show a sketch of the curve with equation

a $y = \mathrm{f}(x + 1)$, **b** $y = |\mathrm{f}(x)|$, **c** $y = \mathrm{f}(|x|)$,

marking on each sketch the coordinates of points at which the curve

i has a turning point,

ii meets the x-axis. [EDEXCEL]

15 Sketch the graph of the function $y = |2x - 1|$, and state the coordinates of any points where the graph meets the axes. [MEI]

16 Solve the equation $|2x + 1| = 6$. [MEI]

17 Solve the inequality $|3x - 2| > 4$. [MEI]

18 The functions f and g are defined as follows:

$$\text{f}: x \mapsto 5e^{\frac{1}{2}x} - 6, x \in \mathbb{R},$$
$$\text{g}: x \mapsto |x - 2|, x \in \mathbb{R}.$$

i Determine the range of f and the range of g.

ii Find an expression for $\text{f}^{-1}(x)$.

iii Find the solutions of the equation $\text{gf}(x) = 7$, giving each in an exact form. [OCR]

19 The function f is defined for all real values of x by

$$\text{f}(x) = |2x - 3| - 1.$$

a Sketch the graph of $y = \text{f}(x)$. Indicate the coordinates of the points where the graph crosses the x-axis and the coordinates of the point where the graph crosses the y-axis.

b State the range of f.

c Find the values of x for which $\text{f}(x) = x$. [AQA]

20 Find the exact solution of the equation

$$|7x - 3| = |7x + 6|.$$ [OCR]

21 Describe, in each of the following cases, a single transformation which maps the graph of $y = e^x$ onto the graph of the function given.

a $y = e^{3x}$.

b $y = e^{x-3}$.

c $y = \ln x$. [AQA]

22 i Given that $\text{f}(x) = x^2$, sketch the graph of $y = \text{f}(x)$.

The graph of $y = \text{g}(x)$ is obtained by reflecting the graph of $y = \text{f}(x)$ in the x-axis. The graph of $y = \text{h}(x)$ is obtained by translating the graph of $y = \text{g}(x)$ by $+2$ units parallel to the y-axis.

ii Sketch and label the graphs of $y = \text{g}(x)$ and $y = \text{h}(x)$ on a single diagram.

iii Write down expressions for $\text{g}(x)$ and $\text{h}(x)$ in terms of x. [OCR]

23 The diagram shows the graphs of $y = x$ and $y = \mathrm{f}(x)$.

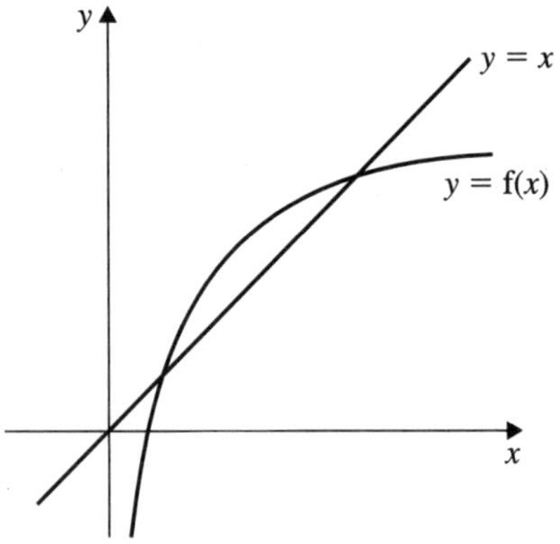

a **i** Describe the geometrical transformation by which the graph of $y = \mathrm{f}^{-1}(x)$ can be obtained from the graph of $y = \mathrm{f}(x)$.

ii Copy the diagram and sketch on the same axes the graph of

$$y = \mathrm{f}^{-1}(x).$$

b The function f is defined for $x > 0$ by

$$\mathrm{f}(x) = 3 \ln x.$$

i Describe the geometrical transformation by which the graph of $y = \mathrm{f}(x)$ can be obtained from the graph of $y = \ln x$.

ii Find an expression for $\mathrm{f}^{-1}(x)$. [AQA]

Part 2

Trigonometry

2.1 Secant, cosecant and cotangent

- In addition to sine, cosine and tangent, it is helpful to define three further trigonometric ratios.

$$\sec x = \frac{1}{\cos x}$$

$$\operatorname{cosec} x = \frac{1}{\sin x}$$

$$\cot x = \frac{1}{\tan x}\left(= \frac{\cos x}{\sin x}\right)$$

> Hint: To remember which ratio you have, look at the *third* letter:
>
> | sec | cosec | cot |
> |---|---|---|
> | ↑ | ↑ | ↑ |
> | cos | sin | tan |

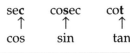

Notice that 'sec x' is written for 'secant x' and so on.

To draw the graphs of $\sec x$, $\operatorname{cosec} x$ and $\cot x$ we refer to the graphs of $\cos x$, $\sin x$ and $\tan x$ respectively. It is not worth remembering what they look like as it is easier to work from $\cos x$, $\sin x$ or $\tan x$ which you should remember. Here is the graph of $\sin x$ and $\operatorname{cosec} x$ for $-360° \leqslant x \leqslant 360°$.

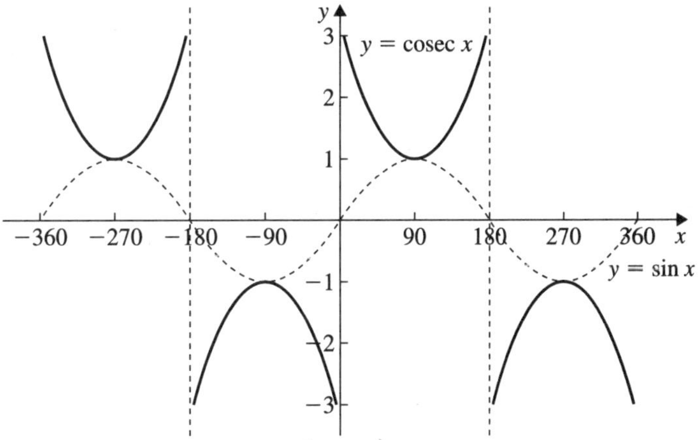

- We are already familiar with the identity $\cos^2\theta + \sin^2\theta \equiv 1$.

If we divide both sides by $\cos^2\theta$ we get $\dfrac{\cos^2\theta}{\cos^2\theta} + \dfrac{\sin^2\theta}{\cos^2\theta} \equiv \dfrac{1}{\cos^2\theta}$

and hence we see that:

$$\boxed{1 + \tan^2\theta \equiv \sec^2\theta}$$

If we divide both sides by $\sin^2\theta$ we get $\dfrac{\cos^2\theta}{\sin^2\theta} + \dfrac{\sin^2\theta}{\sin^2\theta} \equiv \dfrac{1}{\sin^2\theta}$

and hence we see that:

$$\boxed{1 + \cot^2\theta \equiv \operatorname{cosec}^2\theta}$$

Example 1

Evaluate **a)** sec 60° **b)** cot 40°

a) $\sec 60° = \dfrac{1}{\cos 60°} = \dfrac{1}{\frac{1}{2}} = 2$

b) $\cot 40° = \dfrac{1}{\tan 40°} = 1.192$ (3 dp)

Example 2

Solve the equation $\operatorname{cosec} \theta = \sqrt{2}$ for $0 \leqslant \theta \leqslant 360°$.

$\operatorname{cosec} \theta = \sqrt{2}$

$\dfrac{1}{\sin \theta} = \sqrt{2} \Rightarrow \sin \theta = \dfrac{1}{\sqrt{2}}$

$\theta = 45°, 135°$

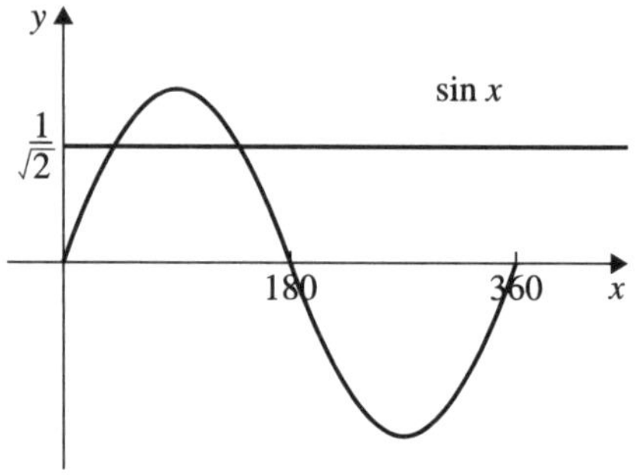

Example 3

Solve the equation $\cot x = \sqrt{3}$ for $0 \leqslant x \leqslant 2\pi$.

$\cot x = \sqrt{3}$

$\dfrac{1}{\tan x} = \sqrt{3} \Rightarrow \tan x = \dfrac{1}{\sqrt{3}}$

From the graph of $y = \tan x$, we obtain

$x = \dfrac{\pi}{6}$ or $\dfrac{7\pi}{6}$.

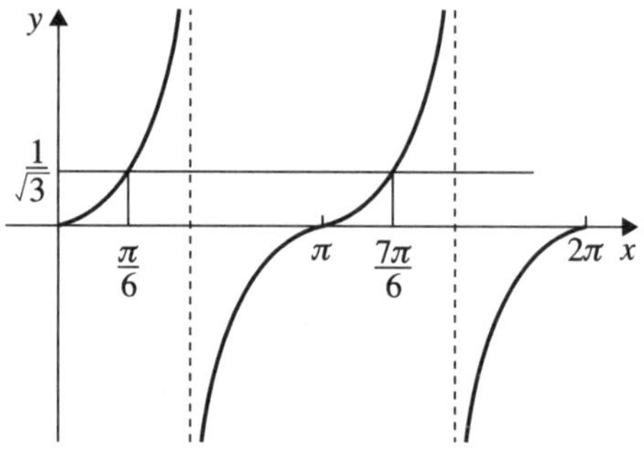

In Examples 2 & 3 notice that we choose to work with one of the 'primary' ratios: $\sin \theta$, $\cos \theta$ or $\tan \theta$ as appropriate.

Example 4

Prove the identity $\sec^2 \theta - \cos^2 \theta = \tan^2 \theta + \sin^2 \theta$.

Consider the left hand side (L.H.S.):

$$
\begin{aligned}
\text{L.H.S.} &= \sec^2 \theta - \cos^2 \theta \\
&= (1 + \tan^2 \theta) - (1 - \sin^2 \theta) \\
&= \tan^2 \theta + \sin^2 \theta \text{ as required.}
\end{aligned}
$$

When proving a trigonometric identity:

1. Do *not* assume that the identity to be proved, is in fact true.
2. Start with one side, generally the more complicated, and simplify it until you can show that both sides are identical.

3 The identity sign ≡ is sometimes used but normally we use an ordinary '=' sign when proving identities. Remember that an identity is true for *all* values of the variable.

For example $\tan\theta \equiv \dfrac{\sin\theta}{\cos\theta}$ for all values of θ.

Example 5

Prove the identity $\dfrac{\cos^2\theta}{1-\sin\theta} = 1 + \sin\theta$.

$$\text{L.H.S.} = \frac{\cos^2\theta}{1-\sin\theta} = \frac{1-\sin^2\theta}{1-\sin\theta}$$

$$= \frac{(1-\sin\theta)(1+\sin\theta)}{(1-\sin\theta)} \quad \text{[Difference of two squares.]}$$

$$= 1 + \sin\theta \text{ as required.}$$

Example 6

Prove the identity $\dfrac{\cos\theta}{1+\cos\theta} = \cot\theta(\operatorname{cosec}\theta - \cot\theta)$.

In this example we could start with either side.

$$\text{L.H.S.} = \frac{\cos\theta}{1+\cos\theta} = \frac{\cos\theta(1-\cos\theta)}{(1+\cos\theta)(1-\cos\theta)} \quad \text{[Multiply numerator and denominator by } (1-\cos\theta).]$$

$$= \frac{\cos\theta(1-\cos\theta)}{1-\cos^2\theta}$$

$$= \frac{\cos\theta(1-\cos\theta)}{\sin^2\theta}$$

$$= \frac{\cos\theta}{\sin\theta}\left(\frac{1-\cos\theta}{\sin\theta}\right)$$

$$= \cot\theta(\operatorname{cosec}\theta - \cot\theta) \text{ as required.}$$

[See for yourself that you can prove this identity just as well by starting with the right hand side.]

EXERCISE 1

1 Evaluate the following. Give answers correct to 3 dp where necessary.

a $\cot 30°$ **b** $\operatorname{cosec} 60°$

c $\sec 10°$ **d** $\operatorname{cosec} 45°$

e $\cot 72°$ **f** $\sec^2 30°$

2 Simplify the expressions.

a $1 + \tan^2 x$ **b** $1 + \tan^2 2x$

c $1 - \cos^2 x$ **d** $\operatorname{cosec} x \tan x$

e $\cot\theta(1 - \cos^2\theta)$ **f** $1 + \cot^2\theta$

3 Prove the following identities.

a $(\sin\theta + \cos\theta)^2 - 2\sin\theta\cos\theta = 1$

b $\tan\theta \operatorname{cosec}\theta = \sec\theta$

c $\operatorname{cosec}\theta(1 - \cos\theta)(1 + \cos\theta) = \sin\theta$

d $\cot\theta \sec\theta \tan\theta\sqrt{(1 - \sin^2\theta)} = 1$

e $\dfrac{1}{\sec^2\theta} + \dfrac{1}{\operatorname{cosec}^2\theta} = 1$

f $\sec\theta - \cos\theta = \sin\theta\tan\theta$

g $\sin^2\theta + 2\cos^2\theta = 2 - \sin^2\theta$

h $\tan\theta + \cot\theta = \dfrac{1}{\sin\theta\cos\theta}$

4 Solve the equations for $0 \leqslant x \leqslant 360°$.

a $\sec x = 2$

b $\cot x = \sqrt{3}$

c $\operatorname{cosec} x = \sqrt{2}$

d $\sec x = 1.2$

e $\cot x = 3$

f $\operatorname{cosec} x = 1$

5 Prove the identities.

a $\operatorname{cosec}\theta - \sin\theta = \cot\theta\cos\theta$

b $\cos^2\theta - \sin^2\theta = 2\cos^2\theta - 1$

c $\operatorname{cosec}^2\theta + \sec^2\theta = \operatorname{cosec}^2\theta\sec^2\theta$

d $\dfrac{\sin^2\theta}{1 - \cos\theta} = 1 + \cos\theta$

e $\dfrac{\sin\theta}{1 + \sin\theta} = \tan\theta(\sec\theta - \tan\theta)$

f $\operatorname{cosec}^2\theta(\tan^2\theta - \sin^2\theta) = \tan^2\theta$

g $\dfrac{\sec\theta}{\tan\theta + \cot\theta} = \sin\theta$

h $\dfrac{\cos\theta}{\sin\theta + 1} + \dfrac{\sin\theta + 1}{\cos\theta} = 2\sec\theta$

6 Sketch the curve $y = \sec x$ for $0 \leqslant x < 90°$.

7 Sketch the curve $y = \operatorname{cosec} x$ for $0 \leqslant x \leqslant 180°$.

8 Solve the equations.

a $2\cot^2\theta - 3\cot\theta + 1 = 0$ for $0 \leqslant \theta \leqslant 360°$

b $\sec^2\theta - \tan\theta = 1$ for $-180° \leqslant \theta \leqslant 180°$

c $\cot^2\theta - 3\operatorname{cosec}\theta + 3 = 0 \quad 0 \leqslant \theta \leqslant 180°$

9 Prove the identities.

a $\sin^4 x - \cos^4 x = \sin^2 x - \cos^2 x$

b $\sec^4 x - \tan^4 x = 1 + 2\tan^2 x$

c $\dfrac{\sin\theta}{1 - \cos\theta} + \dfrac{\sin\theta}{1 + \cos\theta} = 2\operatorname{cosec}\theta$

d $\dfrac{\cot^2\theta}{1 + \cot^2\theta} = \cos^2\theta$

e $\dfrac{1 - \tan^2\theta}{1 + \tan^2\theta} = 1 - 2\sin^2\theta$

f $\dfrac{2\tan\theta}{1 + \tan^2\theta} = 2\sin\theta\cos\theta$

g $\dfrac{\cot\theta + \tan\theta}{\operatorname{cosec}\theta + \sec\theta} = \dfrac{1}{\cos\theta + \sin\theta}$

h $\sin^3 x + \cos^3 x = (\sin x + \cos x)(1 - \sin x\cos x)$

10 Solve the inequality cosec $\theta < 2$ for values of θ between 0 and 180°.

11 **a** Simplify the expression $(1 - \sin\theta)(1 + \sin\theta)\sec\theta$.

b Solve the inequality $(1 - \sin\theta)(1 + \sin\theta)\sec\theta > \frac{1}{2}$ for values of θ between $-90°$ and $90°$.

12 Solve the equations

a $\sec 2x = 3$ for $0 \leqslant x \leqslant 360°$

b $3\,\text{cosec}^2\,2x = 4$ for $0 \leqslant x \leqslant 180°$

13 Solve the simultaneous equations

$$\text{cosec}\,2x = \sqrt{2}$$
$$\cot(x + y) = \sqrt{3}$$

for values of x and y from 0° to 180°.

2.2 Inverse trigonometric functions (arcsin *x*, arccos *x* and arctan *x*)

We know from the work with functions that only one-to-one functions have inverses. We also know that sin, cos and tan are all many-to-one functions unless we restrict the domain.

By considering the graph of $y = \sin x$ we can see that it is one-to-one if we restrict the domain to $-\frac{\pi}{2} \leqslant x \leqslant \frac{\pi}{2}$.
If we restrict the domain in this way then there is an inverse, called arcsin. We write $x = \arcsin y$ or $x = \sin^{-1} y$.

The graph of $y = \arcsin x$ is obtained by reflecting $y = \sin x$ about the line $y = x$. We have the following graphs.

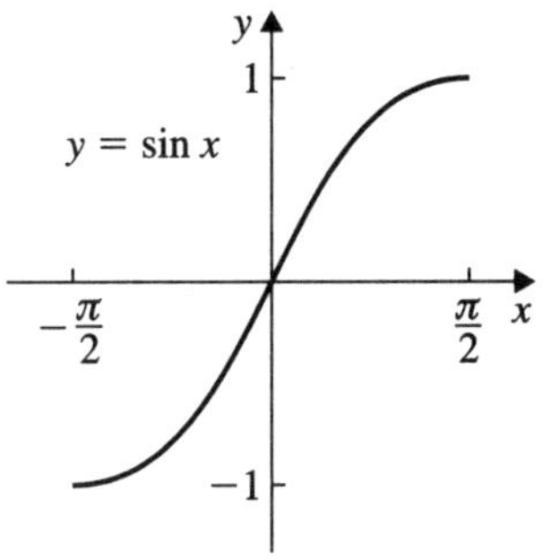

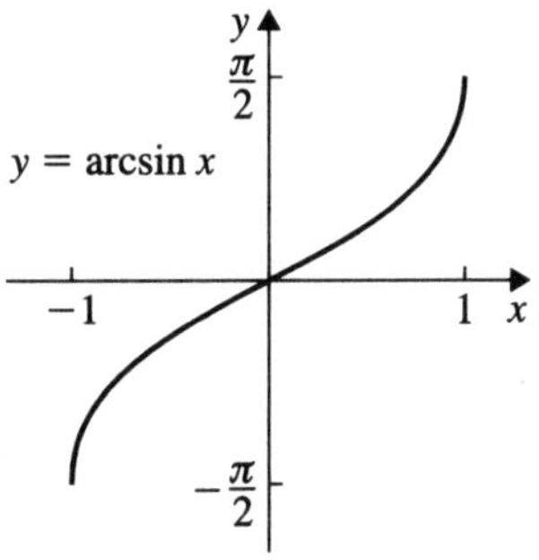

The domain of $y = \arcsin x$ is $-1 \leqslant x \leqslant 1$ and its range is $-\frac{\pi}{2} \leqslant \arcsin x \leqslant \frac{\pi}{2}$
By considering the graph of $y = \cos x$ we can see that it is one-to-one if we restrict the domain to $0 \leqslant x \leqslant \pi$.
If we restrict the domain in this way then there is an inverse, called arccos. We write $x = \arccos y$ or $x = \cos^{-1} y$.

The graph of $y = \arccos x$ is obtained by reflecting $y = \cos x$ about the line $y = x$. We have the following graphs.

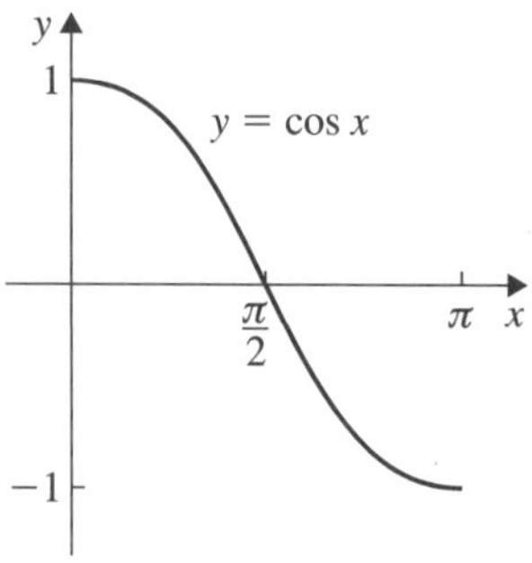

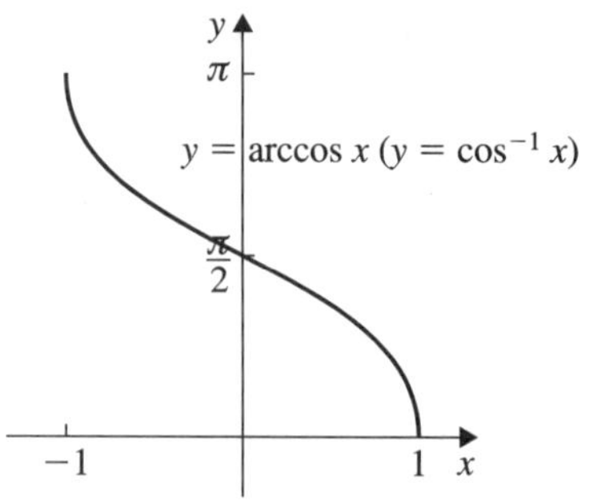

The domain of $y = \arccos x$ is $-1 \leqslant x \leqslant 1$ and its range is $0 \leqslant \arccos x \leqslant \pi$

By considering the graph of $y = \tan x$ we can see that it is one-to-one if we restrict the domain to $-\dfrac{\pi}{2} < x < \dfrac{\pi}{2}$.

If we restrict the domain in this way then there is an inverse, called arctan. We write $x = \arctan y$.

The graph of $y = \arctan x$ is obtained by reflecting $y = \tan x$ about the line $y = x$. We have the following graphs.

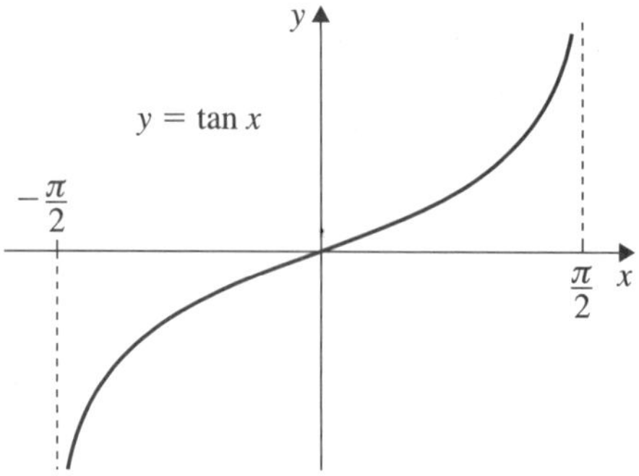

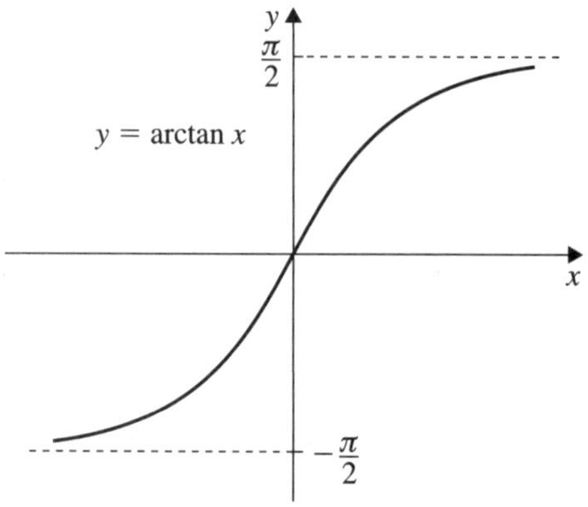

The domain of $y = \arctan x$ is the real numbers and its range is $-\dfrac{\pi}{2} < \arctan x < \dfrac{\pi}{2}$

Note that an alternative notation for $\arcsin x$, $\arccos x$, $\arctan x$ is $\sin^{-1} x$, $\cos^{-1} x$, $\tan^{-1} x$ respectively.

Example 1

a Find, in degrees, the value of $\arcsin\left(\dfrac{1}{\sqrt{2}}\right)$

b Find, in radians in terms of π, the value of $\arctan(-\sqrt{3})$

a $\arcsin\left(\dfrac{1}{\sqrt{2}}\right)$ means 'the angle whose sine is $\dfrac{1}{\sqrt{2}}$'

We know that $\sin 45° = \dfrac{1}{\sqrt{2}}$.

$\therefore \quad \arcsin\left(\dfrac{1}{\sqrt{2}}\right) = 45°$

b $\arctan(-\sqrt{3})$ means 'the angle whose tan is $-\sqrt{3}$'

We know that $\tan\left(-\dfrac{\pi}{3}\right) = -\sqrt{3}$.

$\therefore \quad \arctan(-\sqrt{3}) = -\dfrac{\pi}{3}$

EXERCISE 2

In Questions 1 and 2 do not use a calculator.

1 Evaluate the following in degrees.

a $\arcsin \frac{1}{2}$ **b** $\arccos 0$ **c** $\arctan 1$

d $\arccos(-\frac{1}{2})$ **e** $\arctan\left(-\dfrac{1}{\sqrt{3}}\right)$ **f** $\arcsin\left(\dfrac{\sqrt{3}}{2}\right)$

2 Evaluate the following in radians in terms of π.

a $\arctan(-1)$ **b** $\arccos(\frac{1}{2})$ **c** $\arcsin\left(\dfrac{1}{\sqrt{2}}\right)$

d $\arccos(1)$ **e** $\arctan(\sqrt{3})$ **f** $\arcsin(-1)$

3 Sketch the graph of $f(x) = \arcsin x$ for $-1 \leqslant x \leqslant 1$.

4 Sketch the graph of $f(x) = \arccos x$ for $-1 \leqslant x \leqslant 1$.

5 Solve the equations, for $0 \leqslant x \leqslant 360°$.

a $\sin x = \cos(\arcsin \frac{1}{2})$

b $\tan x = 2\sin\left(\arccos \dfrac{\sqrt{3}}{2}\right)$

c $\cos x = \sin\left(\arctan \dfrac{1}{\sqrt{3}} + \arccos \frac{1}{2}\right)$

6 Simplify the following.

a $\sin(\arcsin \frac{1}{2})$ **b** $\tan\left(\arccos \dfrac{1}{\sqrt{2}}\right) - \sin(\arccos 0)$

c $\cos(\arccos \theta)$ **d** $3\theta - 2\tan(\arctan \theta)$

2.3 Compound angles; sin, cos and tan of (A ± B)

In the diagram $\hat{A} = P\hat{R}X$

$= 90° - X\hat{R}Q$

$= R\hat{Q}X$

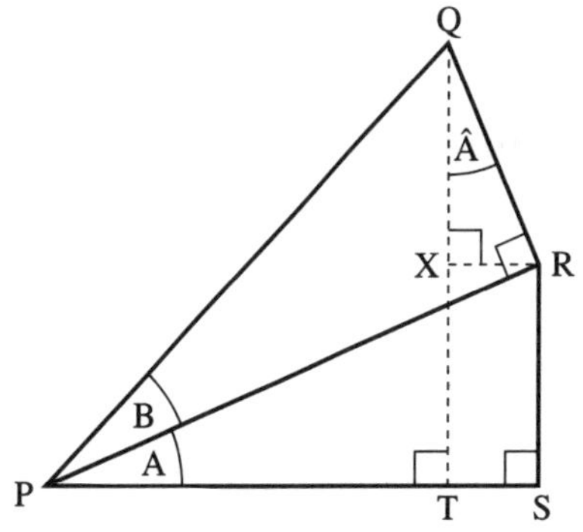

$$\sin(A+B) = \frac{QT}{PQ} = \frac{TX + XQ}{PQ}$$

$$= \frac{TX}{PQ} + \frac{XQ}{PQ} = \frac{RS}{PQ} + \frac{XQ}{PQ}$$

$$= \frac{RS}{PR} \times \frac{PR}{PQ} + \frac{XQ}{QR} \times \frac{QR}{PQ}$$

$$\mathbf{\sin(A + B) = \sin A \cos B + \cos A \sin B} \quad [1]$$

$$\cos(A + B) = \frac{PT}{PQ} = \frac{PS - TS}{PQ} = \frac{PS - RX}{PQ}$$
$$= \frac{PS}{PR} \times \frac{PR}{PQ} - \frac{RX}{QR} \times \frac{QR}{PQ}$$

$$\mathbf{\cos(A + B) = \cos A \cos B - \sin A \sin B} \quad [2]$$

By replacing B by $-B$ in [1] and [2], we obtain:

$$\sin(A - B) = \sin A \cos(-B) + \cos A \sin(-B)$$
$$\mathbf{\sin(A - B) = \sin A \cos B - \cos A \sin B} \quad [3]$$

and $\cos(A - B) = \cos A \cos(-B) - \sin A \sin(-B)$

$$\mathbf{\cos(A - B) = \cos A \cos B + \sin A \sin B} \quad [4]$$

Finally $\tan(A + B) = \dfrac{\sin(A + B)}{\cos(A + B)} = \dfrac{\sin A \cos B + \cos A \sin B}{\cos A \cos B - \sin A \sin B}$

$$= \frac{\dfrac{\sin A \cos B}{\cos A \cos B} + \dfrac{\cos A \sin B}{\cos A \cos B}}{\dfrac{\cos A \cos B}{\cos A \cos B} - \dfrac{\sin A \sin B}{\cos A \cos B}}$$ [Dividing each term by $\cos A \cos B$]

$$\mathbf{\tan(A + B) = \frac{\tan A + \tan B}{1 - \tan A \tan B}}$$

And similarly $\mathbf{\tan A - B = \dfrac{\tan A - \tan B}{1 + \tan A \tan B}}$

In summary:

$$\sin(A \pm B) \equiv \sin A \cos B \pm \cos A \sin B$$
$$\cos(A \pm B) \equiv \cos A \cos B \mp \sin A \sin B$$
$$\tan(A \pm B) \equiv \frac{\tan A \pm \tan B}{1 \mp \tan A \tan B};$$

Example 1

Answer 'true' or 'false': $\sin(30° + 30°) = \sin 30° + \sin 30°$.

False!! This is sadly a common error. If you are still not convinced:

$$\sin(30 + 30) = \sin 60 = \frac{\sqrt{3}}{2}$$

$$\sin 30 + \sin 30 = \tfrac{1}{2} + \tfrac{1}{2} = 1$$

Example 2

Find the exact values of

a $\sin 75°$ **b** $\tan 15°$

a
$$\begin{aligned}\sin 75° &= \sin(45° + 30°)\\ &= \sin 45° \cos 30° + \cos 45° \sin 30°\\ &= \left(\frac{1}{\sqrt{2}} \times \frac{\sqrt{3}}{2}\right) + \left(\frac{1}{\sqrt{2}} \times \frac{1}{2}\right)\\ &= \frac{\sqrt{3}+1}{2\sqrt{2}}\\ &= \frac{(\sqrt{3}+1)\sqrt{2}}{2 \times \sqrt{2} \times \sqrt{2}}\\ &= \frac{\sqrt{6}+\sqrt{2}}{4}\end{aligned}$$

b
$$\begin{aligned}\tan 15° &= \tan(60° - 45°)\\ &= \frac{\tan 60° - \tan 45°}{1 + \tan 60° \tan 45°}\\ &= \frac{\sqrt{3}-1}{1+\sqrt{3}}\\ &= \frac{(\sqrt{3}-1)(1-\sqrt{3})}{(1+\sqrt{3})(1-\sqrt{3})}\\ &= \frac{-4+2\sqrt{3}}{1-3}\\ &= 2-\sqrt{3}\end{aligned}$$

Example 3

Given that A and B are acute angles with $\sin A = \frac{3}{5}$ and $\cos B = \frac{12}{13}$, find the exact value of $\cos(A - B)$.

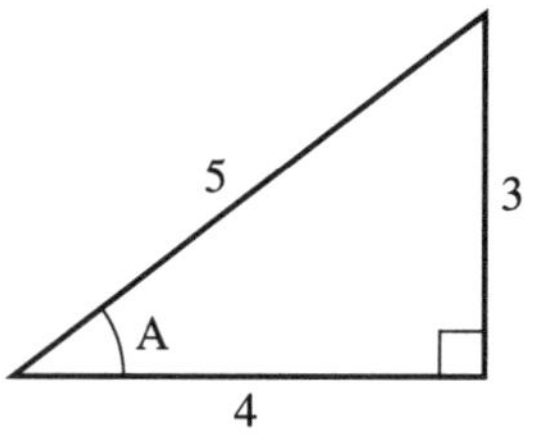

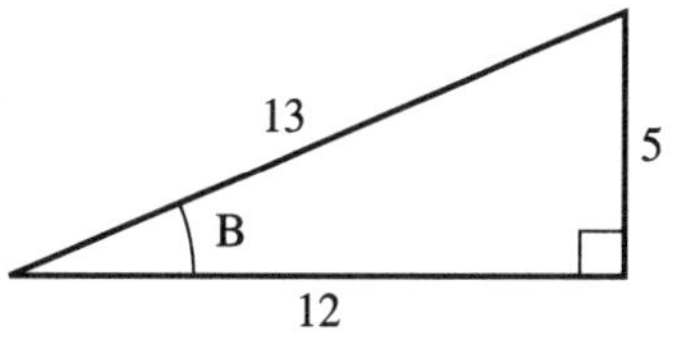

$\sin A = \frac{3}{5}$, $\cos A = \frac{4}{5}$ $\cos B = \frac{12}{13}$, $\sin B = \frac{5}{13}$

$$\begin{aligned}\cos(A - B) &= \cos A \cos B + \sin A \sin B\\ &= \tfrac{4}{5} \times \tfrac{12}{13} + \tfrac{3}{5} \times \tfrac{5}{13}\\ &= \tfrac{63}{65}\end{aligned}$$

Example 4

Solve the equation
$\sin\theta \cos 10° + \cos\theta \sin 10° = 0.8$,
for $0 \leqslant \theta \leqslant 360°$.

$$\sin\theta \cos 10° + \cos\theta \sin 10° = 0.8$$

which is $\sin(\theta + 10°) = 0.8$

$$\theta + 10° = 53.1° \text{ or } 126.9°$$
$$\theta = 43.1° \text{ or } 116.9°$$

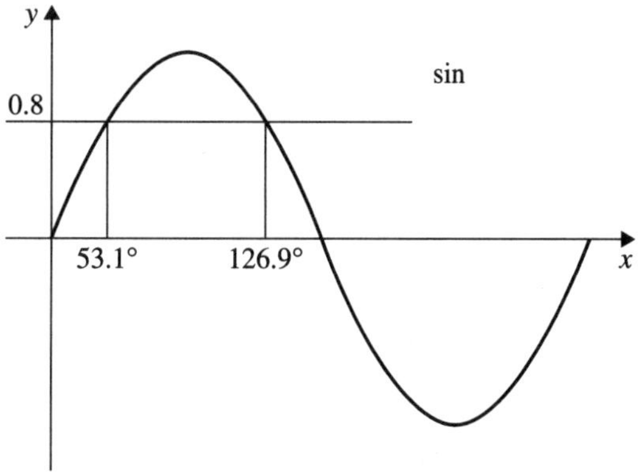

EXERCISE 3

1 **a** Write down the formula for $\sin(A + B)$.

b Evaluate $\sin 80° \cos 10° + \cos 80° \sin 10°$

2 **a** Write down the formula for $\cos(A + B)$.

b Evaluate $\cos 50° \cos 40° - \sin 50° \sin 40°$

3 Answer 'true' or 'false':

a $\tan(A + 45°) = \tan A + 1$ **b** $\tan(A + 45°) = \dfrac{\tan A + 1}{1 - \tan A}$

4 Simplify the following

a $\sin \theta \cos 2\theta + \cos \theta \sin 2\theta$ **b** $\cos 3A \cos A + \sin 3A \sin A$

c $\sin 5x \cos 2x - \cos 5x \sin 2x$ **d** $\cos B \cos B + \sin B \sin B$

5 Find the exact values of the following (using surds):

a $\sin(60° + 45°)$ **b** $\sin 105°$ **c** $\cos 75°$

d $\cos 15°$ **e** $\tan(-15°)$

6 Given that A and B are acute angles such that $\sin A = \frac{5}{13}$ and $\cos B = \frac{3}{5}$, find the exact values of:

a $\sin(A + B)$ **b** $\cos(A - B)$ **c** $\tan(A + B)$

7 Solve the equation $\sin(x + 60°) + \cos(x + 30°) = \frac{1}{2}$ in the range $0° \leqslant x < 360°$.

8 Express the following as single trigonometric expressions:

a $\cos(x + 90°)$ **b** $\sin(x - 90°)$ **c** $\sin(x + 180°)$ **d** $\cos(x - 180°)$

9 Given that $\tan(x + 45°) = 2$, find the value of $\tan x$, without using a calculator.

10 Given that $\tan(A + B) = 3$ and that $\tan B = \frac{1}{2}$, find the exact value of $\tan A$.

11 Solve the equation $\sin(\theta + 45°) = \sqrt{2} \cos \theta$, for $0° \leqslant \theta \leqslant 360°$.

12 Find the greatest value that the following expressions can have and state the value of θ between 0° and 360°, for which these values occur.

a $\sin \theta \cos 40° + \cos \theta \sin 40°$ **b** $\sin \theta \cos 15° - \cos \theta \sin 15°$

13 Prove the following identities:

a $\sin(\theta + 90°) \equiv \cos \theta$ **b** $\sin(A + B) + \sin(A - B) \equiv 2 \sin A \cos B$

c $\cos x \cos 2x + \sin x \sin 2x \equiv \cos x$ **d** $\sin 2x \cos x - \cos 2x \sin x \equiv \sin x$

e $\dfrac{\sin(A + B)}{\sin(A - B)} \equiv \dfrac{\tan A + \tan B}{\tan A - \tan B}$ **f** $\tan A + \tan B \equiv \dfrac{\sin(A + B)}{\cos A \cos B}$

g $\sin(A + B) \sin(A - B) \equiv \sin^2 A - \sin^2 B$

14 Using $\tan(x + \alpha)$ for a suitable value of α, express $\dfrac{1 + \tan x}{1 - \tan x}$ as a single trigonometric expression.

15 a Find an expression for $\tan 2A$ in terms of $\tan A$.

b Given that $\tan \theta = \frac{4}{3}$ and that θ is acute find the exact value of $\tan\left(\dfrac{\theta}{2}\right)$.

16 If $\sin(x - A) = \cos(x + A)$, show that $\tan x = 1$.

17 Find the exact values of the following:

a $\cot 15°$ (in the form $a + \sqrt{b}$)

b $\sec 75°$ (in the form $\sqrt{a} + \sqrt{b}$)

c $\operatorname{cosec} 105°$ (in the form $\sqrt{a} - \sqrt{b}$)

18 Show that $\operatorname{cosec}(A + B) \equiv \dfrac{\sec A \sec B}{\tan A + \tan B}$.

19 Show that $\cot(A + B) \equiv \dfrac{\cot A \cot B - 1}{\cot A + \cot B}$.

20 Prove that if the sum of A, B and C is 180° then $\dfrac{\tan A + \tan B + \tan C}{\tan A \tan B \tan C} \equiv 1$.

21 Show that $\tan 15° = 2 - \sqrt{3}$.

22 Given $\sin(A + B) = k \sin(A - B)$, show that $\tan A = \dfrac{k+1}{k-1} \tan B$.

23 Given that $\tan(A + B) = 1$ and that $\tan A = \frac{1}{2}$, find $\tan B$.

24 a Given that $\sin(x + \alpha) = 2 \sin(x - \alpha)$ show that $\tan x = 3 \tan \alpha$.

b Use the result of part **a** to solve the equation $\sin\left(2y + \dfrac{\pi}{4}\right) = 2 \sin\left(2y - \dfrac{\pi}{4}\right)$ for $0 < y < \pi$.

2.4 The double angle formulae

- We know that $\sin(A + B) = \sin A \cos B + \cos A \sin B$
  Put $B = A$ in this identity.
  $\sin(A + A) = \sin A \cos A + \cos A \sin A$

  or $\mathbf{\sin 2A = 2 \sin A \cos A}$ [1]

- In a similar way we can prove that

  $\mathbf{\cos 2A = \cos^2 A - \sin^2 A}$ [2]

  Using $\sin^2 A + \cos^2 A = 1$, we can show further that

  $\mathbf{\cos 2A = 2 \cos^2 A - 1}$ [2a]
  $\mathbf{\cos 2A = 1 - 2 \sin^2 A}$ [2b]

- Finally $\mathbf{\tan 2A = \dfrac{2 \tan A}{1 - \tan^2 A}}$ [3]

Example 1

Express more simply

a $2 \sin 15° \cos 15°$ **b** $\dfrac{2 \tan 60°}{1 - \tan^2 60°}$ **c** $2 \cos^2 25° - 1$

a From $2 \sin A \cos A = \sin 2A$, we have

$$\begin{aligned} 2 \sin 15° \cos 15° &= \sin 30° \\ &= \tfrac{1}{2} \end{aligned}$$

b From $\dfrac{2 \tan \theta}{1 - \tan^2 \theta} = \tan 2\theta$,

$$\begin{aligned} \frac{2 \tan 60°}{1 - \tan^2 60°} &= \tan 120° \\ &= -\sqrt{3} \end{aligned}$$

c $2 \cos^2 25° - 1 = \cos 50°$

Example 2

You are given that θ is acute and that $\sin \theta = \frac{4}{5}$. Find, without a calculator, the exact values of $\sin 2\theta$ and $\cos 2\theta$.

Draw a right angled triangle and show $\sin \theta = \frac{4}{5}$.

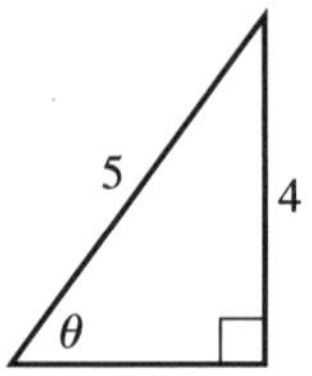

This is a 3, 4, 5 triangle so $\cos \theta = \frac{3}{5}$.

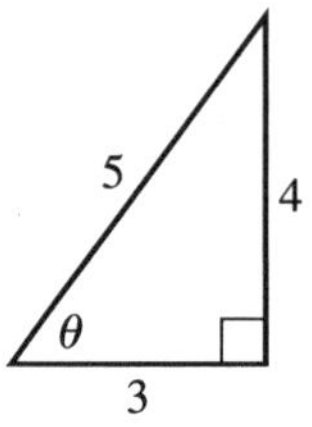

$$\begin{aligned} \text{Now } \sin 2\theta &= 2 \sin \theta \cos \theta \\ &= 2 \times \tfrac{4}{5} \times \tfrac{3}{5} \\ &= \tfrac{24}{25} \end{aligned}$$

$$\begin{aligned} \text{And} \quad \cos 2\theta &= 2 \cos^2 \theta - 1 \\ &= 2 \times \left(\tfrac{3}{5}\right)^2 - 1 \\ &= -\tfrac{7}{25} \end{aligned}$$

Example 3

Solve the equation $\sin 2\theta = \cos \theta$ for $0 \leqslant \theta \leqslant 360°$.

Using $\sin 2\theta = 2 \sin \theta \cos \theta$, we have

$$2 \sin \theta \cos \theta = \cos \theta \qquad [1]$$
$$2 \sin \theta \cos \theta - \cos \theta = 0$$

Factorising, $\cos\theta(2\sin\theta - 1) = 0$

Either $\cos\theta = 0 \quad \Rightarrow \quad \theta = 90°$ or $270°$
Or $\quad 2\sin\theta - 1 = 0$
$\qquad \sin\theta = \frac{1}{2} \quad \Rightarrow \quad \theta = 30°$ or $150°$

So finally $\theta = 30°, 90°, 150°, 270°$

Notice that in line [1] we did not divide through by $\cos\theta$. If we did that we would lose all the solutions we obtain from $\cos\theta = 0$. The correct method is to factorise as above.

Example 4

Prove the identity $\dfrac{\cos 2A}{\cos A + \sin A} = \cos A - \sin A$

Start with the more complicated side and rearrange it until we obtain the expression on the other side. In this case the more complicated side is the left hand side (L.H.S.).

$$\begin{aligned}
\text{L.H.S.} &= \frac{\cos 2A}{\cos A + \sin A} \\
&= \frac{\cos^2 A - \sin^2 A}{\cos A + \sin A} \\
&= \frac{(\cos A - \sin A)(\cos A + \sin A)}{\cos A + \sin A} \qquad \text{[Difference of two squares.]} \\
&= \cos A - \sin A, \text{ as required.}
\end{aligned}$$

Example 5

Prove the identity $\cos 3A = 4\cos^3 A - 3\cos A$

$$\begin{aligned}
\text{Write } \cos 3A &= \cos(2A + A) \\
&= \cos 2A \cos A - \sin 2A \sin A \\
&= (2\cos^2 A - 1)\cos A - (2\sin A\cos A)\sin A \\
&= 2\cos^3 A - \cos A - 2\sin^2 A\cos A \\
&= 2\cos^3 A - \cos A - 2(1 - \cos^2 A)\cos A \\
&= 2\cos^3 A - \cos A - 2\cos A + 2\cos^3 A \\
&= 4\cos^3 A - 3\cos A \text{ as required.}
\end{aligned}$$

EXERCISE 4

1 Evaluate the following, without using a calculator.

a $2\sin 45° \cos 45°$

b $2\cos^2 15° - 1$

c $1 - 2\sin^2 30°$

d $\dfrac{2\tan 22\frac{1}{2}°}{1 - \tan^2 22\frac{1}{2}°}$

2 Express in a more simple form.

a $2\sin 10^\circ \cos 10^\circ$ **b** $2\cos^2 17^\circ - 1$ **c** $1 - 2\sin^2 35^\circ$ **d** $\dfrac{2\tan 11^\circ}{1 - \tan^2 11^\circ}$

e $\sin\theta\cos\theta$ **f** $\dfrac{\tan 2\theta}{1 - \tan^2 2\theta}$ **g** $\dfrac{\sin 2A}{\cos A}$ **h** $2\sin^2 A + \cos 2A$

3 Find the values of $\sin 2\theta$ and $\cos 2\theta$, given that θ is acute and that $\sin\theta = \frac{3}{5}$.

4 Find the values of $\sin 2\theta$ and $\cos 2\theta$, given that θ is acute and that $\cos\theta = \frac{5}{13}$.

5 Find $\tan 2\theta$ given θ is acute and,

a $\tan\theta = \frac{1}{2}$ **b** $\sin\theta = \frac{3}{5}$

6 Find $\sin x$, if $\cos 2x = \frac{1}{8}$

7 Find $\cos x$, if $\cos 2x = \frac{7}{25}$

8 Find the possible values of $\tan\frac{1}{2}\theta$, if $\tan\theta = \frac{3}{4}$

In Questions **9** to **18** solve the equations, for $0 \leqslant \theta \leqslant 360^\circ$.

9 $\sin 2\theta - \sin\theta = 0$

10 $2\sin 2\theta + \cos\theta = 0$

11 $4\cos 2\theta - 6\cos\theta + 5 = 0$

12 $\sin 2\theta = 2\cos 2\theta$

13 $\cos 2\theta + \sin\theta - 1 = 0$

14 $3\cos 2\theta - \cos\theta + 2 = 0$

15 $\tan 2\theta + \tan\theta = 0$

16 $\sin\theta + \sin\dfrac{\theta}{2} = 0$

17 $\sin\dfrac{\theta}{2} = \sin\theta$

18 $2\cos 2x = 5 - 13\sin x$

19 Prove the identity $2\operatorname{cosec} 2\theta = \sec\theta\operatorname{cosec}\theta$.

$\left[\text{Hint: Start with 'L.H.S.} = \dfrac{2}{\sin 2\theta}\text{'}\right]$

20 Prove the identity $\tan 2\theta\sec\theta = 2\sin\theta\sec 2\theta$.

$\left[\text{Hint: start with 'L.H.S.} = \dfrac{\sin 2\theta}{\cos 2\theta} \times \dfrac{1}{\cos\theta}\text{'}\right]$

In Questions **21** to **35** prove the identities.

21 $\dfrac{\cos 2A}{\cos A - \sin A} = \cos A + \sin A$

22 $\dfrac{\sin A}{\sin B} + \dfrac{\cos A}{\cos B} = \dfrac{2\sin(A + B)}{\sin 2B}$

23 $\dfrac{\sin A}{\sin B} - \dfrac{\cos A}{\cos B} = \dfrac{2\sin(A - B)}{\sin 2B}$

24 $\tan A + \cot A = 2\operatorname{cosec} 2A$

25 $\cot A - \tan A = 2\cot 2A$

26 $\dfrac{\sin 2A}{1 + \cos 2A} = \tan A$

27 $\sin 2A = \dfrac{2\tan A}{1+\tan^2 A}$

28 $\cos 2A = \dfrac{1-\tan^2 A}{1+\tan^2 A}$

29 $\sin 3A = 3\sin A - 4\sin^3 A$

30 $\cos 3A = 4\cos^3 A - 3\cos A$

31 $\cot\theta = \dfrac{\sin 2\theta}{1-\cos 2\theta}$

32 $\cos^4\theta - \sin^4\theta = \cos 2\theta$

33 $\cot 2A + \operatorname{cosec} 2A = \cot A$

34 $\sin^4\theta + \cos^4\theta = \frac{1}{4}(\cos 4\theta + 3)$

35 $\tan 3\theta = \dfrac{3\tan\theta - \tan^3\theta}{1-3\tan^2\theta}$

36 Use the double angle formula for cosine to show that $\cos x = 2\cos^2\left(\dfrac{x}{2}\right) - 1$.

37 Express $\sin x$ in terms of $\sin\left(\dfrac{x}{2}\right)$ and $\cos\left(\dfrac{x}{2}\right)$.

38 Express $\tan x$ in terms of $\tan\left(\dfrac{x}{2}\right)$.

39 **a** Show that $\sin^2\left(\dfrac{x}{2}\right) = \dfrac{1}{2}(1-\cos x)$.

b Show that $\cos^2\left(\dfrac{x}{2}\right) = \dfrac{1}{2}(1+\cos x)$.

40 Prove that $\tan 22\frac{1}{2}^\circ = \sqrt{2} - 1$.

41 **a** Show that $\operatorname{cosec} 2x - \cot 2x = \tan x$

b Hence find the exact value of $\tan 75^\circ$, giving your answer in the form $a + b\sqrt{3}$.

42 Solve the equation $4\tan 2\theta = \cot\theta$, for $0^\circ < \theta < 360^\circ$, giving your answers correct to 1 decimal place.

43 Express $8\cos^4\theta$ in the form $A\cos 4\theta + B\cos 2\theta + C$, giving the numerical values of the constants A, B and C.

2.5 The form $a\cos\theta + b\sin\theta$

A special technique is used to express $a\cos\theta + b\sin\theta$ in the form $R\cos(\theta - \alpha)$, where α is an acute angle.

We write $a\cos\theta + b\sin\theta = \sqrt{a^2+b^2}\left[\dfrac{a}{\sqrt{a^2+b^2}}\cos\theta + \dfrac{b}{\sqrt{a^2+b^2}}\sin\theta\right]$

Let $\cos\alpha = \dfrac{a}{\sqrt{a^2+b^2}}$

$\sin\alpha = \dfrac{b}{\sqrt{a^2+b^2}}$

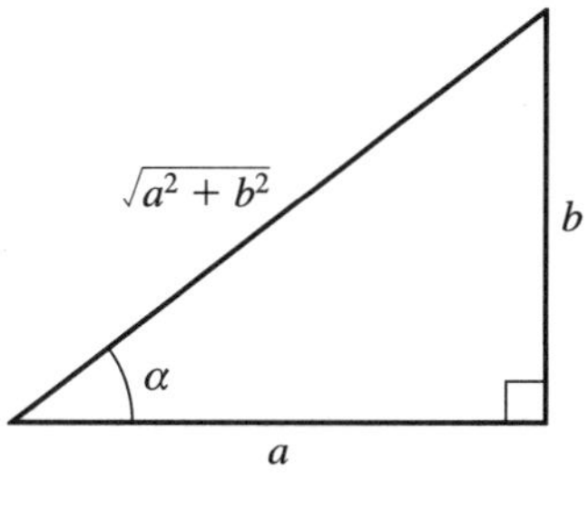

So $a\cos\theta + b\sin\theta = \sqrt{a^2+b^2}\,[\cos\theta\cos\alpha + \sin\theta\sin\alpha]$
$= \sqrt{a^2+b^2}\cos(\theta - \alpha)$

$R = \sqrt{a^2+b^2}$ and the value of α can be obtained from the triangle shown.

The same method, with slight modification, is used for an expression of the form $a\sin\theta - b\cos\theta$ and we can choose to write the expression in the form $R\sin(\theta - \alpha)$ if preferred. The examples below will illustrate the method.

Example 1

Write the expression $f(\theta) = 3\cos\theta + 4\sin\theta$ in the form $R\cos(\theta - \alpha)$.

$$f(\theta) = \sqrt{3^2+4^2}\left[\frac{3}{\sqrt{3^2+4^2}}\cos\theta + \frac{4}{\sqrt{3^2+4^2}}\sin\theta\right]$$
$$= 5(\tfrac{3}{5}\cos\theta + \tfrac{4}{5}\sin\theta)$$

Let $\cos\alpha = \frac{3}{5}$ and $\sin\alpha = \frac{4}{5}$

So $f(\theta) = 5(\cos\alpha\cos\theta + \sin\alpha\sin\theta)$
$= 5\cos(\theta - \alpha)$

Since $\cos\alpha = \frac{3}{5}$, we have $\alpha = \cos^{-1}\frac{3}{5} = 53.1°$ (to 1 decimal place)

Finally $f(\theta) = 5\cos(\theta - 53.1°)$

Example 2

a Write $f(\theta) = 2\sin\theta - 3\cos\theta$ in the form $R\sin(\theta - \alpha)$.

b Hence write down the largest possible value of $f(\theta)$ and state the smallest positive value of θ for which this value occurs.

a $f(\theta) = 2\sin\theta - 3\cos\theta$

$$= \sqrt{2^2+(-3)^2}\left[\frac{2}{\sqrt{2^2+3^2}}\sin\theta - \frac{3}{\sqrt{2^2+3^2}}\cos\theta\right]$$
$$= \sqrt{13}\left[\underset{\substack{\uparrow\\ \cos\alpha}}{\frac{2}{\sqrt{13}}}\sin\theta - \underset{\substack{\uparrow\\ \sin\alpha}}{\frac{3}{\sqrt{13}}}\cos\theta\right]$$

$f(\theta) = \sqrt{13}\sin(\theta - \alpha)$ where $\alpha = \cos^{-1}\dfrac{2}{\sqrt{13}} = 56.3°$ (to 1 d.p.)

$f(\theta) = \sqrt{13}\sin(\theta - 56.3°)$

b The largest value for the sine of any angle is one. The largest value for $f(\theta)$ is therefore $\sqrt{13}$ and this value occurs when $\theta - 56.3° = 90°$, ie when $\theta = 146.3°$ (correct to one decimal place).

Example 3

Solve the equation $4\sin\theta - \cos\theta = 2$, for $0 \leqslant \theta \leqslant 360°$.

Since this is an *equation*, rather than an expression, we divide each term on both sides by $\sqrt{4^2+(-1)^2}$.

$$\frac{4}{\sqrt{17}}\sin\theta - \frac{1}{\sqrt{17}}\cos\theta = \frac{2}{\sqrt{17}}$$

Let $\cos\alpha = \frac{4}{\sqrt{17}}$, $\sin\alpha = \frac{1}{\sqrt{17}}$. $\alpha = 14.04°$ (2 decimal places)

We have $\sin\theta\cos\alpha - \cos\theta\sin\alpha = \frac{2}{\sqrt{17}}$

$$\sin(\theta - 14.04°) = \frac{2}{\sqrt{17}}$$

$\theta - 14.04° = 29.02°, 150.98°$

$\theta = 43.1°$ or $165.0°$

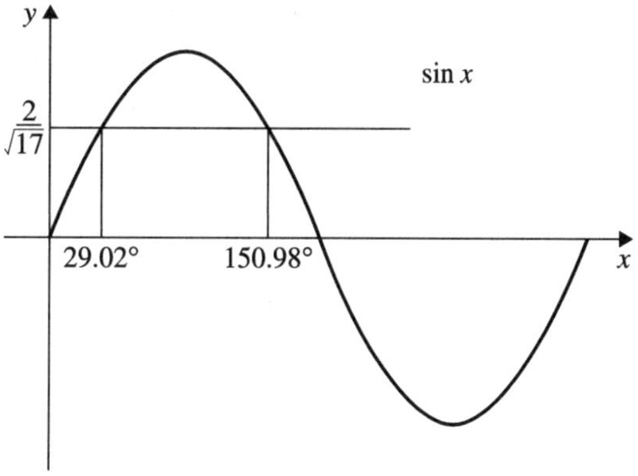

Notice that we chose to work with α correct to 2 decimal places in the working and then gave the final answer correct to 1 decimal place.
You can of course retain the value of α in the memory of your calculator.

Example 4

Solve the equation $3\cos\theta + 2\sin\theta = 1$, for $0 \leqslant \theta \leqslant 2\pi$, giving your answers in radians correct to 3 decimal places.

Divide both sides by $\sqrt{3^2 + 2^2}$.

$$\frac{3}{\sqrt{13}}\cos\theta + \frac{2}{\sqrt{13}}\sin\theta = \frac{1}{\sqrt{13}}$$

Let $\cos\alpha = \frac{3}{\sqrt{13}}$, $\sin\alpha = \frac{2}{\sqrt{13}}$, so $\alpha = \cos^{-1}\frac{3}{\sqrt{13}} = 0.588\,00$ (to 5 d.p.)

We have $\cos\theta\cos\alpha + \sin\theta\sin\alpha = \frac{1}{\sqrt{13}}$

$$\cos(\theta - 0.588\,00) = \frac{1}{\sqrt{13}}$$

From a calculator $\cos^{-1}\frac{1}{\sqrt{13}} = 1.289\,76$ radians (correct to 5 d.p.)

$$\theta - 0.588\,00 = 1.289\,76 \text{ or } (2\pi - 1.289\,76)$$
$$\theta - 0.588\,00 = 1.289\,76, 4.993\,42$$

$\theta = 1.878$ rad or 5.581 rad (correct to 3 decimal places).

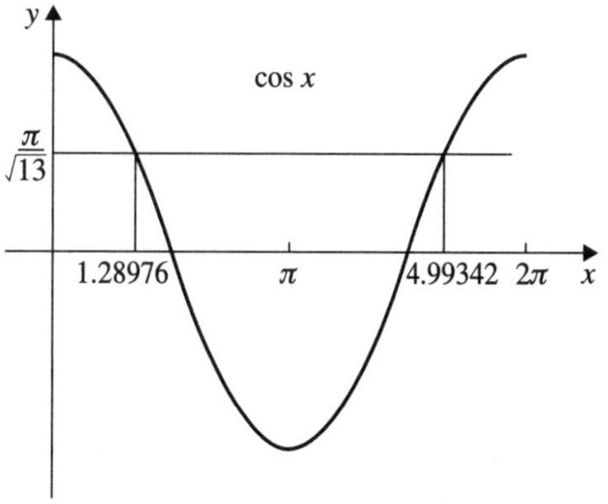

EXERCISE 5

1 Write each expression in the form $R\cos(\theta - \alpha)$, where $0 < \alpha < 90°$.

a $4\cos\theta + 3\sin\theta$ **b** $5\cos\theta + 12\sin\theta$ **c** $\cos\theta + 2\sin\theta$

2 Find the value of R and α in each of the following.

a $3\sin\theta + 4\cos\theta = R\sin(\theta + \alpha)$

b $4\cos\theta - 3\sin\theta = R\cos(\theta + \alpha)$

c $2\cos\theta + \sin\theta = R\cos(\theta - \alpha)$

d $8\sin\theta - 15\cos\theta = R\sin(\theta - \alpha)$

e $8\sin\theta + 6\cos\theta = R\sin(\theta + \alpha)$

f $\cos\theta + \sin\theta = R\cos(\theta - \alpha)$

3 **a** Express $f(\theta) = \sin\theta + \sqrt{3}\cos\theta$ in the form $R\sin(\theta + \alpha)$.

b Find the maximum value of $f(\theta)$ and state the smallest positive value of θ that gives this maximum value.

4 Find the maximum value of each expression and state the smallest positive value of θ that gives this value.

a $3\cos\theta + 4\sin\theta$

b $5\sin\theta - 12\cos\theta$

c $\cos\theta + \sin\theta$

d $8\cos\theta - 15\sin\theta$

5 **a** Express $4\cos\theta + 3\sin\theta$ in the form $R\cos(\theta - \alpha)$, where $R > 0$ and $0 < \alpha < 90°$.

b Solve the equation $4\cos\theta + 3\sin\theta = 2$, for $0 \leqslant \theta \leqslant 360°$.

6 **a** Express $\sin\theta + \sqrt{8}\cos\theta$ in the form $R\sin(\theta + \alpha)$.

b Solve the equation $\sin\theta + \sqrt{8}\cos\theta = 2$, for $0 \leqslant \theta \leqslant 360°$.

7 Solve the following equations, correct to one decimal place, for $0 \leqslant \theta \leqslant 360°$.

a $8\cos\theta + 15\sin\theta = 8.5$

b $5\sin\theta + 12\cos\theta = 7$

c $\sin\theta + \sqrt{3}\cos\theta = 1$

d $4\sin\theta - 3\cos\theta = 2$

e $4\cos\theta - 6\sin\theta = 5$

f $5\sin\theta - 3\cos\theta = 1$

g $8\sin\theta - 24\cos\theta = 3$

h $5\cos 2x + 2\sin 2x = 3$

8 Solve the equations for $0 < \theta < 2\pi$, giving your answers in radians correct to two decimal places.

a $\cos(\theta - 1.2^c) = \frac{1}{2}$

b $\sin(\theta + 0.4^c) = 0.7$

9 **a** Express $2\cos x + \sqrt{3}\sin x$ in the form $R\cos(x - \alpha)$, where $R > 0$ and $0 < \alpha < \dfrac{\pi}{2}$.

b Find the smallest positive root of the equation $2\cos x + \sqrt{3}\sin x = 2$, giving your answer in radians correct to two decimal places.

10 Solve the equations for $0 \leqslant \theta \leqslant 2\pi$, giving your answers in radians correct to two decimal places.

a $3\sin\theta + \cos\theta = 2$

b $\cos\theta - 2\sin\theta = 1$

c $\cos\theta - \sin\theta = 1$

d $3\sin\theta - 4\cos\theta = 3.5$

11 **a** Express $\sin\theta - 3\cos\theta$ in the form $R\sin(\theta - \alpha)°$, giving the values of R and α.

b Explain why the equation $\sin\theta - 3\cos\theta = 4$ has no solutions.

12 Find the range of values of the constant k for which the equation $\sin\theta + \sqrt{2}\cos\theta = k$, has real solutions for θ.

13 You are given $f(\theta) = (\cos\theta - \sin\theta)(17\cos\theta - 7\sin\theta)$

a Show that $f(\theta)$ may be written in the form $5\cos 2\theta - 12\sin 2\theta + c$, where c is a constant. State the value of c.

b Write $5\cos 2\theta - 12\sin 2\theta$ in the form $R\cos(2\theta + \alpha)$, where $0 < \alpha < \frac{\pi}{2}$. State the value of R and find α in radians correct to 3 decimal places.

14 a Express $3\cos\theta - 4\sin\theta$ in the form $R\cos(\theta + \alpha)$.

b Find the greatest possible value of $\dfrac{2}{(3\cos\theta - 4\sin\theta + 6)}$.

15 a Express $f(\theta) = 5\cos\theta + 12\sin\theta$ in the form $R\cos(\theta - \alpha)$.

b Find the smallest possible value of $\dfrac{30}{f(\theta) + 2}$.

16 a Write $8\sin\theta\cos\theta - 6\sin^2\theta$ in the form $R\sin(2\theta + \alpha) + k$.

b Hence solve the equation $8\sin\theta\cos\theta - 6\sin^2\theta = 1$, for $0 < \theta < 180°$.

17 a Express $4\cos x + 3\sin x$ in the form $R\cos(x - \alpha)$.

b Find the greatest and least values of $\dfrac{1}{(4\cos x + 3\sin x + 7)}$.

c Find, in radians correct to 3 decimal places, the solution of the equation $(4\cos x + 3\sin x)^2 = 12.5$, for $0 \leqslant x \leqslant \pi$.

REVIEW EXERCISE 2

1 Solve the following trigonometric equations in the given intervals (to 1 dp where necessary).

a $\sin x = \frac{2}{3}$, for $0 \leqslant x \leqslant 360°$

b $\cos x = -\frac{3}{4}$, for $0 \leqslant x \leqslant 360°$

c $\tan x = -0.7$, for $0 \leqslant x \leqslant 360°$

d $\sin x = -\frac{1}{2}$, for $0 \leqslant x \leqslant 360°$

e $\cos x = \frac{1}{2}$, for $0 \leqslant x \leqslant 2\pi$

f $\tan x = 1$, for $-\pi \leqslant x \leqslant \pi$

g $2\sin^2 x = 3\cos x$, for $0 \leqslant x \leqslant 360°$

h $3\tan x = 2\sin x$, for $0 \leqslant x \leqslant 360°$

2 Solve these equations, giving all values of x to the nearest degree in the interval $0 \leqslant x \leqslant 360°$.

a $2\tan^2 x + \sec x - 1 = 0$

b $\operatorname{cosec}^2 x + \cot x - 2 = 0$

3 Find, in terms of π, the solutions of the equation $\tan^4 x - 4\tan^2 x + 3 = 0$, for $0 \leqslant x \leqslant \pi$.

4 Evaluate the following in degrees.

a $\sin^{-1}\dfrac{1}{\sqrt{2}}$

b $\cos^{-1} 1$

c $\tan^{-1}\sqrt{3}$

5 Evaluate the following in radians, in terms of π.

a $\cos^{-1}\dfrac{\sqrt{3}}{2}$

b $\tan^{-1}\dfrac{1}{\sqrt{3}}$

c $\sin^{-1}\dfrac{1}{2}$

6 Prove the following identities.

a $\operatorname{cosec}\theta \tan\theta \equiv \sec\theta$

b $\operatorname{cosec}\theta - \sin\theta \equiv \cot\theta \cos\theta$

c $(\sin\theta + \operatorname{cosec}\theta)^2 \equiv \sin^2\theta + \cot^2\theta + 3$

d $\sec^2\theta(\cot^2\theta - \cos^2\theta) \equiv \cot^2\theta$

7 Solve the following equations in the range $0° \leqslant \theta \leqslant 360°$.

a $6\sec^2\theta = 5(\tan\theta + 1)$ **b** $3\cot\theta = 2\sin\theta$ **c** $\cot\theta + 3\cos\theta = 0$

8 If $\cos A = \frac{4}{5}$, $\cos B = \frac{5}{13}$ and A and B are acute angles, find the value of $\cos(A + B)$.

9 The angle θ is obtuse and $\sin\theta = \frac{4}{5}$.

a Find the value of $\cos\theta$.

b Find the value of $\sin 2\theta$ and $\cos 2\theta$, giving your answers as fractions in their lowest terms.

10 Prove the following identities.

a $\dfrac{\sin 2\theta}{1 + \cos 2\theta} \equiv \tan\theta$ **b** $\dfrac{1 - \cos 2\theta}{1 + \cos 2\theta} \equiv \tan^2\theta$

c $\dfrac{\sin 2\theta + \sin\theta}{1 + \cos 2\theta + \cos\theta} \equiv \tan\theta$ **d** $\dfrac{1 - \cos 2\theta + \sin 2\theta}{1 + \cos 2\theta + \sin 2\theta} = \tan\theta$

11 Solve the following equations in the range $0° \leqslant x \leqslant 360°$.

a $\cos 2\theta + \cos\theta = 0$ **b** $3\sin 2\theta = \sin\theta$

c $2\cos 2\theta = 3\sin\theta$ **d** $\sin 2\theta = \cos\theta$

e $\cos 2\theta = \cos\theta$ **f** $2\sin 2\theta = 3\tan\theta$

12 If θ is an acute angle such that $\cos\theta = \frac{3}{5}$ then find the exact values of:

a $\cos 2\theta$ **b** $\sin 2\theta$ **c** $\tan 2\theta$

13 **a** Find the exact values of the following (using surds):

i $\sin 15°$ **ii** $\tan 75°$ **iii** $\cos 105°$

b Hence show that $\sin 15° + \cos 105° = 0$.

14 If θ is an acute angle such that $\cos\theta = \frac{5}{13}$ then find the exact values of:

a $\cos 2\theta$ **b** $\sin\left(\dfrac{\theta}{2}\right)$ **c** $\tan 2\theta$

15 If θ is an acute angle such that $\tan\theta = \frac{3}{4}$ then find the exact value of $\tan\left(\dfrac{\theta}{2}\right)$:

16 Solve the equations in the range $0° \leqslant \theta \leqslant 360°$.

a $\cos 2\theta = \sin\theta$ **b** $3\tan\theta = \tan 2\theta$

c $\cos 2\theta + \cos\theta + 1 = 0$ **d** $2\sin\theta = \tan\frac{1}{2}\theta$

e $\tan\theta = 3\tan\frac{1}{2}\theta$

17 Express $\sin 3\theta$ in terms of $\sin \theta$.

18 a Express $\cos x - \sin x$ in the form $R\cos(x + \alpha)$, $R > 0$, $0 < \alpha < \frac{\pi}{2}$. State the value of R in surd form and find the exact value of α in radians.

b Hence find the exact solutions in the interval $0 \leqslant x \leqslant 2\pi$ of the equation $\cos x - \sin x = 1$.

c State the maximum value of $\cos x - \sin x$.

19 a Express $\sin \theta° + \sqrt{2}\cos \theta°$ in the form $R\sin(\theta + \alpha)°$, where $R > 0$ and $0 < \alpha < 90$.

b Solve the equation $\sin \theta + \sqrt{2}\cos \theta = 1$, for $0° < \theta < 360°$.

20 a Express $f(\theta) = \cos \theta - \sqrt{3}\sin \theta$ in the form $R\cos(\theta + \alpha)$, where $R > 0$ and $0 < \alpha < 90°$.

b Hence find the least value of $\dfrac{1}{[f(\theta)]^2}$.

21 a Express $3\cos \theta + 4\sin \theta$ in the form $R\sin(\theta + \alpha)$, where $R > 0$ and $0 < \alpha < 90°$. Give the value of α correct to one decimal place.

b Solve the equation $3\cos \theta + 4\sin \theta = 2$, for $0 < \theta < 360°$, giving each solution to the nearest degree.

EXAMINATION EXERCISE 2

1 Find the solutions of

$$6\tan \theta - \sec^2 \theta = 7$$

in the interval $0 \leqslant \theta < 2\pi$, giving each answer in radians to one decimal place.

[AQA]

2 i Express $\sec^2 \theta + \operatorname{cosec}^2 \theta$ in terms of $\sin \theta$ and $\cos \theta$, giving your answer as a single fraction as simply as possible.

ii Hence prove that

$$\sec^2 \theta + \operatorname{cosec}^2 \theta = 4\operatorname{cosec}^2 2\theta.$$

iii Find the values of θ, for $0 < \theta < \frac{1}{2}\pi$, such that

$$\sec^2 \theta + \operatorname{cosec}^2 \theta = 10.$$

[OCR]

3 i Prove, by counter-example, that the statement

"$\sec(A + B) \equiv \sec A + \sec B$, for all A and B"

is false

ii Prove that

$$\tan \theta + \cot \theta \equiv 2\operatorname{cosec} 2\theta,\ \theta \neq \frac{n\pi}{2},\ n \in \mathbb{Z}.$$

[EDEXCEL]

4 a Prove that

$$\frac{2\tan x}{1+\tan^2 x} \equiv \sin 2x.$$

b Hence or otherwise find the exact value of $\tan 15°$ in the form $a + b\sqrt{3}$, where a and b are integers. [AQA]

5 a Prove that

$$\frac{1-\cos 2\theta}{\sin 2\theta} \equiv \tan\theta, \quad \theta \neq \frac{n\pi}{2}, \quad n \in \mathbb{Z}.$$

b Solve, giving exact answers in terms of π,

$$2(1-\cos 2\theta) = \tan\theta, \quad 0 < \theta < \pi.$$ [EDEXCEL]

6 a Show that $\dfrac{\cot^2\theta}{1+\cot^2\theta} \equiv \cos^2\theta$.

b Hence solve $\dfrac{\cot^2\theta}{1+\cot^2\theta} = 2\sin 2\theta$ for $0° \leqslant \theta \leqslant 360°$. [AQA]

7 Without using a calculator, find the exact value of

$$(\sin 22\tfrac{1}{2}° + \cos 22\tfrac{1}{2}°)^2.$$ [OCR]

8 a Express $6\sin^2\theta$ in the form $a + b\cos 2\theta$, where a and b are constants.

b Find the exact value of $\displaystyle\int_0^{\frac{\pi}{12}} 6\sin^2\theta\,\mathrm{d}\theta$.

c Solve the equation

$$3 - 3\cos 2\theta = 2\operatorname{cosec}\theta$$

giving all solutions in radians in the interval $0 < \theta < 2\pi$. [AQA]

9 i Given that $\sin(\theta + 45°) = 2\sin\theta$, show that

$$\tan\theta = \frac{1}{2\sqrt{2}-1}.$$

ii Hence solve the equation

$$\sin(\theta + 45°) = 2\sin\theta,$$

for values of θ between 0° and 360°, giving your answers correct to the nearest degree. [OCR]

10 a Show that $\tan(45° + \theta) = \dfrac{1+\tan\theta}{1-\tan\theta}$.

b Hence obtain the exact value of $\tan 105°$ in the form $a + b\sqrt{3}$, where a and b are integers to be found. [AQA]

11 **i** Write down the formula for $\tan 2x$ in terms of $\tan x$.

ii By letting $\tan x = t$, show that the equation

$$4 \tan 2x + 3 \cot x \sec^2 x = 0$$

becomes

$$3t^4 - 8t^2 - 3 = 0.$$

iii Find all the solutions of the equation

$$4 \tan 2x + 3 \cot x \sec^2 x = 0$$

which lie in the range $0 \leqslant x \leqslant 2\pi$. [OCR]

12 **a** **i** By expanding $\cos(\frac{1}{4}\pi + \frac{1}{6}\pi)$, express $\cos \frac{5}{12}\pi$ in surd form.

ii Express $\tan(\frac{1}{4}\pi + x)$ in terms of $\tan x$.

iii Hence or otherwise show that

$$\int_0^{\frac{1}{6}\pi} \frac{1 + \tan x}{1 - \tan x} \, dx = \ln\left(\frac{2}{\sqrt{3} - 1}\right).$$

[OCR]

13 **a** Express $\sin x + \sqrt{3} \cos x$ in the form $R \sin(x + \alpha)$, where $R > 0$ and $0 < \alpha < 90°$.

b Show that the equation $\sec x + \sqrt{3} \operatorname{cosec} x = 4$ can be written in the form

$$\sin x + \sqrt{3} \cos x = 2 \sin 2x.$$

c Deduce from parts **a** and **b** that $\sec x + \sqrt{3} \operatorname{cosec} x = 4$ can be written in the form

$$\sin 2x - \sin(x + 60°) = 0.$$

[EDEXCEL]

14 **a** Given that $\tan x \neq 1$, show that

$$\frac{\cos 2x}{\cos x - \sin x} \equiv \cos x + \sin x.$$

b By expressing $\cos x + \sin x$ in the form $R \sin(x + a)$, solve, for $0° \leqslant x \leqslant 360°$,

$$\frac{\cos 2x}{\cos x - \sin x} = \frac{1}{2}.$$

[AQA]

15 The diagram shows a rectangle $OABC$ in which B has coordinates $(\sin\theta, 2\cos\theta)$, where $0 \leqslant \theta \leqslant \frac{\pi}{2}$. The perimeter of the rectangle is of length L.

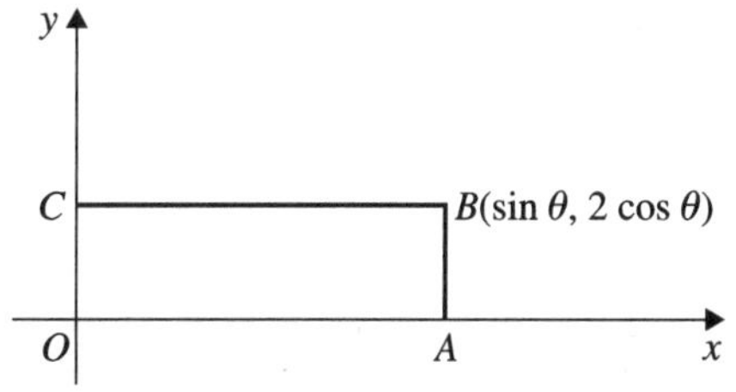

a **i** Write down the length L in terms of θ.

ii Hence obtain an expression for L in the form $R \sin(\theta + \alpha)$, where $R > 0$ and $0 \leqslant \alpha \leqslant \frac{\pi}{2}$.

Give your answer for α to three decimal places.

b Given that θ varies between 0 and $\frac{\pi}{2}$:

i write down the maximum value of L;

ii find the value of θ, to two decimal places, for which L is maximum.

[AQA]

16 i In the diagram, $ABCD$ represents a rectangular table with sides 3.5 m and 1.5 m. It has been turned so that it wedges in a passage of width 2.5 m. Given that $\theta°$ is the acute angle between the longer side and the passage, as shown in the diagram, show clearly why

$$7\sin\theta° + 3\cos\theta° = 5.$$

ii Express $7\sin\theta° + 3\cos\theta°$ in the form

$$R\sin(\theta + \alpha)°,$$

where $R > 0$ and $0 < \alpha < 90$. [OCR]

iii Find θ.

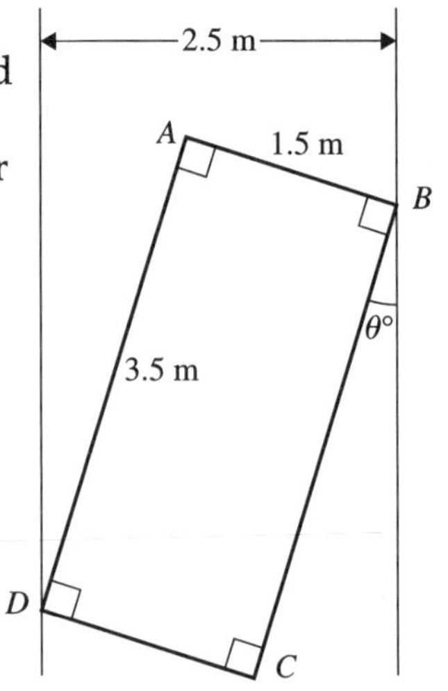

17 a Express $2\cos\theta + 5\sin\theta$ in the form $R\cos(\theta - \alpha)$, where $R > 0$ and $0 < \alpha < \frac{\pi}{2}$. Give the values of R and α to 3 significant figures.

b Find the maximum and minimum values of $2\cos\theta + 5\sin\theta$ and the smallest possible value of θ for which the maximum occurs.

The temperature T°C, of an unheated building is modelled using the equation

$$T = 15 + 2\cos\left(\frac{\pi t}{12}\right) + 5\sin\left(\frac{\pi t}{12}\right), \quad 0 \leqslant t < 24,$$

where t hours is the number of hours after 1200.

c Calculate the maximum temperature predicted by this model and the value of t when this maximum occurs.

d Calculate, to the nearest half hour, the times when the temperature is predicted to be 12°C. [EDEXCEL]

18 The diagram shows the graphs of $y = \cos^{-1} x$ and $y = \tan^{-1} x$, for $-1 \leqslant x \leqslant 1$ in each case. The graphs intersect at the point with coordinates (p, q).

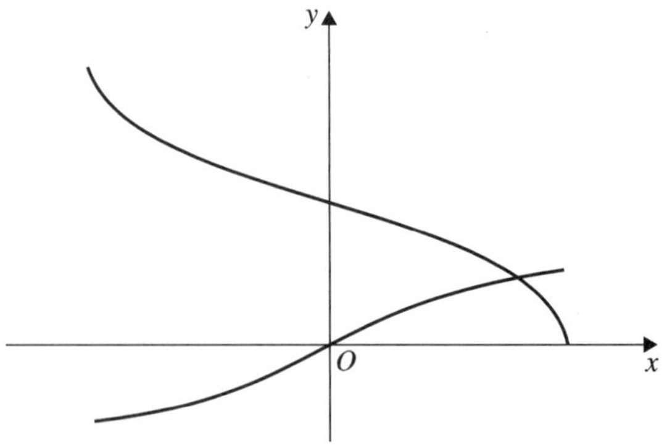

i State the exact values of y for $x = -1, 0$ and 1 on each graph.

ii Write down, in terms of p and q, the coordinates of the corresponding point of intersection of the graphs of $y = \cos x$ and $y = \tan x$, and hence show that $\cos^2 q = \sin q$.

iii Deduce that $p^4 + p^2 - 1 = 0$.

iv Hence find in exact form the solution of the equation

$$\cos^{-1} x = \tan^{-1} x.$$

[OCR]

PART 3

Exponentials and logarithms

3.1 e^x and $\ln x$

An exponential function has the form a^x where a is a constant and x is the index or *exponent*.
So, for example 2^x, 7^x and $(\frac{1}{2})^x$ are all exponential functions.

When mathematicians talk about *the* exponential function they are referring to the function e^x, where e is the constant 2.718 281 correct to six decimal places. The number e is very important in mathematics and is a number like π which is also an *irrational* number. It is a non-repeating, non-recurring decimal. You will find e on your calculator [Press $\boxed{e^x}$ $\boxed{1}$ $\boxed{=}$]

The function e^x is important because it has the unique property that when you differentiate e^x you obtain e^x. We return to differentiate e^x later in Part 4.

The graph of e^x is shown. Notice:

a At $x = 0, y = 1$

b As $x \to -\infty, y \to 0$

c $e^x > 0$ for all values of x.

d The function is one–one.

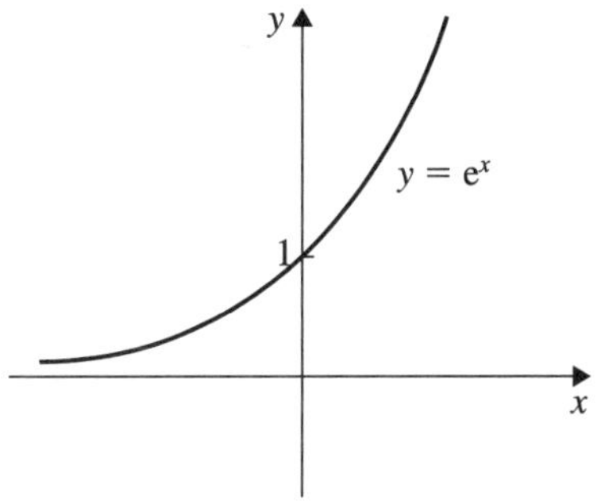

Because e^x is a one–one function it has an inverse. The inverse function can be found by the method discussed in Part 1 of this book.

For the function $f(x) = e^x$, let $y = e^x$.
So $x = \log_e y$

So we can obtain x using the expression $\log_e y$.
So the inverse function $f^{-1}(x)$ is $f^{-1}(x) = \log_e x$.

We usually write $\log_e x$ as $\ln x$ and $\ln x$ is a *natural* logarithm.

The graph of $\ln x$ is the reflection of the graph of e^x in the line $y = x$.

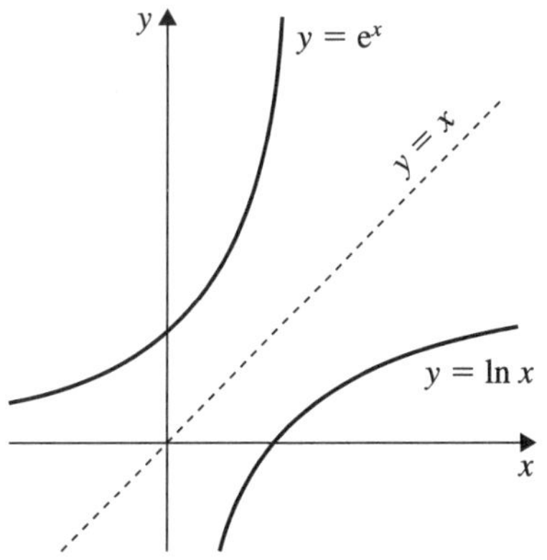

Note that $\ln x$ is not defined for $x \leqslant 0$. In other words we cannot find the logarithm of a negative number.

We can sketch the graphs of transformations of e^x and $\ln x$ using the basic transformations summarised below.

$y = f(x) + a$: Translation by a units parallel to the y-axis.
$y = f(x - a)$: Translation by a units parallel to the x-axis (note the negative sign).
$y = af(x)$: Stretch parallel to the y-axis by a scale factor a.
$y = f(ax)$: Stretch parallel to the x-axis by a scale factor $\frac{1}{a}$. (note the inverse of a).
$y = f(-x)$: Reflection in the y-axis.
$y = -f(x)$: Reflection in the x-axis.

Example 1

Sketch the graphs of

a $y = e^{x+2}$

b $y = e^{2x}$

c $y = \ln(x + 1) + 3$

d $y = \frac{1}{2}e^{-x}$

a

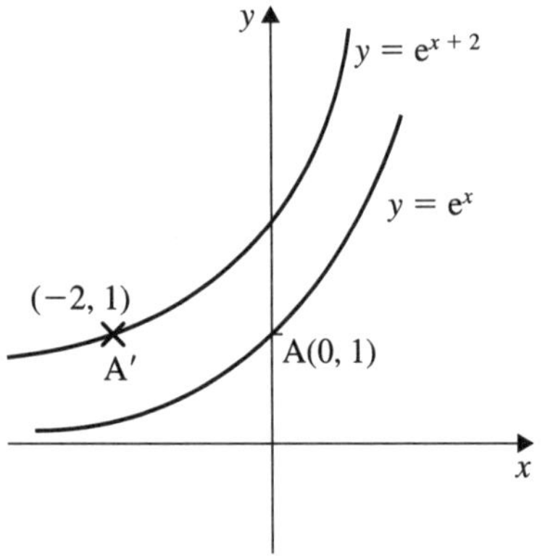

To obtain $y = e^{x+2}$, translate $y = e^x$ 2 units to the left parallel to the x-axis.

b

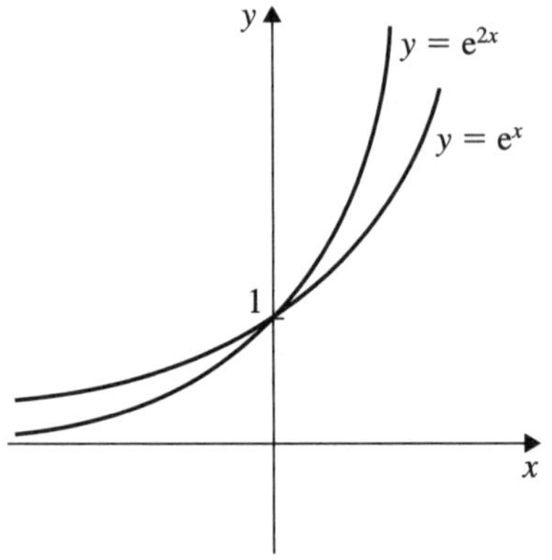

To obtain $y = e^{2x}$ stretch $y = e^x$ parallel to x-axis, scale factor $\frac{1}{2}$.

c

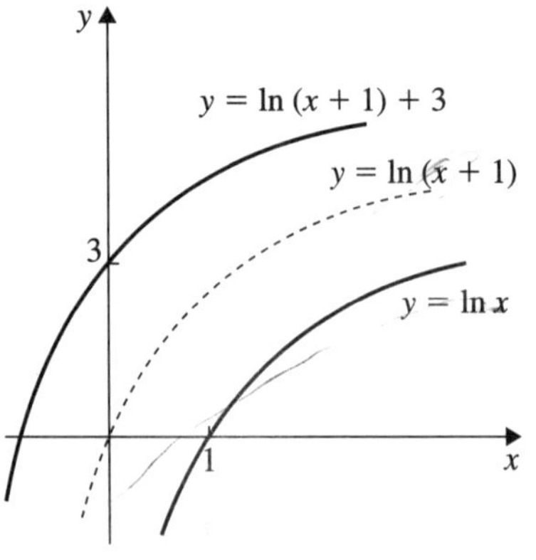

To obtain $y = \ln(x + 1) + 3$ perform the translations $\begin{pmatrix} -1 \\ 0 \end{pmatrix}$ and then $\begin{pmatrix} 0 \\ 3 \end{pmatrix}$.

d

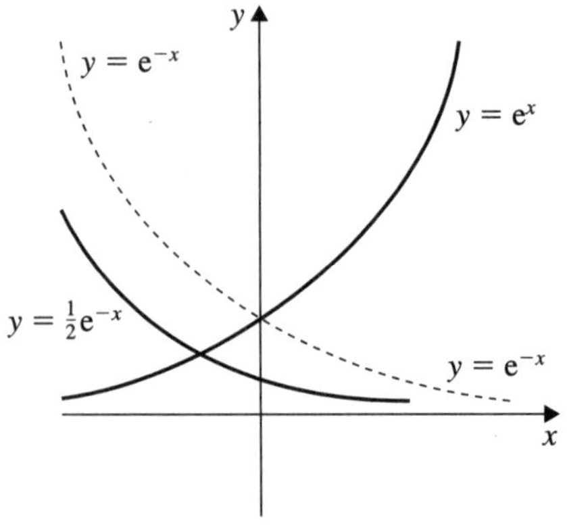

Reflect $y = e^x$ in the y-axis then stretch, scale factor $\frac{1}{2}$, parallel to the y-axis.

Example 2

The function f is defined by

$$f(x) = \ln(3x - 1), \quad x > \tfrac{1}{3}.$$

a Find an expression for $f^{-1}(x)$.

b Write down the domain of $f^{-1}(x)$.

c On the same axes sketch the graphs of $y = f(x)$ and $y = f^{-1}(x)$.

a Let $y = \ln(3x - 1)$

$\therefore \quad e^y = 3x - 1$ [Using the method for finding an inverse function in Part 1.]

$3x = e^y + 1$

$x = \tfrac{1}{3}(e^y + 1)$

$\therefore \quad f^{-1}(x) = \tfrac{1}{3}(e^x + 1)$

b The domain of $f^{-1}(x)$ is the range of $f(x)$.

$\therefore$ The domain of $f^{-1}(x)$ is all real values of x.

c For $f(x) = \ln(3x - 1)$

$f(\tfrac{2}{3}) = \ln 1 = 0$

f^{-1} is the reflection of f in the line $y = x$.

$f(x)$ and $y = f^{-1}(x)$ are shown.

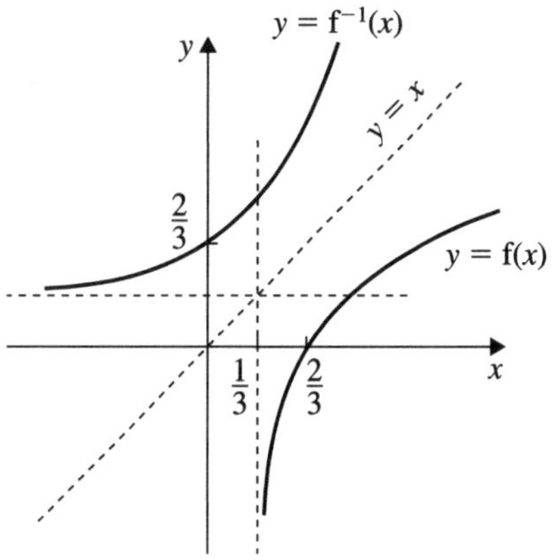

Example 3

The function $f(x)$ is defined by $f(x) = 4e^{2x}, x \in \mathbb{R}$.

a Describe a series of geometrical transformations that maps the graph of $y = e^x$ onto the graph of $y = 4e^{2x}$.

b Obtain an expression for $f^{-1}(x)$.

a $e^x \to e^{2x}$ Stretch, scale factor $\tfrac{1}{2}$, parallel to the x-axis

$e^{2x} \to 4e^{2x}$ Stretch, scale factor 4, parallel to the y-axis.

b Let $y = 4e^{2x}$

$\therefore \quad e^{2x} = \dfrac{y}{4}$

$2x = \ln\left(\dfrac{y}{4}\right)$

$x = \tfrac{1}{2}\ln\left(\dfrac{y}{4}\right)$

$\therefore \quad f^{-1}(x) = \tfrac{1}{2}\ln\left(\dfrac{x}{4}\right)$

EXERCISE 1

Draw sketch graphs of the following:

1 $y = e^x$ **2** $y = -e^x$ **3** $y = \ln x$

4 $y = e^{-x}$ **5** $y = 2\ln x$ **6** $y = \ln x + 2$

7 $y = e^{\frac{x}{2}}$ **8** $y = e^{x+2}$ **9** $y = \ln(x - 3)$

10 $y = 2\ln(x + 1)$ **11** $y = e^{2x} - 1$ **12** $y = -e^{-x}$

13 **a** Using the same axes sketch the graphs of $y = e^{-x}$ and $y = x(2 - x)$.

b Use symmetry to write down the coordinates of the turning point of the curve $y = x(2 - x)$.

c How many solutions are there to the equation $e^{-x} = x(2 - x)$?

14 The function $f(x) = \ln 2x, \quad x > 0$.

a Sketch $y = f(x)$, showing the point where the graph cuts the x-axis.

b Find an expression for $f^{-1}(x)$.

c Sketch the graph of $y = f^{-1}(x)$ and describe the relationship between the graphs of $y = f(x)$ and $y = f^{-1}(x)$.

15 On the same axes draw a sketch of the graph of $y = \ln x$ and a sketch of the graph of $y = \ln(x^3)$. Label your two curves clearly.

16 The functions $f(x)$ and $g(x)$ are defined by

$$f(x) = \ln x, \quad x > 0 \quad \text{and } g(x) = x^2, \quad x \in \mathbb{R}.$$

a Write down expressions for $fg(x)$ and $gf(x)$.

b Show that $fg(x) = kf(x)$, for some constant k, and find k.

17 The function $f(x)$ is defined by

$$f(x) = 3e^x \text{ for all } x.$$

a Sketch on the same axes $y = f(x)$ and $y = f^{-1}(x)$.

b Find an expression for $f^{-1}(x)$.

c State the domain and range of f^{-1}.

18 The function f is defined by $f(x) = 3 + \ln x, \quad x > 0$.

a State the range of the function f.

b State the domain and range of the inverse function f^{-1}.

c Obtain an expression for $f^{-1}(x)$.

d The function g is defined for all x by

$$g(x) = ex^4.$$

Show that $fg(x) = 4(1 + \ln x)$

e Solve the equation $fg(x) = 16$.

19 The functions f and g are defined for all real numbers by

$$f(x) = 3e^{2x},$$
$$g(x) = x - 4$$

a Evaluate fg(4).

b Find an expression for $f^{-1}x$.

c Sketch the graph of $y = |gf(x)|$ and obtain the x coordinate of the point where the graph meets the x-axis.

3.2 Solving equations with exponential and logarithmic functions

We solve equations involving exponential and logarithmic functions by working from first principles and applying the basic laws of indices and logarithms which are summarised below:

$e^a \times e^b = e^{a+b}$ $\qquad$ $\ln a + \ln b = \ln ab$

$e^a \div e^b = e^{a-b}$ $\qquad$ $\ln a - \ln b = \ln \dfrac{a}{b}$

$(e^a)^n = e^{an}$ $\qquad$ $\ln a^n = n \ln a$

$e^{-a} = \dfrac{1}{e^a}$

$e^0 = 1$ $\qquad$ $\ln e = \log_e e = 1$

$\ln 1 = 0$

Example 1

Solve the equation $e^x = 7$

$$e^x = 7$$
$$\log_e e^x = \log_e 7 \quad \text{[We have written } \log_e e \text{ for clarity.]}$$
$$x \log_e e = \log_e 7$$
$$x = \ln 7 \ (= 1.946 \text{ to 3 decimal places})$$

Example 2

Solve the equation $e^{2x+1} = 3$.

Take $\log_e$ of both sides.

$$\log_e e^{2x+1} = \log_e 3$$
$$(2x + 1) \ln e = \ln 3$$
$$2x + 1 = \ln 3 \qquad (\ln e = 1)$$
$$x = \frac{\ln 3 - 1}{2} \quad (= 0.0493 \text{ to 3 significant figures})$$

Example 3

Solve the equation $\ln(3x - 1) = 4$.

$$\log_e(3x - 1) = 4$$

So $\quad 3x - 1 = e^4$

$$x = \frac{e^4 + 1}{3} \qquad (= 18.53 \text{ to 2 decimal places})$$

Example 4

Express $\log_{10}(x + 2) - \log_{10} x$ as a single logarithm.
Hence solve the equation $\log_{10}(x + 2) - \log_{10} x = 2$.

$$\log_{10}(x + 2) - \log_{10} x = \log_{10}\left(\frac{x+2}{x}\right)$$

$$\log_{10}(x + 2) - \log_{10} x = 2$$

$$\log_{10}\left(\frac{x+2}{x}\right) = 2$$

$$10^2 = \frac{x+2}{x}$$

$$100x = x + 2$$

$$x = \tfrac{2}{99}.$$

EXERCISE 2

1 Simplify the following.

a $e^x \times e^x$ **b** $\dfrac{e^{8x}}{e^{2x}}$ **c** $(e^x)^3$ **d** e^0

e $\ln 2 + \ln x$ **f** $\ln 6 - \ln 2$ **g** $\ln x^3$ **h** $5 \ln 1$

i $\ln e$ **j** $e^2 \times e^x$ **k** $\ln \sqrt{e}$ **l** $\ln e^x$

m $\ln e^a$ **n** $(e^0) \times 7$ **o** $\dfrac{e^3 \times e^{-1}}{e}$ **p** $\ln(\ln e)$

2 Use a calculator to evaluate the following, giving the answer correct to two decimal places.

a $e^{2.5}$ **b** $\ln 10$ **c** $\dfrac{e^3 - 4}{7}$

d $\ln 12 - \ln 2$ **e** $\dfrac{\ln 18}{e^2}$ **f** $e^{-1} \times \ln 8$

3 Solve the equation $e^x = 7$, giving your answer correct to two decimal places.

In Questions **4** to **14** solve the equations.

4 $e^{2x} = 8$ **5** $2e^x = 9$ **6** $e^{x-1} = 11$

7 $e^{2x+1} = 4$ **8** $e^{3x-1} = 8$ **9** $e^{5x+2} = 100$

10 $e^{3x-1} = 1$ **11** $\ln x - \ln 4 = 4$ **12** $\ln x + \ln 2x = 12$

13 $\ln x + \ln 3 - \ln 6 = \ln 10$

14 $\ln 2 + \ln x^2 = \ln 18$

15 Solve the equation $2 \ln 2x = 1 + \ln 3$. Give your answer correct to two decimal places.

16 **a** Express $\ln(x + 2) - \ln x$ as a single logarithm.
b Solve the equation $\ln(x + 2) - \ln x = 4$.
Give your answer in terms of e.

17 Simplify the following.

a $\ln x^2 - \ln xy$

b $\ln x + \ln \frac{1}{x}$

c $\ln(\ln x^2) - \ln(\ln x)$

d $e^{\ln x}$

18 Solve the equations,

a $e^x \times e^{x+1} = 10$

b $\left(\frac{1}{e}\right)^{x+1} = 4$

c $e^{2x} - 3e^x + 2 = 0$

d $e^{\ln x} + \ln e^x = 8$

19 Solve the simultaneous equations $\ln(xy) = 5$

$[x > 0, y > 0]$ $\ln\left(\frac{x}{y}\right) = 1$

20 Find the value, or values, of x for which

$\ln(x + 6) = 2 \ln(x - 6)$.

21 At time t minutes the temperature of a body in °C is

$5e^{-t} + 20$.

a Write down the temperature of the body at $t = 0$.
b Calculate the value of t when the temperature of the body reaches 22°C.
Give your answer correct to one decimal place.

22 Establish the formula $\log_a x = \frac{1}{\log_x a}$.

23 The positive numbers x, y and z are in geometric progression. Show that $\log x$, $\log y$ and $\log z$ are in arithmetic progression.

Review Exercise 3

1 Draw sketch graphs of the following:

a $y = e^x + 1$

b $y = \ln x$

c $y = 2e^{-x}$

d $y = 3 \ln x$

e $y = e^{x-1}$

f $y = \ln(x - 2), x > 2$

2 **a** If $f(x) = \ln x$, $x > 0$, write down an expression for $f^{-1}(x)$.
b Solve the equation $f^{-1}(x) = e$.

3 The function f is defined by $f(x) = 4 + \ln x, \quad x > 0$. Obtain an expression for $f^{-1}(x)$.

4 The functions f and g are defined for all real values of x by $f(x) = 2x - 100$
$g(x) = e^{-2x}$.

a State the range of g.

b Find an expression for $g^{-1}(x)$.

c On the same axes sketch the graphs of $y = g(x)$ and $y = g^{-1}(x)$.

d Evaluate $fg(-2)$, giving your answer correct to 2 decimal places.

5 Given $f(x) = \ln(3 - 2x)$ and $g(x) = e^x$, evaluate $gf(1)$.

6 Sketch the graph of $y = \ln e^{x^2} - x, \quad x \in \mathbb{R}$.

7 Show that the equation $x = \ln(x + 5)$ has a root between $x = 1$ and $x = 2$.

8 Solve the equations, giving your answers correct to 2 decimal places where necessary.

a $e^{3x} = 7$ **b** $3e^x = 2$ **c** $e^{x+1} = 5$

9 Solve the equations, giving your answers correct to 2 decimal places where necessary.

a $\ln(x + 1) = 2$ **b** $\ln x - \ln 2 = 3$

c $2 \ln 3x = 1$ **d** $\ln x = \ln e^3$

10 The function f is defined by $f(x) = 5(2^{-x}) - 1$ for $x \geqslant 0$.
The graph of $y = f(x)$ meets the y-axis at $(0, a)$ and the x-axis at $(b, 0)$.

a State the value of a.

b Find the value of b, correct to 2 decimal places.

c Sketch the graph of $y = f(x)$.

d Copy and complete the following statement:
'As $x \to \infty$, $f(x) \to \square$.'

11 Solve the equation $2 \ln 2x = 1 + \ln 3$. Give your answer in terms of e.

12 Solve the equation $\ln(x + 1) - \ln x = e$. Give your answer correct to 4 decimal places.

13 The temperature, θ°C, of an oven at time t minutes is given by $\theta = 300 + 50\,e^{-0.04t}$.
Find the value of θ when $t = 10$.

14 **a** By substituting $y = 2^x$ into the equation

$$2^{2x} - 5 \times 2^x + 4 = 0$$

show that $y^2 - 5y + 4 = 0$.

b Hence find the values of x which satisfy the equation $2^{2x} - 5 \times 2^x + 4 = 0$.

Examination Exercise 3

1 Solve the equation $e^{\ln x} + \ln e^x = 8$. [EDEXCEL]

2 Simplify:

a $\log x^2 - \log xy$ **b** $\log z + \log \frac{1}{z}$ **c** $\frac{\log x^3 - \log x}{\log x^2 - \log x}$ [OCR]

3 Express $\log(2\sqrt{10}) - \frac{1}{3}\log 0.8 - \log(\frac{10}{3})$ in the form $c + \log d$ where c and d are rational numbers and the logarithms are to base 10. [OCR]

4 Solve the equation

$$\ln(5x + 6) = 2\ln(5x - 6)$$ [EDEXCEL]

5 It is given that $\ln x = p + 2$ and $\ln y = 3p$.

i Express each of the following in terms of p:

a $\ln(xy)$, **b** $\ln(x^3)$, **c** $\ln\left(\frac{y}{e}\right)$.

ii Express y in terms of x and e, simplifying your answer. [OCR]

6 The function f is defined by $f : x \mapsto e^x + k$, $x \in \mathbb{R}$ and k is a positive constant.

a State the range of f.

b Find $f(\ln k)$, simplifying your answer.

c Find f^{-1}, the inverse function of f, in the form $f^{-1} : x \mapsto \ldots$, stating its domain.

d On the same axes, sketch the curves with equations $y = f(x)$, and $y = f^{-1}(x)$, giving the coordinates of all points where the graphs cuts the axes. [EDEXCEL]

7 The function f is defined by

$$f : x \mapsto \ln(5x - 2),\ x > \tfrac{2}{5}$$

a Find an expression for $f^{-1}(x)$.

b Write down the domain of f^{-1}.

c Solve, giving your answer to 3 decimal places,

$$\ln(5x - 2) = 2$$ [EDEXCEL]

8 The diagram shows the graph of $y = f(x)$, where f is defined for all real numbers by

$$f(x) = 2e^{-x}.$$

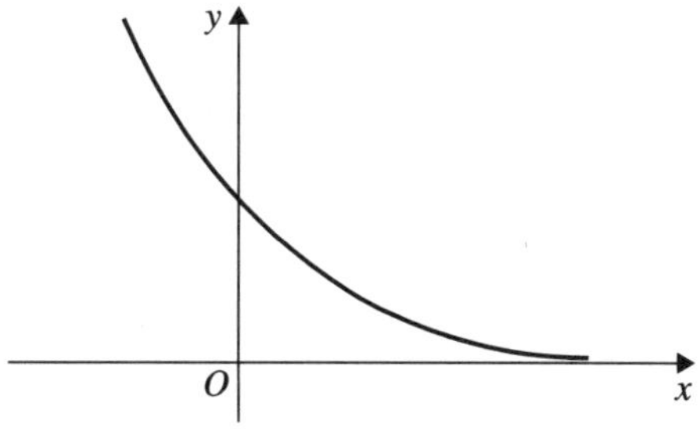

a Describe a sequence of geometrical transformations by which the above graph can be obtained from the graph of $y = e^x$.

b Copy the above diagram and sketch on the same axes the graph of

$$y = f^{-1}(x).$$

c Find an expression for $f^{-1}(x)$.

d State the domain and range of f^{-1}. [AQA]

9 a

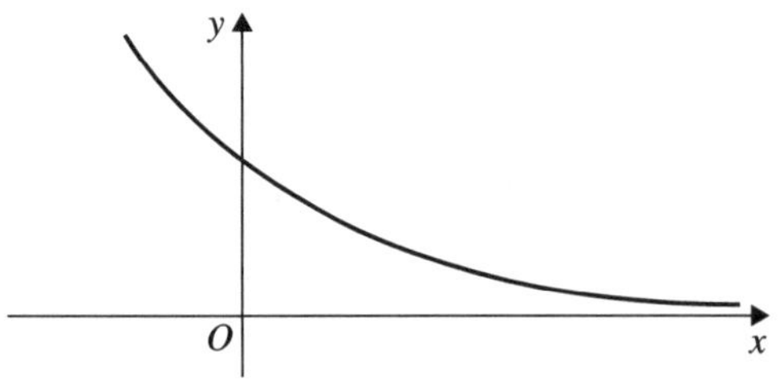

The diagram shows the graph of $y = \text{f}(x)$, where the function f is defined for all values of x by

$$\text{f}(x) = 5\text{e}^{-x}.$$

i Write down the coordinates of the point where the graph intersects the y-axis.

ii State the range of the function f.

iii Find the value of $\text{f}(\ln 6)$, giving your answer as a fraction.

b The function g is defined for all values of x by

$$\text{g}(x) = x + 10.$$

i Show that $\text{gf}(x) = 5(\text{e}^{-x} + 2)$.

ii State the range of the function gf.

iii Sketch the graph of $y = \text{gf}(x)$.

iv Show that $\text{gf}(x) = 11 \Rightarrow x = \ln 5$. [AQA]

10 The functions f and g are defined for all real numbers by

$$\text{f} : x \mapsto 2\text{e}^{3x},$$
$$\text{g} : x \mapsto x - k,$$

where k is a positive constant.

i Evaluate $\text{fg}(k)$.

ii Find an expression for $\text{f}^{-1}(x)$.

iii Sketch on the same diagram the graphs of $y = \text{f}(x)$ and $y = \text{f}^{-1}(x)$, indicating the relationship between the graphs.

iv

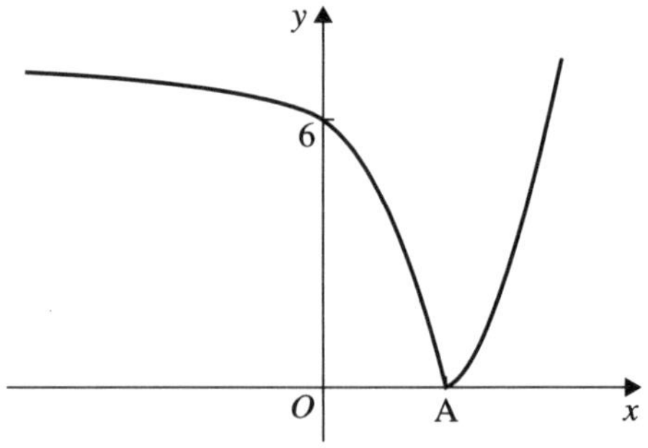

The diagram shows the graph of $y = |\text{gf}(x)|$, which meets the x-axis at A and the y-axis at (0, 6). Find k and determine the exact value of the x-coordinate of A. [OCR]

11 The diagram shows the graph of $y = \mathrm{f}(x)$, where f is defined for $x > 0$ by

$$\mathrm{f}(x) = 2 + \ln x.$$

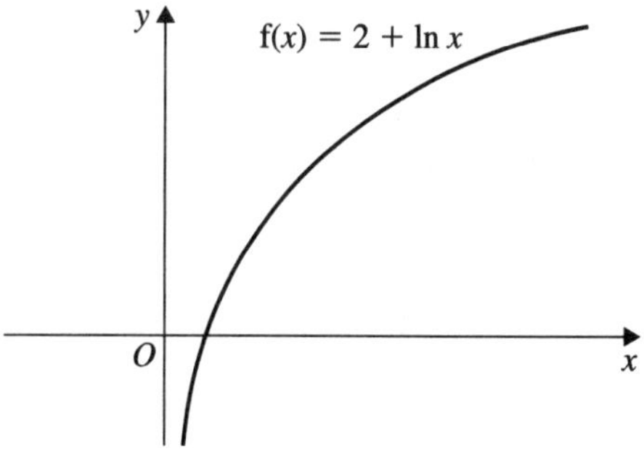

a **i** Differentiate $\mathrm{f}(x)$ to find $\mathrm{f}'(x)$.

ii Find the gradient of the curve at the point where $x = \mathrm{e}$.

b Describe the geometrical transformation by which the graph of $y = 2 + \ln x$ can be obtained from the graph of $y = \ln x$.

c **i** State the range of the function f.

ii State the domain and range of the inverse function f^{-1}.

iii Find an expression for $\mathrm{f}^{-1}(x)$.

d The function g is defined for all x by

$$\mathrm{g}(x) = \mathrm{e}\,x^3$$

Show that:

i $\mathrm{fg}(x) = 3(1 + \ln x)$;

ii $\mathrm{fg}(x) = 9 \Rightarrow x = \mathrm{e}^2$. [AQA]

Part 4

Differentiation 1

4.1 Function of a function

Consider the function $y = (x + 3)^7$. Suppose we let $u = x + 3$, so that $y = u^7$. Now y is a function of u and u is a function of x. So y 'is a function of a function' of x.

The Chain Rule

> If y is a function of u and u is a function of x,
>
> $$\frac{dy}{dx} = \frac{dy}{du} \times \frac{du}{dx}$$

Note that the chain rule is easy to remember because it *appears* that the term du cancels. Remember that cancellation in this context is in fact meaningless.

In the above example $y = (x + 3)^7$ and $u = x + 3$ so that $y = u^7$.

$$\frac{dy}{du} = 7u^6 \quad \text{and} \quad \frac{du}{dx} = 1$$

By the chain rule,

$$\frac{dy}{dx} = \frac{dy}{du} \times \frac{du}{dx} = 7u^6 \times 1$$
$$= 7(x + 3)^6$$

Example 1

If $y = (4x - 1)^3$, find $\frac{dy}{dx}$

Method 1: Let $u = 4x - 1$, so $y = u^3$

$$\frac{du}{dx} = 4 \quad \text{and} \quad \frac{dy}{du} = 3u^2$$

$$\frac{dy}{dx} = \frac{dy}{du} \times \frac{du}{dx} = 3u^2 \times 4$$
$$= 12u^2$$
$$\frac{dy}{dx} = 12(4x - 1)^2$$

Method 2: When you have used the above method several times you will see that a quicker method can be used.

We have $y = (4x - 1)^3$

$$\frac{dy}{dx} = 3(4x - 1)^2 \times \frac{d}{dx}(4x - 1)$$
$$= 3(4x - 1)^2 \times 4$$
$$= 12(4x - 1)^2.$$

Example 2

Find $\frac{dy}{dx}$ in each case.

a $y = (5x + 1)^4$

$$\frac{dy}{dx} = 4(5x + 1)^3 \times \frac{d}{dx}(5x + 1)$$
$$= 4(5x + 1)^3 \times 5$$
$$= 20(5x + 1)^3$$

b $y = 3(1 + x^2)^7$

$$\frac{dy}{dx} = 3 \times 7(1 + x^2)^6 \times \frac{d}{dx}(1 + x^2)$$
$$= 21(1 + x^2)^6 \times 2x$$
$$= 42x(1 + x^2)^6$$

c $y = \sqrt{2x + 3}$

$$= (2x + 3)^{\frac{1}{2}}$$
$$\frac{dy}{dx} = \frac{1}{2}(2x + 3)^{-\frac{1}{2}} \times \frac{d}{dx}(2x + 3)$$
$$= \tfrac{1}{2}(2x + 3)^{-\frac{1}{2}} \times 2$$
$$= (2x + 3)^{-\frac{1}{2}}$$

d $y = \frac{1}{(3x^3 + 1)^2} = (3x^3 + 1)^{-2}$

$$\frac{dy}{dx} = -2(3x^3 + 1)^{-3} \times \frac{d}{dx}(3x^3 + 1)$$
$$= -2(3x^3 + 1)^{-3} \times 9x^2$$
$$= \frac{-18x^2}{(3x^3 + 1)^3}$$

EXERCISE 1

1 Find $\frac{dy}{dx}$.

a $y = (2x + 5)^3$, let $u = 2x + 5$.

b $y = (x^2 + 7)^4$, let $u = x^2 + 7$.

2 Find $\frac{dy}{dx}$.

a $y = (3x - 4)^4$ **b** $y = (8x + 11)^5$ **c** $y = (x^2 - 3)^2$

d $y = (3x^3 + 1)^3$ **e** $y = (1 - 3x)^4$ **f** $y = (3 - x)^{10}$

3 Differentiate the following.

a $y = (1 + 3x)^{-1}$ **b** $y = \frac{1}{(2x + 1)^2}$ **c** $y = \frac{1}{(5x + 2)^3}$

d $y = \frac{3}{(4x - 1)^2}$ **e** $y = \frac{5}{(x^2 + 1)}$ **f** $y = \frac{10}{(1 - x)^4}$

4 a If $f(x) = \sqrt{2x^2 - 1}$, show that $f'(x) = \frac{2x}{\sqrt{2x^2 - 1}}$.

b Find $f'(2)$.

5 Find $f'(x)$.

a $f(x) = \sqrt{5x^2 + 3}$ **b** $f(x) = (3x + 1)^{\frac{1}{3}}$

c $f(x) = \frac{1}{\sqrt{x^2 - 3}}$ **d** $f(x) = \frac{1}{x^3 + 1}$

e $f(x) = \frac{1}{\sqrt{8x + 7}}$ **f** $f(x) = \frac{1}{\sqrt{x} + 2}$

6 Find the equation of the tangent to the curve $y = (x^2 + 1)^4$ at the point (1, 16).

7 Find the equation of the normal to the curve $y = (3x + 1)^{\frac{3}{2}}$ at the point where $x = 5$.

8 For the curve $y = (2 + x^2)^3$, find the coordinates of the point where the gradient is zero.

9 Given $y = (3x + 1)^2 + (3x + 1)^3$, find the value of $\frac{dy}{dx}$ at $x = \frac{1}{3}$.

10 Find the equation of the tangent to the curve $y = \sqrt{x^2 + 3}$ at the point (1, 2).

11 Differentiate the following

a $y = (2\sqrt{x} - 3x)^3$ **b** $y = \sqrt{1 - \frac{1}{x}}$ **c** $y = \frac{2}{x^{\frac{3}{2}} + 2}$

d $y = \left(\frac{1}{1 + \sqrt{x}}\right)^2$ **e** $y = \sqrt{x^2 - \frac{1}{x}}$ **f** $y = 10x^2 + \sqrt{2x - 3}$

12 **a** Find the coordinates of the two stationary points of $y = \frac{8}{(x + 1)} + 2x + 1$.

b Show that one is a local maximum and the other is a local minimum.

13 **a** Find the coordinates of the stationary point of $y = \frac{32}{(3x + 1)^2} + 3x - 3$.

b Show that this is a local minimum.

4.2 Product Rule

Consider the product uv where u and v are functions of x.

$$y = uv \qquad [1]$$

Let x increase by a small amount δx and let the corresponding increase in u by δu, in v be δv and in y be δy.

So $\quad y + \delta y = (u + \delta u)(v + \delta v) \qquad [2]$

$$[2] - [1] \quad y + \delta y - y = (u + \delta u)(v + \delta v) - uv$$

$$\delta y = u\,\delta v + v\,\delta u + \delta u\,\delta v$$

$$\frac{\delta y}{\delta x} = u\frac{\delta v}{\delta x} + v\frac{\delta u}{\delta x} + \delta u\frac{\delta v}{\delta x}$$

As $\delta x \to 0$ $\frac{\delta y}{\delta x} \to \frac{dy}{dx}$, $\frac{\delta v}{\delta x} \to \frac{dv}{dx}$ and $\frac{\delta v}{\delta x} \to \frac{dv}{dx}$.

And as $\delta x \to 0$ δu and δv both tend towards zero.

$$\frac{dy}{dx} = u\frac{dv}{dx} + v\frac{du}{dx}$$

Remember this important result.

Example 1

Find $\frac{dy}{dx}$:

a $y = x(x+2)^2$ **b** $y = (x+3)^2(2x+1)^3$ **c** $y = (2x+1)^2\sqrt{4x-3}$

a We have $y = uv$, where $u = x$ and $v = (x+2)^2$

$$\left[\therefore \quad \frac{du}{dx} = 1 \quad \text{and} \quad \frac{dv}{dx} = 2(x+2)\right]$$

$$\begin{aligned}\frac{dy}{dx} &= u\frac{dv}{dx} + v\frac{du}{dx}\\ &= x \times 2(x+2)^1 + (x+2)^2 \times 1\\ &= (x+2)(2x+x+2)\\ &= (x+2)(3x+2)\end{aligned}$$

b $y = (x+3)^2(2x+1)^3$

We have $y = uv$, where $u = (x+3)^2$ and $v = (2x+1)^3$

$$\begin{aligned}\frac{dy}{dx} &= u\frac{dv}{dx} + v\frac{du}{dx}\\ &= (x+3)^2 \times 3(2x+1)^2 \times 2 + (2x+1)^3 \times 2(x+3)^1\\ &= (x+3)(2x+1)^2[6(x+3) + (2x+1) \times 2]\\ &= (x+3)(2x+1)^2(10x+20)\\ &= 10(x+3)(2x+1)^2(x+2)\end{aligned}$$

c $y = (2x+1)^2\sqrt{4x-3}$

We have $y = uv$, where $u = (2x+1)^2$ and $v = (4x-3)^{\frac{1}{2}}$

$$\begin{aligned}\frac{dy}{dx} &= (2x+1)^2 \times \frac{1}{2}(4x-3)^{-\frac{1}{2}} \times 4 + (4x-3)^{\frac{1}{2}} \times 2(2x+1)^1 \times 2\\ &= \frac{(2x+1)}{(4x-3)^{\frac{1}{2}}}[2(2x+1) + (4x-3) \times 4]\\ &= \frac{2x+1}{\sqrt{4x-3}}[20x-10]\\ &= \frac{10(2x+1)(2x-1)}{\sqrt{4x-3}}\end{aligned}$$

Exercise 2

Differentiate the following with respect to x and simplify your answers.

1 $y = x(2x+1)^2$

2 $y = x^2(3x-1)^2$

3 $y = x^3(x-1)^2$

4 $y = (x+2)^2(x+3)^2$

5 $y = 2x(1-x)^3$

6 $y = (4x^3+1)(x-1)^2$

7 $y = 5x(1+2x)^4$

8 $y = (3x+1)^4(x-3)$

9 $y = (x + 2)x^{\frac{3}{2}}$

10 $y = (4x + 1)^2 \sqrt{x}$

11 $y = x^2 \sqrt{2x + 1}$

12 $y = x^3 \sqrt{4x - 1}$

13 Find the equation of the tangent to the curve $y = 6x\sqrt{x + 8}$ at the point (1, 18).

14 Find the gradient of the curve $y = x^3(x + 1)^2$ at the point (1, 4).

15 Find the x coordinates of the two points where the gradient of the curve $y = \sqrt{x}(x - 2)^2$ is zero.

16 **a** Find the x coordinates of the turning points of the curve $y = (x + 3)^2(x - 1)$.
b Determine the nature of each turning point.

17 **a** For the curve $y = (x + 2)(x - 4)^2$, find
i the points where it meets the axes,
ii the turning points.
b Sketch the curve.

18 The curve $y = x^2(x - 3)$ has a minimum value at the point A.

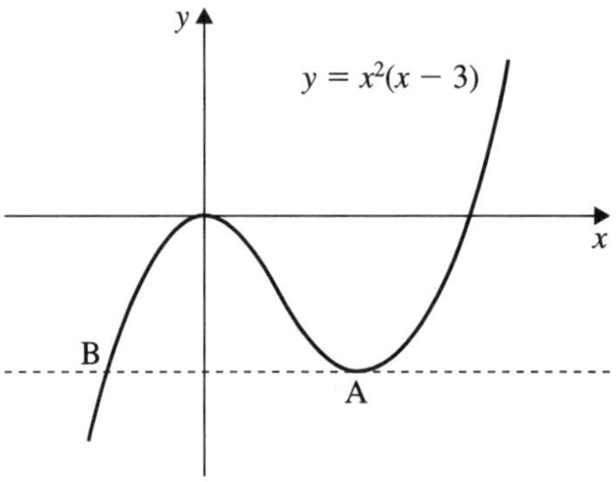

a Find the coordinates of A.
b The tangent to the curve at A meets the curve again at B.
Find the coordinates of B.

19 **i** Given that $y = (x + 1)\sqrt{2x - 1}$, show that $\frac{dy}{dx}$ can be written in the form $\frac{kx}{\sqrt{2x - 1}}$ and state the value of k.
ii Hence evaluate

$$\int_1^5 \frac{x}{\sqrt{2x - 1}}\,dx$$

4.3 The quotient rule

Consider the function $y = \frac{u}{v}$, where u and v are both functions of x.

Write $y = uv^{-1}$ and then differentiate using the product rule.

$$\begin{aligned}\frac{dy}{dx} &= u\frac{d}{dx}(v^{-1}) + v^{-1}\frac{d}{dx}(u)\\ &= u \times (-1)v^{-2}\frac{dv}{dx} + v^{-1}\frac{du}{dx}\\ &= -\frac{u}{v^2}\frac{dv}{dx} + \frac{1}{v}\frac{du}{dx}\\ &= \frac{-u\frac{dv}{dx} + v\frac{du}{dx}}{v^2}\end{aligned}$$

or, finally

$$\boldsymbol{\frac{d}{dx}\left(\frac{u}{v}\right) = \frac{v\frac{du}{dx} - u\frac{dv}{dx}}{v^2}}$$

Remember this result or perhaps this verse!
'low D-high minus high D-low over the square of what's below'.

Example 1

Find $\frac{dy}{dx}$ in each case.

a $y = \frac{x^2}{2x + 3}$

This is of the form $y = \frac{u}{v}$, where $u = x^2$ and $v = 2x + 3$.

By the quotient rule,

$$\begin{aligned}\frac{dy}{dx} &= \frac{(2x + 3) \times 2x - x^2 \times 2}{(2x + 3)^2}\\ &= \frac{4x^2 + 6x - 2x^2}{(2x + 3)^2}\\ &= \frac{2x^2 + 6x}{(2x + 3)^2} = \frac{2x(x + 3)}{(2x + 3)^2}\end{aligned}$$

b $y = \frac{3x^2 - 1}{(2x - 1)^3}$

This is of the form $y = \frac{u}{v}$, where $u = 3x^2 - 1$ and $v = (2x - 1)^3$.

$$\frac{du}{dx} = 6x \quad \text{and} \quad \frac{dv}{dx} = 3(2x - 1)^2 \times 2$$
$$= 6(2x - 1)^2$$

$$\frac{dy}{dx} = \frac{(2x-1)^3 \times 6x - (3x^2-1) \times 6(2x-1)^2}{(2x-1)^6}$$

$$= \frac{(2x-1)^2[6x(2x-1) - 6(3x^2-1)]}{(2x-1)^6}$$

$$= \frac{12x^2 - 6x - 18x^2 + 6}{(2x-1)^4} \qquad \text{[cancel } (2x-1)^2\text{]}$$

$$= \frac{6 - 6x - 6x^2}{(2x-1)^4} = \frac{6(1 - x - x^2)}{(2x-1)^4}$$

Example 3

a Find $\frac{dy}{dx}$ if $y = \frac{x}{\sqrt{2x-1}}, x > \frac{1}{2}$.

b Find any turning points and determine their nature.

a $y = \frac{x}{(2x-1)^{\frac{1}{2}}}$

So $y = \frac{u}{v}$, where $u = x$ and $v = (2x-1)^{\frac{1}{2}}$

Differentiating u and v, $\frac{du}{dx} = 1$, $\frac{dv}{dx} = \frac{1}{2}(2x-1)^{-\frac{1}{2}} \times 2 = (2x-1)^{-\frac{1}{2}}$

$$\therefore \quad \frac{dy}{dx} = \frac{(2x-1)^{\frac{1}{2}} \times 1 - x \times \frac{1}{(2x-1)^{\frac{1}{2}}}}{(2x-1)}$$

Multiply numerator and denominator by $(2x-1)^{\frac{1}{2}}$.

$$\frac{dy}{dx} = \frac{(2x-1) - x}{(2x-1)^{\frac{3}{2}}}$$

$$\frac{dy}{dx} = \frac{x-1}{(2x-1)^{\frac{3}{2}}}$$

b At turning points $\frac{dy}{dx} = 0 \quad \Rightarrow \quad \frac{x-1}{(2x-1)^{\frac{3}{2}}} = 0$

$$\Rightarrow \quad x = 1$$

When $x = 1$, $y = \frac{1}{\sqrt{2-1}} = 1$.

There is a turning point at (1, 1).

To determine the nature of the turning point find $\frac{d^2y}{dx^2}$.

$$\frac{d^2y}{dx^2} = \frac{(2x-1)^{\frac{3}{2}} \times 1 - (x-1) \times \frac{3}{\cancel{2}}(2x-1)^{\frac{1}{2}} \times \cancel{2}}{(2x-1)^3}$$

$$= \frac{(2x-1)^{\frac{1}{2}}[(2x-1) - 3(x-1)]}{(2x-1)^3}$$

$$= \frac{-x+2}{(2x-1)^{\frac{5}{2}}}$$

When $x = 1$, $\dfrac{d^2y}{dx^2} = \dfrac{-1 + 2}{1} = 1$ which is greater than zero.

This tells us that there is a minimum value at (1, 1).

Here is a sketch of the curve.

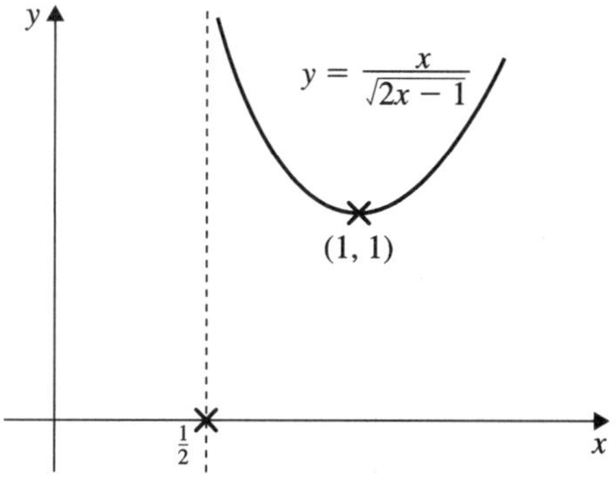

Notice that as $x \to \frac{1}{2}$, $y \to \infty$.

EXERCISE 3

Find $\dfrac{dy}{dx}$.

1 $y = \dfrac{2x}{x + 1}$ **2** $y = \dfrac{x + 7}{2x - 1}$ **3** $y = \dfrac{x^2}{3x + 1}$

4 $y = \dfrac{4x + 5}{x^2 + 1}$ **5** $y = \dfrac{(x + 1)^2}{x^2}$ **6** $y = \dfrac{2x + 1}{(x + 1)^3}$

7 $y = \dfrac{(3x + 2)^2}{(4x + 1)^3}$ **8** $y = \dfrac{x^3}{(x - 3)^4}$ **9** $y = \dfrac{(3x - 4)^3}{(2x + 1)^2}$

10 $y = \dfrac{\sqrt{x + 1}}{x}$ **11** $y = \dfrac{x}{\sqrt{(4x + 3)}}$ **12** $y = \dfrac{3x + 1}{\sqrt{2x - 1}}$

13 Find the equation of the tangent to the curve $y = \dfrac{x^2}{(x + 1)^3}$ at the point $(1, \frac{1}{8})$.

14 Find the equation of the tangent to the curve $y = \dfrac{x + 1}{2x - 1}$ at the point $(0, -1)$.

15 Find the equation of the normal to the curve $y = \dfrac{x + 4}{x^2 + 1}$ at the point $(0, 4)$.

16 Find the x coordinate of the point on the curve $y = \dfrac{2x}{\sqrt{4x - 3}}$ where the gradient is zero.

17 Find the coordinates of the turning points of $y = \dfrac{1 - x}{x^2 + 3}$. Determine their nature.

18 Show that the gradient of the curve $y = \dfrac{x^3}{\sqrt{x^2+1}}$ is positive for all values of x except $x = 0$.

19 Show that if $y = \dfrac{x^2}{\sqrt{x+1}}$ then $\dfrac{dy}{dx} = \dfrac{x(3x+4)}{2(x+1)\sqrt{x+1}}$.

4.4 Differentiating e^x

In Part 3 of this book we stated that when we differentiate the function e^x we obtain e^x. It is not necessary at this level to understand a proof of this result but we indicate below how we can 'justify' the result by a numerical method.

At this stage remember: $\mathbf{\dfrac{d}{dx}(e^x) = e^x}$

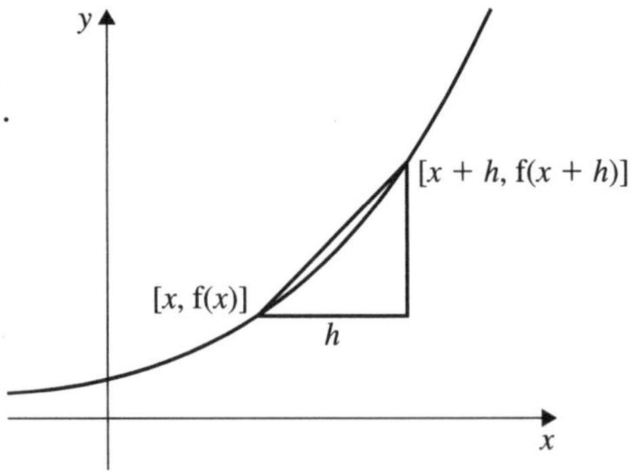

For a function, f(x), $f'(x) = \lim\limits_{h \to 0}\left(\dfrac{f(x+h) - f(x)}{h}\right)$.

So if $f(x) = a^x$ we see that

$$f'(x) = \lim_{h \to 0}\left(\frac{a^{x+h} - a^x}{h}\right) = \lim_{h \to 0}\left(\frac{a^x.a^h - a^x}{h}\right)$$

$$= a^x \lim_{h \to 0}\left(\frac{a^h - 1}{h}\right).$$

We require the value of a such that $f'(a^x) = a^x$.

We need $\lim\limits_{h \to 0}\left(\dfrac{a^h - 1}{h}\right) = 1$.

Try $a = 2$, $h = 0.001$: $\dfrac{2^{0.001} - 1}{0.001} \approx 0.693$. So $a = 2$ is too small.

Try $a = 3$, $h = 0.001$: $\dfrac{3^{0.001} - 1}{0.001} \approx 1.099$. So $a = 3$ is too large.

Try $a = 2.71$, $h = 0.001$: $\dfrac{2.71^{0.001} - 1}{0.001} \approx 0.997$. So $a = 2.71$ is too small.

Try $a = 2.72$, $h = 0.001$: $\dfrac{2.72^{0.001} - 1}{0.001} \approx 1.001$. So $a = 2.72$ is too large.

When this process is continued we obtain the number 2.718 281 828 correct to 9 decimal places. Note that this is an *approximate* value for e. The number e is in fact an irrational number which means that it cannot be written in the form $\dfrac{a}{b}$ where a and b are integers.

A frequently asked question is 'What is e?'
Answer: e is just a *number* which has the unique property that $\dfrac{d}{dx}(e^x) = e^x$.

Example 1

Find the gradient of the curve $y = e^x$ at the point $(2, e^2)$.

$$y = e^x$$
$$\frac{dy}{dx} = e^x$$

When $x = 2$, $\frac{dy}{dx} = e^2$.

At the point $(2, e^2)$ the gradient of the curve is e^2.
[Normally the answer would be given as 'e^2'. As a decimal the gradient is 7.39 correct to 2 decimal places.]

Example 2

Find $\frac{dy}{dx}$, if $y = e^{3x}$.

$$y = e^{3x}$$

Let $u = 3x$, so $y = e^u$

Using $\frac{dy}{dx} = \frac{dy}{du} \times \frac{du}{dx}$, $\frac{dy}{dx} = e^u \times 3$

$\therefore \quad \frac{dy}{dx} = 3e^{3x}$.

Example 3

Find $\frac{dy}{dx}$ in each of the following.

a $y = e^{x^2}$

Let $u = x^2$

$\therefore \quad y = e^u$

$$\frac{dy}{dx} = \frac{dy}{du} \times \frac{du}{dx}$$
$$= e^u \times 2x$$
$$\frac{dy}{dx} = 2x\, e^{x^2}$$

b $y = e^{4x+1}$

Let $u = 4x + 1$

$\therefore \quad y = e^u$

$$\frac{dy}{dx} = \frac{dy}{du} \times \frac{du}{dx}$$
$$= e^u \times 4$$
$$\frac{dy}{dx} = 4\, e^{4x+1}$$

Notice that in each case, for $y = e^{f(x)}$

$$\frac{dy}{dx} = f'(x)\, e^{f(x)}$$

This is a quick way of finding the derivative of $e^{f(x)}$.

Example 4

Find $\frac{dy}{dx}$ in each of the following.

a $y = e^{3x^2}$

$\frac{dy}{dx} = 6x\, e^{3x^2}$ (using the quick method above)

b $y = x^2 e^x$

We have $y = uv$, where $u = x^2$, $v = e^x$

$$\frac{dy}{dx} = u\frac{dv}{dx} + v\frac{du}{dx}$$
$$= (x^2 e^x) + (e^x \times 2x)$$
$$= x\, e^x(x + 2)$$

c $y = \dfrac{e^{2x}}{3x+1}$

We have $y = \dfrac{u}{v}$, where $u = e^{2x}$ and $v = 3x + 1$.

$$\frac{dy}{dx} = \frac{v\dfrac{du}{dx} - u\dfrac{dv}{dx}}{v^2} = \frac{(3x+1)\,2e^{2x} - e^{2x} \times 3}{(3x+1)^2}$$
$$= \frac{e^{2x}(6x + 2 - 3)}{(3x+1)^2} = \frac{e^{2x}(6x-1)}{(3x+1)^2}$$

Example 5

Find the equation of the tangent to the curve $y = x + e^x$ at the point $(1, 1 + e)$.

$$y = x + e^x$$
$$\frac{dy}{dx} = 1 + e^x$$

At $x = 1$, $\dfrac{dy}{dx} = 1 + e$.

The gradient of the tangent at $x = 1$ is $1 + e$.

The equation of the tangent is $y - (1 + e) = (1 + e)(x - 1)$

$$y - 1 - e = (1 + e)x - 1 - e$$

The equation of the tangent is $y = (1 + e)x$.

Example 6

Find the coordinates of the turning point of the curve $y = x\,e^{2x}$ and determine its nature.

$$y = x\,e^{2x}$$
$$\frac{dy}{dx} = x\,2e^{2x} + e^{2x} \times 1 \quad \text{[product rule]}$$
$$= e^{2x}(2x + 1)$$

At turning points $\quad e^{2x}(2x + 1) = 0$

$$\Rightarrow \quad 2x + 1 = 0 \quad [\text{since } e^{2x} \text{ is never zero}]$$
$$x = -\tfrac{1}{2}$$

When $x = -\dfrac{1}{2}$, $y = -\dfrac{1}{2} \times e^{-1} = -\dfrac{1}{2e}$

$$\frac{d^2y}{dx^2} = e^{2x} \times 2 + (2x + 1) \times 2e^{2x} \quad \text{[product rule]}$$
$$= 2e^{2x}[1 + 2x + 1]$$
$$= 4e^{2x}(x + 1)$$

When $x = -\dfrac{1}{2}$, $\dfrac{d^2y}{dx^2} = 2e^{-1}$ which is >0.

So there is a minimum point at $\left(-\dfrac{1}{2}, -\dfrac{1}{2e}\right)$

Here is a sketch of the curve.

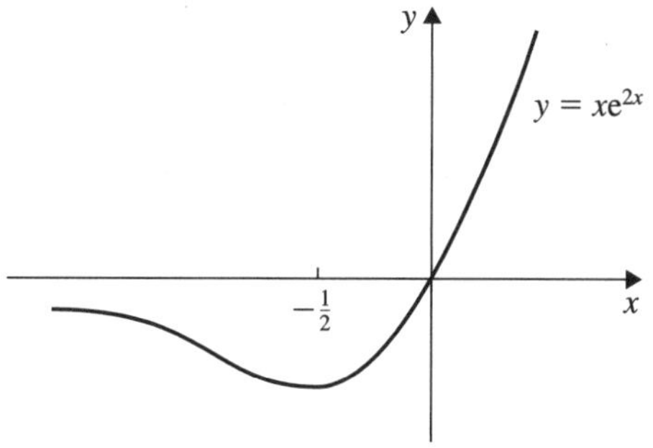

EXERCISE 4

1 Find $\frac{dy}{dx}$ for each of the following.

a $y = e^x$ **b** $y = e^{3x}$ **c** $y = 2e^x$
d $y = e^{x^2}$ **e** $y = e^{2x+1}$ **f** $y = e^{-x}$
g $y = 5e^{4x}$ **h** $y = e^x + e^{2x}$ **i** $y = x^2 - e^{-x}$
j $y = \frac{1}{e^{2x}}$ **k** $y = \frac{4}{e^x}$ **l** $y = e^{x^3} + x^3$

2 Differentiate with respect to x.

a $x\,e^x$ **b** $x^2\,e^x$ **c** $2x\,e^{2x}$
d $(x^2 + 1)e^x$ **e** $4x^3\,e^{2x}$ **f** $x^2(e^{3x} - x)$
g $\frac{e^{3x}}{x}$ **h** $\frac{e^x + 1}{x^2}$ **i** $\frac{x^2}{e^x}$
j $e^x(1 + x)^3$ **k** $e^{-x}(x^3 + 1)$ **l** $\frac{e^{2x}}{1 - e^x}$

3 Find the gradient of each curve at the given point.

a $y = x\,e^x$ at $x = 2$
b $y = x\,e^{x-2}$ at (2, 2)
c $y = e^x + x^2$ at (0, 1)
d $y = \frac{e^{2x} + 1}{e^{2x} - 1}$ at $x = 1$.

4 Find the equation of the tangent to the curve $y = e^x$ at the point where $x = 2$.

5 Find the equation of the tangent to the curve $y = x\,e^x$ at the point (0, 0).

6 a Sketch the curves $y = e^x$ and $y = e^{x-2}$.
b Find the equation of the normal to the curve $y = e^{x-2}$ at the point where the curve meets the y-axis.
c Find the coordinates of the point where the normal meets the x-axis.

7 Find the coordinates of the turning point on the curve $y = x\,e^x$ and show whether it is a maximum or minimum point.

8 Find the coordinates and nature of any turning points on the curve $y = \dfrac{e^x}{x}$.

9 a Find the stationary point of the curve $y = e^x + e^{-x}$.
b Show that this is a local minimum.

10 a Find the coordinates and nature of any turning points on the curve $y = x^2 e^x$.
b Sketch the curve.

11 Given that $y = e^x + \dfrac{1}{e^x}$, show that $y\dfrac{dy}{dx} = e^{2x} - \dfrac{1}{e^{2x}}$.

12 Given that $y = e^x - e^{-x}$, show that $\left(\dfrac{dy}{dx}\right)^2 = y^2 + 4$.

13 Show that if $y = e^{2x} + e^{-2x}$ then $\dfrac{d^2y}{dx^2} = 4y$.

14 Given that $y = A + B\,e^{-4x}$, where A and B are constants, show that $\dfrac{d^2y}{dx^2} + 4\dfrac{dy}{dx} = 0$.

15 Given that $y = e^{2t}(A + Bt)$, where A and B are constants, show that $\dfrac{d^2y}{dx^2} - 4\dfrac{dy}{dx} + 4y = 0$.

16 Find the coordinates of the turning point on the curve $y = 2e^{3x} + 8e^{-3x}$, and determine the nature of this turning point.

4.5 Differentiating ln x

To find $\dfrac{d}{dx}(\ln x)$ we use the result $\dfrac{dy}{dx} = \dfrac{1}{\left(\dfrac{dx}{dy}\right)}$.

This result is justified below.

Now if $y = \ln x$, [Remember $\ln x = \log_e x$]
then $x = e^y$ [1]

Differentiate with respect to y.

$$\frac{dx}{dy} = e^y$$

$$\frac{dy}{dx} = \frac{1}{\left(\dfrac{dx}{dy}\right)} = \frac{1}{e^y}$$

$$\therefore \quad \frac{dy}{dx} = \frac{1}{x} \quad \text{[from [1]]}$$

So if $y = \ln x$, $\dfrac{dy}{dx} = \dfrac{1}{x}$ Remember this result.

To show that $\dfrac{dy}{dx} = \dfrac{1}{\left(\dfrac{dx}{dy}\right)}$:

$$\frac{dy}{dx} = \lim_{\delta x \to 0}\left(\frac{\delta y}{\delta x}\right) = \lim_{\delta x \to 0}\left[\frac{1}{\left(\dfrac{\delta x}{\delta y}\right)}\right] \qquad [2]$$

Now as $\delta x \to 0$, $\delta y \to 0$ also and so [2] becomes

$$\frac{dy}{dx} = \lim_{\delta y \to 0}\left[\frac{1}{\left(\dfrac{\delta x}{\delta y}\right)}\right]$$

and $\boldsymbol{\dfrac{dy}{dx} = \dfrac{1}{\left(\dfrac{dx}{dy}\right)}}$ Remember this result.

Example 1

a Sketch the curve $y = \ln x$.

b Find the gradient of the curve at the point where the curve cuts the x-axis.

a The curve is shown.
Remember that $\ln x$ is defined for $x > 0$ and that the curve cuts the x-axis at (1, 0).

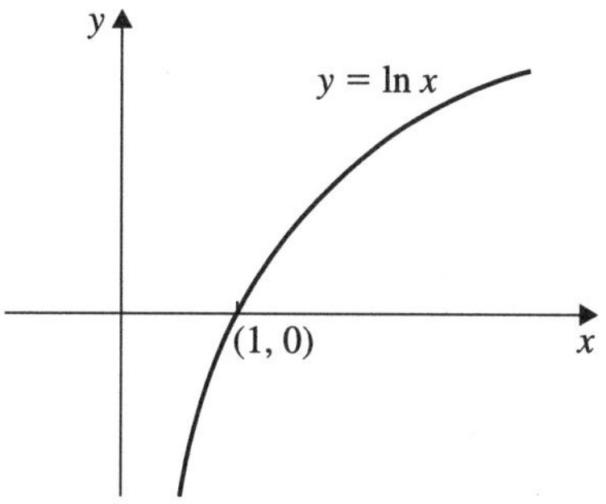

b

$$y = \ln x$$

$$\frac{dy}{dx} = \frac{1}{x}$$

At $x = 1$, $\dfrac{dy}{dx} = \dfrac{1}{1} = 1$.

The gradient of the curve is 1 at the point where it cuts the x-axis.

Example 2

Find $\dfrac{dy}{dx}$ in each of the following:

a $y = \ln 3x^2$

Let $u = 3x^2$

$\therefore\ y = \ln u$

$$\frac{dy}{dx} = \frac{dy}{du} \times \frac{du}{dx}$$

$$= \frac{1}{u} \times 6x$$

$$\frac{dy}{dx} = \frac{1}{3x^2} \times 6x = \frac{2}{x}$$

b $y = \ln(4x - 1)$

Let $u = 4x - 1$

$\therefore\ y = \ln u$

$$\frac{dy}{dx} = \frac{dy}{du} \times \frac{du}{dx}$$

$$= \frac{1}{u} \times 4$$

$$= \frac{4}{4x - 1}$$

Notice that in each case we have $y = \ln[\mathrm{f}(x)]$

$$\text{and } \frac{dy}{dx} = \frac{1}{\mathrm{f}(x)} \times \mathrm{f}'(x).$$

This is a quick way of differentiating $\ln[\mathrm{f}(x)]$.

Example 3

Find $\frac{dy}{dx}$ in each case.

a $y = (x^2 + 1)\ln 2x$

We have $y = uv$, where $u = x^2 + 1$ and $v = \ln 2x$.

Differentiating, $\frac{du}{dx} = 2x$, $\frac{dv}{dx} = \frac{1}{2x} \times 2 = \frac{1}{x}$.

$$\frac{dy}{dx} = u\frac{dv}{dx} + v\frac{du}{dx}$$
$$= (x^2 + 1)\frac{1}{x} + (\ln 2x)\, 2x$$
$$= \frac{x^2 + 1}{x} + 2x \ln 2x$$

b $y = e^x \ln x^2$

We have $y = uv$, where $u = e^x$ and $v = \ln x^2$.

Differentiating, $\frac{du}{dx} = e^x$, $\frac{dv}{dx} = \frac{1}{x^2} \times 2x = \frac{2}{x}$.

$$\frac{dy}{dx} = u\frac{dv}{dx} + v\frac{du}{dx}$$
$$= e^x \frac{2}{x} + (\ln x^2)\, e^x$$
$$= \frac{2e^x}{x} + e^x \ln x^2$$

EXERCISE 5

1 Find $\frac{dy}{dx}$ for each of the following.

a $y = \ln 4x$ **b** $y = \ln x^3$ **c** $y = 6 \ln x$

d $y = \ln(3x - 1)$ **e** $y = \ln(1 - 2x)$ **f** $y = \ln(x^3 + x)$

g $y = \ln\frac{x + 1}{2}$ **h** $y = \ln 4x^2$ **i** $y = 3\ln(x + 2)$

j $y = \ln\frac{1}{x}$ **k** $y = \ln\sqrt{x}$ **l** $y = \ln(x^2 + x - 2)$

2 **a** Write $\ln\left(\frac{x + 4}{x - 2}\right)$ as the difference of 2 logarithms.

b Hence find $\frac{d}{dx}\left[\ln\left(\frac{x + 4}{x - 2}\right)\right]$

3 Differentiate with respect to x.

a $y = \ln\left(\dfrac{3x}{x+1}\right)$ **b** $y = \ln\left(\dfrac{2x+3}{4x-1}\right)$

4 Find $\dfrac{dy}{dx}$ for each of the following.

a $y = x \ln x$ **b** $y = x^2 \ln x$ **c** $y = x \ln(1 + x)$

d $y = \dfrac{\ln x}{x}$ **e** $y = \dfrac{\ln x}{(x+1)^2}$ **f** $y = \dfrac{x}{\ln x}$

g $y = x^2 + \ln x$ **h** $y = 3 \ln x + \dfrac{1}{x}$ **i** $y = \ln \sqrt{x} + \sqrt{x}$

j $y = \dfrac{1 + \ln x}{x}$ **k** $y = x^2 \ln(1 + x^2)$ **l** $y = \ln\left(\dfrac{x+1}{x+2}\right)$

5 If $y = x^2 \ln x$, find $\dfrac{dy}{dx}$ and $\dfrac{d^2y}{dx^2}$.

6 Given that $y = \ln(\ln x)$, show that $x \ln x \dfrac{dy}{dx} = 1$.

7 Given that $y = \ln(1 + e^x)$, find $\dfrac{dy}{dx}$ and show that $(1 + e^x)\dfrac{d^2y}{dx^2} = \dfrac{dy}{dx}$.

8 **a** Sketch the curves $y = \ln x$ and $y = \ln(x + 1)$.

b Find the equation of the tangent to the curve $y = \ln(x + 1)$ at the point $(2, \ln 3)$.

9 Find the equation of the normal to the curve $y = \ln(1 + x^2)$ at the point where $x = 1$.

10 **a** Find the equation of the tangent to the curve $y = \ln(x - 1)$ at the point where the curve meets the x-axis.

b The tangent above meets the x and y axes at the points A and B respectively. Find the area of triangle OAB where O is the origin.

11 The diagram shows the curve $y = \dfrac{\ln x}{x}$.

The curve crosses the x-axis at A and B is the maximum point on the curve. Find the coordinates of A and B.

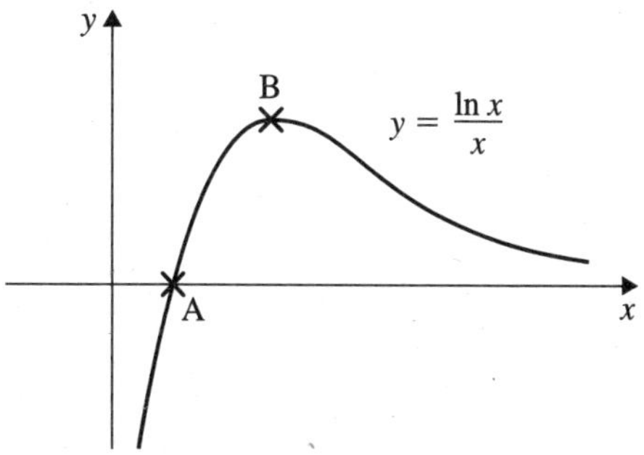

12 Find the coordinates of the point on the curve $y = x \ln x$ where the gradient is zero.

13 Find the x-coordinate of the point on the curve $y = x \ln 2x$ where the gradient is zero.

14 **a** Find the stationary point of the curve $y = \ln(3x - 8) - x - 2$.

b Show that this is a local maximum.

15 **a** If $y = \ln(f(x))$ then find $\frac{dy}{dx}$ in terms of $f(x)$ and $f'(x)$.

b Hence find $\int \frac{3x^2}{x^3 + 1}\,dx$.

16 Find the equation of the tangent to the curve $y = e^x \ln x$ at the point where $x = 1$.

17 Find the coordinates of the point on the curve $y = \ln(e^x + x)$ where the gradient of the curve is 1.

4.6 Differentiating $\sin x$, $\cos x$, $\tan x$

When we differentiate the trigonometric functions we must measure angles in *radians*. You will see why below.

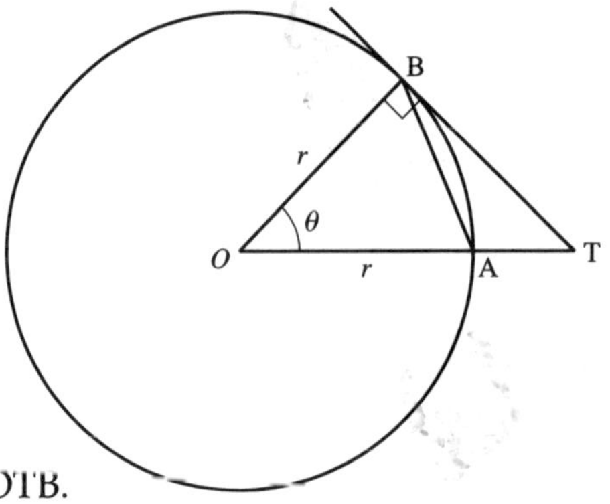

In the diagram, O is the centre of a circle of radius r, AB is a chord and BT is a tangent to the circle.
Angle AOB $= \theta$ radians.

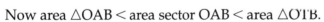

By trigonometry, $\tan\theta = \frac{BT}{r} \Rightarrow BT = r\tan\theta$.

So the area of $\triangle OTB = \frac{1}{2}r \times r\tan\theta$.

$= \frac{1}{2}r^2 \tan\theta$.

Now area $\triangle OAB <$ area sector OAB $<$ area $\triangle OTB$.

$$\tfrac{1}{2}r^2 \sin\theta < \tfrac{1}{2}r^2\,\theta < \tfrac{1}{2}r^2 \tan\theta$$

Divide each term by $\frac{1}{2}r^2$, $\quad \sin\theta < \theta < \frac{\sin\theta}{\cos\theta}$.

θ is acute, so we can divide by $\sin\theta$ which is positive.

$$\frac{\sin\theta}{\sin\theta} < \frac{\theta}{\sin\theta} < \frac{1}{\cos\theta}$$

As $\theta \to 0$, $\cos\theta \to 1$ and so $\frac{\theta}{\sin\theta} \to 1$.

We have $\lim_{\theta \to 0}\left(\frac{\theta}{\sin\theta}\right) = 1$.

This is an interesting and useful result. You can easily test it with your calculator in radian mode. Find the sine of various small angles in radians (0.1, 0.01, 0.001 etc.)

Now consider $y = \sin x$.

$$y + \delta y = \sin(x + \delta x)$$
$$= \sin x \cos\delta x + \cos x \sin\delta x \quad [\text{using } \sin(A + B)]$$

δx is very small, $\cos \delta x \approx 1$ and $\sin \delta x \approx \delta x$.

$$\begin{aligned} y + \delta y &\approx \sin x + \cos \delta x \\ \delta y &\approx \sin x + \cos x\, \delta x - \sin x \\ \delta y &\approx \cos x\, \delta x \\ \frac{\delta y}{\delta x} &\approx \cos x \end{aligned}$$

$$\lim_{\delta x \to 0}\left(\frac{\delta y}{\delta x}\right) = \frac{\mathrm{d}y}{\mathrm{d}x} = \cos x$$

Remember: $\mathbf{\dfrac{d}{dx}(\sin x) = \cos x}$

We can find $\dfrac{\mathrm{d}}{\mathrm{d}x}(\cos x)$ by writing $\cos x = \sin\left(\dfrac{\pi}{2} - x\right)$ and then $\dfrac{\mathrm{d}}{\mathrm{d}x}(\tan x)$ by differentiating $\dfrac{\sin x}{\cos x}$ using the quotient rule.

It is left to the reader to obtain the next two results.

$$\mathbf{\frac{d}{dx}(\cos x) = -\sin x \qquad \frac{d}{dx}(\tan x) = \sec^2 x}$$

Remember these three results and remember also that the angle x must be measured in radians. In the rest of this book whenever we use calculus with trigonometric functions the assumption is made that the angles are given in radians.

Example 1

Find $\dfrac{\mathrm{d}y}{\mathrm{d}x}$.

a $y = \cos x$

$$\frac{\mathrm{d}y}{\mathrm{d}x} = -\sin x$$

b $y = 3 \tan x$

$$\frac{\mathrm{d}y}{\mathrm{d}x} = 3 \sec^2 x$$

c $y = 2 \sin x + \cos x$

$$\frac{\mathrm{d}y}{\mathrm{d}x} = 2 \cos x - \sin x$$

d $y = \sin 4x$

Let $u = 4x$

So $y = \sin u$

$$\begin{aligned} \frac{\mathrm{d}y}{\mathrm{d}x} &= \frac{\mathrm{d}y}{\mathrm{d}u} \times \frac{\mathrm{d}u}{\mathrm{d}x} \\ &= \cos u \times 4 \\ \frac{\mathrm{d}y}{\mathrm{d}x} &= 4 \cos 4x \end{aligned}$$

e $y = \cos 3x^2$

Let $u = 3x^2$

So $y = \cos u$

$$\begin{aligned} \frac{\mathrm{d}y}{\mathrm{d}x} &= \frac{\mathrm{d}y}{\mathrm{d}u} \times \frac{\mathrm{d}u}{\mathrm{d}x} \\ &= -\sin u \times 6x \\ \frac{\mathrm{d}y}{\mathrm{d}x} &= -6x \sin 3x^2 \end{aligned}$$

In **d** and **e** observe that $\dfrac{\mathrm{d}}{\mathrm{d}x}[\sin f(x)] = f'(x) \cos f(x)$,

$$\frac{\mathrm{d}}{\mathrm{d}x}[\cos f(x)] = -f'(x) \sin f(x).$$

and similarly $\dfrac{\mathrm{d}}{\mathrm{d}x}[\tan f(x)] = f'(x) \sec^2 f(x)$

We use this as a quick method to avoid the use of the substitution $u = f(x)$.

Example 2

Find $\frac{dy}{dx}$.

a $y = \sin 4x^2$

$\frac{dy}{dx} = 8x \cos 4x^2$

b $y = 2 \tan 3x$

$\frac{dy}{dx} = 2 \times 3 \times \sec^2 3x$

$= 6 \sec^2 3x$

c $y = \cos\left(\frac{x}{5}\right)$

$\frac{dy}{dx} = -\frac{1}{5}\sin\left(\frac{x}{5}\right)$

Example 3

Find $\frac{dy}{dx}$.

a $y = \sin^2 x$

$y = (\sin x)^2$

$\frac{dy}{dx} = 2(\sin x)^1 \times \cos x$

$= 2 \sin x \cos x$

[Function of a function]

b $y = \cos^3 4x$

$y = (\cos 4x)^3$

$\frac{dy}{dx} = 3(\cos 4x)^2 \times (-4 \sin 4x)$

$= -12 \cos^2 4x \sin 4x$

c $y = x^2 \tan 7x$

$\frac{dy}{dx} = x^2 \times 7 \sec^2 7x + (\tan 7x) \times 2x$

$= 7x^2 \sec^2 7x + 2x \tan 7x$

d $y = \sqrt{\sin 4x}$

$y = (\sin 4x)^{\frac{1}{2}}$

$\frac{dy}{dx} = \frac{1}{2}(\sin 4x)^{-\frac{1}{2}} \times (\cos 4x) \times 4$

$= 2\,\frac{\cos 4x}{\sqrt{\sin 4x}}$

Example 4

Find the equation of the tangent to the curve $y = \cos x$ at the point $\left(\frac{\pi}{6}, \frac{\sqrt{3}}{2}\right)$.

$y = \cos x$

$\frac{dy}{dx} = -\sin x$

At $x = \frac{\pi}{6}$, $\frac{dy}{dx} = -\sin\frac{\pi}{6} = -\frac{1}{2}$

The tangent passes through $\left(\frac{\pi}{6}, \frac{\sqrt{3}}{2}\right)$.

The equation of the tangent is

$y - \frac{\sqrt{3}}{2} = -\frac{1}{2}\left(x - \frac{\pi}{6}\right)$

$2y - \sqrt{3} = -x + \frac{\pi}{6}$

$2y + x = \sqrt{3} + \frac{\pi}{6}$.

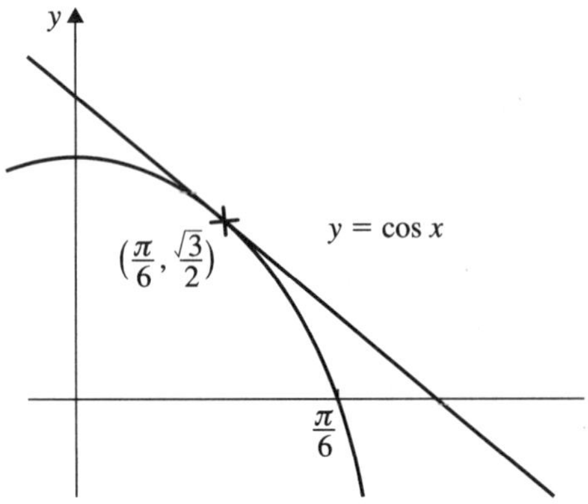

EXERCISE 6

1 Differentiate the following with respect to x.

a $\sin x$ **b** $\cos 3x$ **c** $4 \tan x$

d $\sin 6x$ **e** $\cos \frac{3}{2}x$ **f** $5 \sin 2x$

g $\tan \frac{1}{2}x$ **h** $10 \cos x + \sin x$ **i** $\tan x - \cos x$

j $\sin(x + 1)$ **k** $\tan(2x - 1)$ **l** $-\cos \frac{1}{2}x$

2 Find $\dfrac{dy}{dx}$ in the following.

a $y = (\sin x)^2$ **b** $y = 3(\cos x)^2$ **c** $y = \sin^3 x$

d $y = 2\cos^3 x$ **e** $y = \sqrt{\cos x}$ **f** $y = \cos^2 4x$

g $y = \tan^2 9x$ **h** $y = x + \sqrt{\sin x}$ **i** $y = (\sin 2x)^{\frac{3}{2}}$

3 Differentiate the following with respect to x.

a $x \sin x$ **b** $x \cos 2x$ **c** $x^2 \sin x$

d $\sin x \cos x$ **e** $\dfrac{\cos x}{x}$ **f** $\dfrac{\sin 2x}{x^2}$

g $\dfrac{x}{\sin x}$ **h** $\dfrac{x^3}{\cos x}$ **i** $\dfrac{\cos x}{\sin x}$

4 Given $\cos x = \sin\left(\dfrac{\pi}{2} - x\right)$, differentiate $\sin\left(\dfrac{\pi}{2} - x\right)$ to show that $\dfrac{d}{dx}(\cos x) = -\sin x$.

5 Differentiate $\dfrac{\sin x}{\cos x}$ to show that $\dfrac{d}{dx}(\tan x) = \sec^2 x$.

6 Find the gradient of each curve at the point given.

a $y = 3 \sin 2x$ at $\left(\dfrac{\pi}{2}, 0\right)$. **b** $y = 4 \sin x - \cos x$ at $(0, -1)$.

c $y = (\cos 2x)^4$ at $\left(\dfrac{\pi}{8}, \dfrac{1}{4}\right)$. **d** $y = x \tan x$ at $\left(\dfrac{\pi}{4}, \dfrac{\pi}{4}\right)$.

7 Given that $y = \cos 2x$, show that $\dfrac{d^2y}{dx^2} = -4y$.

8 Given that $y = \sin x + \cos x$, show that $\dfrac{d^2y}{dx} + y = 0$.

9 Find the equation of the tangent to the curve $y = \tan x$ at the point $\left(\dfrac{\pi}{4}, 1\right)$.

10 Differentiate with respect to x.

a $\sin \pi x$ **b** $\tan 2\pi x$ **c** $\cos(x - \pi)$

d $\cos \dfrac{\pi}{2}x$ **e** $\sin x^2$ **f** $\tan x^3$

g $\sin(2x - \pi)$ **h** $\sin \dfrac{x}{\pi}$ **i** $\sin^2 x^2$

11 Consider the curve $y = \sin x^2$ for $0 \leqslant x \leqslant \pi$.
Find the x coordinates of the two points on the curve where the gradient is zero.

12 Show that the equation of the tangent to the curve $y = \sin x$ at the point $(\pi, 0)$ is $y + x = \pi$.

13 Find the equation of the normal to the curve $y = x - \cos x$ at the point where $x = \frac{\pi}{2}$.

14 Show that, at $x = \frac{\pi}{4}$, $\frac{\mathrm{d}}{\mathrm{d}x}(\sin^2 x + \cos^3 x) = \frac{2\sqrt{2} - 3}{2\sqrt{2}}$.

15 Find the maximum value of $y = x + \sin 2x$ for $0 < x < \frac{\pi}{2}$.

16 Find the minimum value of $y = \tan x - 8 \sin x$ which is given by a value of x between 0 and $\frac{\pi}{2}$.

17 Show that for $0 < x < \frac{\pi}{2}$, $f(x) = x \sin x + \cos x$ is an increasing function.

18 **a** If $y = \frac{1 + \cot x}{1 - \cot x}$, show that $\frac{\mathrm{d}y}{\mathrm{d}x} = \frac{2}{\sin 2x - 1}$.

b Find the gradient of the curve at $x = \frac{\pi}{12}$.

19 The tangent to the curve $y = \cos 2x$ at the point where $x = -\frac{\pi}{6}$ meets the y-axis at the point Y.
Find the distance OY where O is the origin.

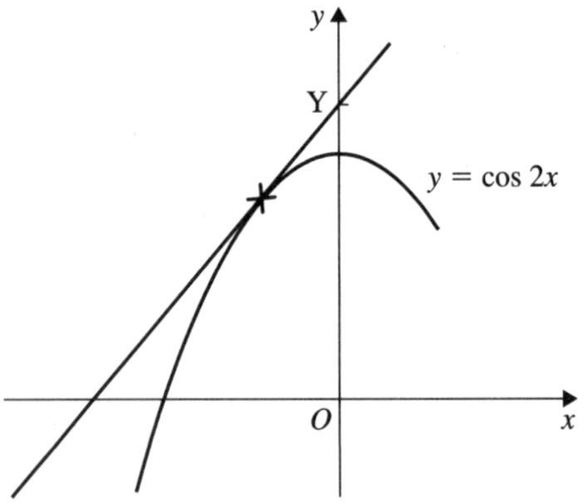

20 Given $f(\theta) = 4 \cos \theta - \sin^2 \theta$, find $f'\left(\frac{\pi}{2}\right)$.

21 Find the coordinates of the stationary points on the curve $y = \sin x + \cos x$ for $0 \leqslant x \leqslant 2\pi$.
Determine the nature of each point.

22 **a** If $y = \sin x(\cos x)^2$, find $\frac{\mathrm{d}y}{\mathrm{d}x}$.

b Find the three values of x between 0 and π radians at which $\frac{\mathrm{d}y}{\mathrm{d}x} = 0$. Give your answers as decimals correct to 2 decimal places.

23 a Find the equation of the tangent to the curve $y = \sin x$ at the point where $x = \frac{4\pi}{3}$.

b Find the coordinates of the point where this tangent meets the y-axis.

24 a Find the gradient of the curve $y = \sin \pi x$ at the point where $x = 1$.

b Find the equation of the normal to the curve $y = \sin \pi x$ at the point where $x = 1$.

Derivatives of cosec x, cot x and sec x

The derivatives of cosec x, cot x and sec x can be found as follows.

a $y = \operatorname{cosec} x = \frac{1}{\sin x} = (\sin x)^{-1}$

$$\frac{dy}{dx} = -(\sin x)^{-2}\cos x$$
$$= -\frac{\cos x}{\sin^2 x}$$
$$= -\frac{1}{\sin x} \times \frac{\cos x}{\sin x}$$
$$= -\operatorname{cosec} x \cot x$$

b $y = \sec x = \frac{1}{\cos x} = (\cos x)^{-1}$

$$\frac{dy}{dx} = -(\cos x)^{-2}(-\sin x)$$
$$= \frac{\sin x}{\cos^2 x} = \frac{1}{\cos x} \times \frac{\sin x}{\cos x}$$
$$= \sec x \tan x$$

c $y = \cot x = (\tan x)^{-1}$

$$\frac{dy}{dx} = -(\tan x)^{-2}\sec^2 x$$
$$= -\frac{\cos^2 x}{\sin^2 x} \times \frac{1}{\cos^2 x}$$
$$= -\operatorname{cosec}^2 x$$

In summary:

$$\mathbf{\frac{d}{dx}(\operatorname{cosec} x) = -\operatorname{cosec} x \cot x}$$
$$\mathbf{\frac{d}{dx}(\sec x) = \sec x \tan x}$$
$$\mathbf{\frac{d}{dx}(\cot x) = -\operatorname{cosec}^2 x}$$

Some people learn these results, while others, including the authors of this book, can never remember them and work them out each time!

EXERCISE 7

1 Differentiate with respect to x.

a $\operatorname{cosec} x$ **b** $\cot x$ **c** $\sec 2x$ **d** $\cot 3x$

e $\operatorname{cosec} 4x$ **f** $2\cot x$ **g** $\sec(2x + 1)$ **h** $x^2 + \sin x + \sec x$

2 Find $\frac{dy}{dx}$.

a $y = (\cot x)^2$ **b** $y = (\sec x)^2$ **c** $y = \tan x \sec x$

d $\sec^2 3x$ **e** $x \sec x$ **f** $x^2 \cot 2x$

g $\frac{\sec x}{x}$ **h** $\frac{x^2}{\cot x}$ **i** $(1 + \sec x)^2$

3 Show that $\frac{d}{dx}\left(\frac{\tan x}{1 + \sec x}\right) = \frac{1}{1 + \cos x}$

4 Given that $y = \cot x$, show that $\dfrac{d^2y}{dx^2} = 2(1 + y^2)y$.

5 If $y = \sqrt{\sec x + \tan x}$, show that $\dfrac{dy}{dx} = \dfrac{1}{2} y \sec x$.

6 Work out $\displaystyle\int \frac{2 + \cos x}{\sin^2 x}\,dx$.

4.7 Further questions on differentiation

Here is a summary of the results used in this section.

$$\frac{dy}{dx} = \frac{dy}{du} \times \frac{du}{dx} \quad \text{(chain rule)}$$

$$\frac{d}{dx}(uv) = u\frac{dv}{dx} + v\frac{du}{dx} \qquad \frac{d}{dx}\left(\frac{u}{v}\right) = \frac{v\dfrac{du}{dx} - u\dfrac{dv}{dx}}{v^2}$$

$$\frac{d}{dx}(e^x) = e^x \qquad \frac{d}{dx}(\ln x) = \frac{1}{x}$$

$$\frac{d}{dx}(\sin x) = \cos x \qquad \frac{d}{dx}(\cos x) = -\sin x \qquad \frac{d}{dx}(\tan x) = \sec^2 x$$

$$\frac{dy}{dx} = \frac{1}{\dfrac{dx}{dy}}$$

In this section you will do questions where more than one of these results may be required.

Example 1

Find $\dfrac{dy}{dx}$.

a $y = e^x \sin x^2$

$$\frac{dy}{dx} = e^x(\cos x^2)2x + \sin x^2\, e^x \quad \text{(product rule)}$$
$$= e^x(2x \cos x^2 + \sin x^2)$$

b $y = e^{2x} \ln(1 + x)$

$$\frac{dy}{dx} = e^{2x} \times \frac{1}{(1 + x)} + \ln(1 + x) \times 2e^{2x} \quad \text{(product rule)}$$
$$= e^{2x}\left[\frac{1}{1 + x} + 2\ln(1 + x)\right]$$

c $y = \dfrac{\sin 4x}{e^x}$

$$\frac{dy}{dx} = \frac{e^x \times 4\cos 4x - \sin 4x \times e^x}{(e^x)^2} \quad \text{(quotient rule)}$$
$$= \frac{4\cos 4x - \sin 4x}{e^x}$$

Example 2

Find the equation of the tangent to the curve $y = e^x \sin x$ at the point $(\pi, 0)$.

$$y = e^x \sin x$$
$$\frac{dy}{dx} = e^x \cos x + \sin x.e^x$$
$$= e^x(\cos x + \sin x)$$

At $x = \pi$, $\dfrac{dy}{dx} = e^{\pi}(\cos \pi + \sin \pi)$
$$= -e^{\pi}$$

Equation of tangent at $(\pi, 0)$ is $y - 0 = -e^{\pi}(x - \pi)$
$$y = e^{\pi}(\pi - x)$$

Example 3

If $x = \sin y$, find $\dfrac{dy}{dx}$ in terms of x.

$$x = \sin y$$

Differentiate both sides with respect to y.

$$\frac{dx}{dy} = \cos y$$
$$\frac{dy}{dx} = \frac{1}{\frac{dx}{dy}} = \frac{1}{\cos y}$$

Now $\sin^2 y + \cos^2 y = 1$
$$\cos^2 y = 1 - \sin^2 y$$
$$\cos y = \sqrt{1 - \sin^2 y}$$
$$= \sqrt{1 - x^2}$$

$$\therefore \quad \frac{dy}{dx} = \frac{1}{\sqrt{1 - x^2}}$$

Example 4

If $x = \ln 5y$, find $\dfrac{dy}{dx}$ in terms of x.

$$x = \ln 5y \qquad [1]$$

Differentiate both sides with respect to y.

$$\frac{dx}{dy} = \frac{1}{5y} \times 5 = \frac{1}{y} \qquad [2]$$

From [1] $5y = e^x$
$$y = \tfrac{1}{5}e^x$$

From [2] $\frac{dy}{dx} = \frac{1}{\frac{dx}{dy}} = \frac{1}{\frac{1}{y}} = y \quad \therefore \quad \frac{dy}{dx} = \frac{1}{5}e^x$

EXERCISE 8

1 Differentiate the following:

a $\ln x$ **b** $\ln(2x)$ **c** $2\ln(3x)$

d $x + \ln(2x)$ **e** $5e^x + \ln(3x)$ **f** $2e^x + 5\ln x$

2 **a** Find the exact coordinates of the stationary point of the curve $y = \ln x - 3x + 1$.

b Show that this is a local maximum.

3 **a** Find the exact coordinates of the stationary point of the curve $y = \ln(4x) - x - 2$.

b Show that this is a local maximum.

4 **a** Show that the curve $y = \ln(2x) + x^2$ passes through $(\frac{1}{2}, \frac{1}{4})$.

b Find the equation of the normal to $y = \ln(2x) + x^2$, at $(\frac{1}{2}, \frac{1}{4})$.

5 Find the equation of the tangent to the curve $y = \ln x + x^2 + 3x$ at the point (1, 4).

6 Find $\frac{dy}{dx}$ in the following:

a $y = 5e^x$ **b** $y = 7e^x$ **c** $y = 5e^{4x}$

d $y = \ln(x + 1)$ **e** $y = \frac{4}{e^x}$ **f** $y = (e^{3x} + 1)^2$

g $y = 2e^x + \ln 3x$ **h** $y = \frac{5e^{3x} + 2e^{2x}}{e^{2x}}$ **i** $y = \frac{e^x + 1}{e^x - 1}$

7 If $y = (e^{\frac{x}{2}} + 1)(e^{\frac{x}{2}} - 1)$ then find $\frac{dy}{dx}$.

8 Differentiate the following:

a $y = \ln(x^2)$ **b** $y = \ln(2x^3)$ **c** $y = \ln(\sqrt{x})$

d $y = \ln\left(\frac{1}{x}\right)$ **e** $y = \ln(\sqrt{9x})$ **f** $y = \ln\left(\frac{1}{\sqrt[3]{x}}\right)$

9 Find $\frac{dy}{dx}$ in the following.

a $y = (x + 1)^2$ **b** $y = (2x + 1)^7$ **c** $y = (3x - 5)^8$

d $y = 5(3x - 7)^3$ **e** $y = (4x + 3)^{-3}$ **f** $y = \frac{1}{(5x - 2)^4}$

g $y = (4x + 11)^{\frac{1}{2}}$ **h** $y = \sqrt[3]{(15x - 17)}$ **i** $y = (ax + b)^n$

10 Show that the equation of the tangent to the curve $y = 2e^x$ at the point where $x = \frac{1}{2}$ is $y = \sqrt{e}(2x + 1)$.

11 Find the turning point of the curve $y = x^2 - 2\ln x$, where $x > 0$, and show that this is a local minimum.

12 Find the turning point of the curve $y = x - e^x$ and show that this is a local maximum.

13 Find $\frac{dy}{dx}$ in the following.

a $y = x \ln x$ **b** $y = x^2 e^x$ **c** $y = \sin x \cos x$

d $y = \sqrt{x}e^x$ **e** $y = e^x \tan x$ **f** $y = x \cot x$

g $y = e^x \operatorname{cosec} x$ **h** $y = x^3 \sec x$ **i** $y = x^3 \cos x$

14 If $y = x^n e^x$ then show that $\frac{dy}{dx} = x^{n-1}e^x(x + n)$.

15 Find $\frac{dy}{dx}$ in the following, simplifying and factorising where possible:

a $y = \frac{\ln x}{x}$ **b** $y = \frac{e^x}{x^2}$ **c** $y = \frac{\sec x}{e^x}$

d $y = \frac{\sin x}{\cos x}$ **e** $y = \frac{\cos x}{\sin x}$ **f** $y = \frac{x^2}{e^x}$

g $y = \frac{x + 1}{\sqrt{x}}$ **h** $y = \frac{x^3}{\cos x}$ **i** $y = \frac{\cot x}{x^2}$

16 a If $y = \frac{2x + 3}{\sqrt{x}}$ then show that $\frac{dy}{dx} = \frac{2x - 3}{2x\sqrt{x}}$.

b Show that the turning point is at $(\frac{3}{2}, 2\sqrt{6})$.

17 The gradient of the curve $y = e^x(ax^2 + bx + c)$ is zero at $x = 1$ and at $x = -2$. The curve passes through the point $(0, -1)$. Find the values of a, b and c.

18 Find $\frac{dy}{dx}$ in the following.

a $y = \sin 2x$ **b** $y = e^{5x}$ **c** $y = e^{\sin x}$

d $y = \cos(x^2)$ **e** $y = e^{\sec x}$ **f** $y = \operatorname{cosec}\left(3x + \frac{\pi}{3}\right)$

g $y = \cot(3x)$ **h** $y = \ln(x^3 + 3)$ **i** $y = \ln(\sin 2x)$

j $y = \ln(\cos x)$ **k** $y = e^{x^3}$ **l** $y = \ln(\sec x)$

m $y = (x^2 + 1)^3$ **n** $y = \sin^3 x$ **o** $y = \tan^3 (2x)$

19 If $x = \cos y$, find $\frac{dy}{dx}$. Give your answer in terms of x.

20 If $x = \sin 2y$, find $\frac{dy}{dx}$ in terms of x.

21 Find $\frac{dy}{dx}$, in terms of x.

a $x = \tan y$ **b** $x = \tan 3y$ **c** $x = e^{2y}$

d $x = e^{-y}$ **e** $x = \ln 5y$ **f** $x = \ln y^2$

g $x = \sin\left(2y - \frac{\pi}{6}\right)$

22 **a** Consider $x = y^2, y \geqslant 0$. Find $\frac{dy}{dx}$, using $\frac{dy}{dx} = \frac{1}{\frac{dx}{dy}}$.

b Write $y = \sqrt{x}$ and find $\frac{dy}{dx}$. Confirm that your result is the same as that obtained in part **a**.

23 Express $\frac{dy}{dx}$ in terms of $f(x)$ and $f'(x)$:

a $y = f(x)^n$ **b** $y = e^{f(x)}$ **c** $y = \sin(f(x))$

24 **a** If $\log_a x = p$, $\log_e a = q$ and $\log_e x = r$ then write down three equations involving p, q and r without logarithms.

b Hence show that $p = \frac{r}{q}$.

c Hence express $\log_a x$ in terms of $\ln x$ and $\ln a$ and so show that if $y = \log_a x$ where a is a constant then $\frac{dy}{dx} = \frac{1}{x \ln a}$.

REVIEW EXERCISE 4

1 Differentiate the following with respect to x.

a $y = (3x - 1)^4$ **b** $y = 5(3x - 4)^3$ **c** $y = (4x + 3)^{-2}$

d $y = \frac{1}{(5x - 1)^3}$ **e** $y = \frac{1}{(2 - 3x)^4}$ **f** $y = (3x + 1)^{\frac{1}{2}}$

2 Differentiate the following with respect to x.

a $y = (x^2 + 3)^4$ **b** $y = (2 - x^3)^5$ **c** $y = (3x^2 + 1)^5$

d $y = (x^2 + x)^4$ **e** $y = \left(1 + \frac{1}{x}\right)^3$ **f** $y = \left(x^2 + \frac{1}{x}\right)^{-3}$

3 Differentiate with respect to x, simplify your results.

a $y = x(x + 1)^3$ **b** $y = (x + 1)(2x - 3)^4$

c $y = (x + 1)^3(x - 1)^2$ **d** $y = x^2(3x - 2)^2$

4 Sketch the curve $y = (x - 1)^2(x + 1)$, showing the coordinates of

a the points where it meets the axes

b the turning points.

5 Find the coordinates of the turning points of the curve $y = (x - 2)(x + 3)^2$. Determine the nature of each turning point.

6 Differentiate with respect to x.

a $\dfrac{x^2}{x + 1}$ **b** $\dfrac{\sqrt{x + 1}}{x}$ **c** $\dfrac{x + 2}{x + 1}$

d $\dfrac{e^{3x}}{x^2}$ **e** $\dfrac{3 - 2x}{x^2}$ **f** $\dfrac{\sin x}{x}$

7 Show that the curve $y = \dfrac{x^3}{(1 + x^4)^{\frac{1}{2}}}$ has a positive gradient for all values of x except at $x = 0$.

8 Differentiate the following with respect to x.

a $y = e^x$ **b** $y = -3e^x$ **c** $y = \frac{1}{2}e^x$

d $y = 2x^3 - 4e^x$ **e** $y = 3(e^x - x^{\frac{1}{2}})$ **f** $y = \frac{1}{2}(x^{\frac{1}{3}} - \frac{1}{2}e^x)$

9 Differentiate the following with respect to x.

a $y = \ln x$ **b** $y = 4 \ln x$ **c** $y = \ln 2x$

d $y = \ln x^2$ **e** $y = \ln x^{\frac{1}{2}}$ **f** $y = \ln\left(\dfrac{x}{2}\right)$

g $y = \ln\left(\dfrac{1}{x}\right)$

10 Find the gradient of the following curves at the given points.

a $y = e^x - x$: $x = 2$

b $y = 3e^x - 2x$: $x = 0$

c $y = \ln 2x - \sqrt{x}$: $x = 4$

d $y = x^3 - 3 \ln x^2$: $x = 2$

11 Find the equation of the tangent to the following curves at the given point.

a $y = e^x$: $x = 0$

b $y = \ln x$: $x = 1$

c $y = 1 - e^x$: $x = 2$

12 Find the equation of the normal to the following curves at the given point.

a $y = e^x$: $x = 1$

b $y = \ln x^2$: $x = 3$

13 Find the coordinates of the stationary point of the following curves and show whether it is a maximum or a minimum.

a $y = e^x - x$

b $y = e^x + e^{-x}$

c $y = -x + \ln x$

d $y = 2x^2 - \ln x$

e $y = \ln x - 8x + 1$

14 The curve $y = \dfrac{\ln x}{x}$ cuts the x-axis at A and has a maximum point at B.
Find the coordinates of A and B.

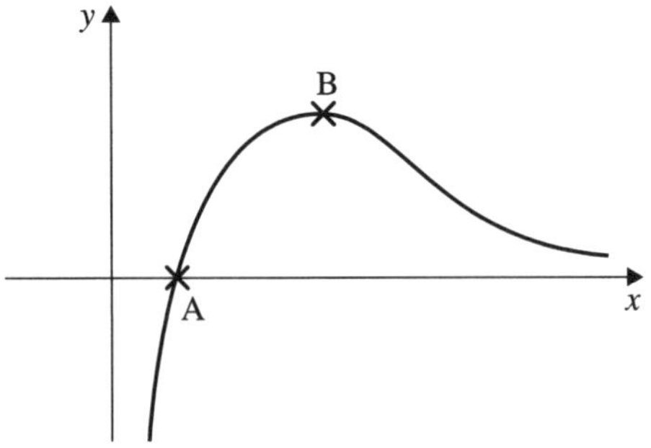

15 If $y = (e^x + 1)^2$, show that $\dfrac{dy}{dx} = 2e^x(e^x + 1)$.

16 Differentiate the following with respect to x.

a $\sin x + 3$ **b** $2\cos x$ **c** $\sin x + \cos x$

d $2\sin x - 3\cos x$ **e** $\tan x + x^2 + 4$ **f** $3\sin x - \tan x$

17 Find the gradient of the following curves at the given point.

a $y = \sin x$: $x = \dfrac{\pi}{3}$

b $y = \cos x$: $x = \dfrac{\pi}{2}$

c $y = \cos x + \sin x$: $x = \dfrac{\pi}{4}$

d $y = x + \cos x$: $x = \dfrac{\pi}{6}$

e $y = 3 - \cos 2x$: $x = \dfrac{\pi}{12}$

18 Find the equation of the tangent to the curve $y = x - \cos x$ where $x = \dfrac{\pi}{2}$

19 Find the equation of the normal to the curve $y = 2\sin x + \cos x$ where $x = \dfrac{\pi}{2}$

20 Differentiate the following with respect to x.

a $y = \sin 3x$ **b** $y = \cos 4x$ **c** $y = \tan 5x$

d $y = \sin^2 x$ **e** $y = \tan^4 x$ **f** $y = \sin\left(2x + \frac{\pi}{4}\right)$

g $y = \sqrt{\sin x}$ **h** $y = \sin^2 3x$ **i** $y = \cos^3 2x$

21 Differentiate the following with respect to x.

a $y = e^{2x}$ **b** $y = e^{-3x}$ **c** $y = e^{x^2}$

d $y = \frac{1}{e^x}$ **e** $y = e^{\sin x}$ **f** $y = \ln(x^2 + 1)$

g $y = \ln(x^3 - 2)$ **h** $y = \ln(x^3 + 2x - 1)$ **i** $y = \ln\left(\frac{x^2+1}{x-1}\right)$

j $y = \ln\left(\frac{1}{x^2+3}\right)$ **k** $y = e^{\ln x}$ **l** $y = e^{\sqrt{x}}$

22 Show that if $y = (e^x + e^{-x})^2$ then $\frac{dy}{dx} = 2(e^{2x} - e^{-2x})$.

23 Given that $y = e^x \ln(1 + \sin x)$, find $\frac{dy}{dx}$ in terms of x.

24 Differentiate with respect to x.

a $y = x e^x$ **b** $y = x \ln x$ **c** $y = e^x \ln x$

d $y = x \sin x$ **e** $y = x^2 \cos x$ **f** $y = e^x \tan x$

25 The diagram shows a right circular cone of slant height ℓ.

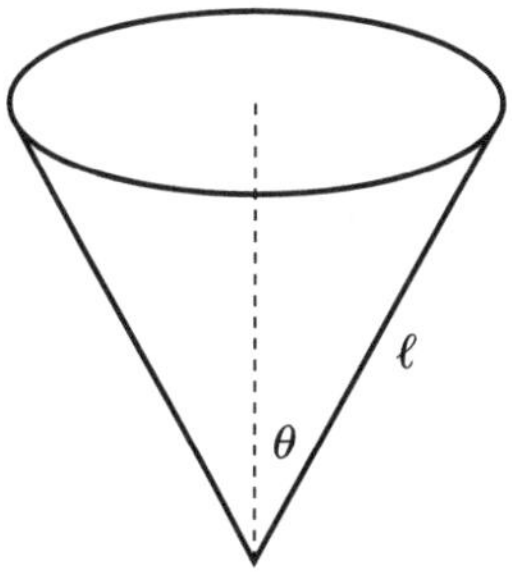

a Show that the volume V of the cone is given by $V = \frac{1}{3}\pi\ell^3 \sin^2\theta \cos\theta$

b Given that ℓ is constant and that θ varies, find the maximum value of V in terms of ℓ and π.

26 Find the coordinates of the points on the curve $y = \sin x \cos^3 x$ where the gradient is zero. Work in the range $0 < x < \pi$.

27 Differentiate the following with respect to x.

a $y = \sin^2 x \cos x$ **b** $y = \ln(2x + 5)$ **c** $y = \sin 2x \cos 3x$

d $y = e^{x^2+2}$ **e** $y = (\ln x)^2$ **f** $y = \sin x^2$

28 Find $\frac{dy}{dx}$.

a $y = \sec x$ **b** $y = \operatorname{cosec} x$ **c** $y = \cot x$

29 Find the gradient of the curve $y = \sec 3x$ at the point where $x = \frac{\pi}{9}$.

30 Find the coordinates of the stationary point on the curve $y = 2x - \ln(x^4)$ and determine the nature of this stationary point.

31 Use $\dfrac{dy}{dx} = \dfrac{1}{\frac{dx}{dy}}$ to find $\dfrac{dy}{dx}$ in terms of x for the following.

a $x = \sin y$ **b** $x = e^{3y}$ **c** $x = \ln y$ **d** $x = \ln 2y$

32 Given that $x = y^3 \ln 2y, \quad y > 0$

a Find $\dfrac{dx}{dy}$

b Use your answer to part (a) to find, in terms of e, the value of $\dfrac{dy}{dx}$ at $y = e$.

33 Given that $y = \dfrac{e^x}{x^2 - 3}$, find the x coordinates of the two points on the curve where the gradient is zero.

Examination Exercise 4

1 Differentiate with respect to x

i $(3 - 2x)^5$,

ii $\ln(4x + 7)$. [OCR]

2 Find $\dfrac{dy}{dx}$ for each of the following cases

a $y = e^{2x} \sin 3x$,

b $y = (2x^2 + 1)^5$. [AQA]

3 Find the equation of the tangent to the curve $y = (3x - 4)^5$ at the point for which $x = 2$, giving your answer in the form $y = mx + c$. [OCR]

4 **a** Differentiate $\sqrt{1 + x^2}$.

b Differentiate $\dfrac{1 + x}{1 - x}$. [MEI]

5 **a** Differentiate:

i $2x^{\frac{1}{2}}$;

ii $\ln(x + 1)$;

b Hence show that $\displaystyle\int_1^4 \left(x^{-\frac{1}{2}} + \frac{1}{x + 1}\right) dx = 2 + \ln\tfrac{5}{2}$. [AQA]

6 The equation of a curve is $y = xe^{-2x}$.

Find the coordinates of the turning point on the curve. [OCR]

7 Fig. 1 shows the graph of $y = x\sqrt{1+x}$. The point P on the curve is on the x-axis.

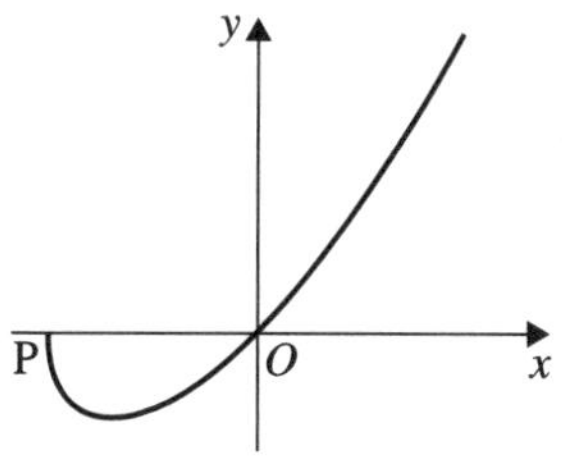

Fig. 1

i Write down the coordinates of P.

ii Show that $\dfrac{dy}{dx} = \dfrac{3x+2}{2\sqrt{1+x}}$.

iii Hence find the coordinates of the turning point on the curve. What can you say about the gradient of the curve at P? [MEI]

8 The curve C has equation $y = 2e^x + 3x^2 + 2$. The point A with coordinates (0, 4) lies on C. Find the equation of the tangent to C at A. [EDEXCEL]

9 Use the derivatives of $\sin x$ and $\cos x$ to prove that the derivative of $\tan x$ is $\sec^2 x$. [AQA]

10 Find the equation of the tangent to the curve $y = \dfrac{2+x}{\cos x}$ at the point on the curve where $x = 0$. [AQA]

11

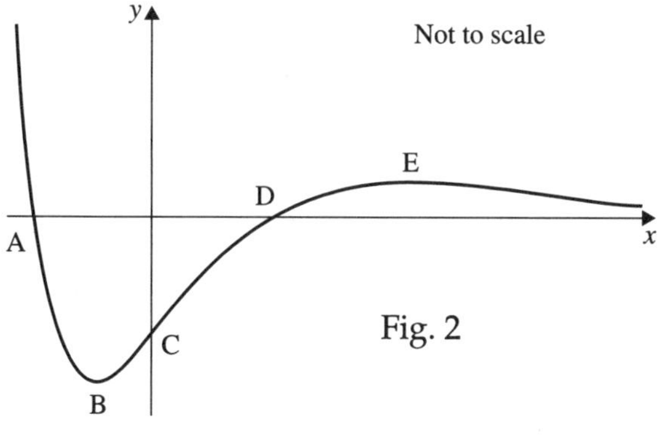

Fig. 2

Fig. 2 shows a sketch of the graph of $y = (x^2 - 3)e^{-x}$. The graph crosses the x-axis at points A and D and the y-axis at C. Points B and E are stationary points on the curve.

i Find the coordinates of the points A, C and D.

ii Show that $\dfrac{dy}{dx} = -(x^2 - 2x - 3)e^{-x}$.

iii Deduce that the x-coordinates of the points B and E are -1 and 3 respectively, and find the corresponding y-coordinates. [MEI]

12 The function f is defined for all $x > 0$ by $f(x) = 4 \ln x - \frac{1}{x}$.

a Find the derivative $f'(x)$.

b Explain why f is an increasing function.

c State, giving a reason, whether f^{-1} exists. [AQA]

13 Fig. 1 shows part of the graph with equation $y = x\sqrt{9 - 2x^2}$. It crosses the x-axis at $(a, 0)$.

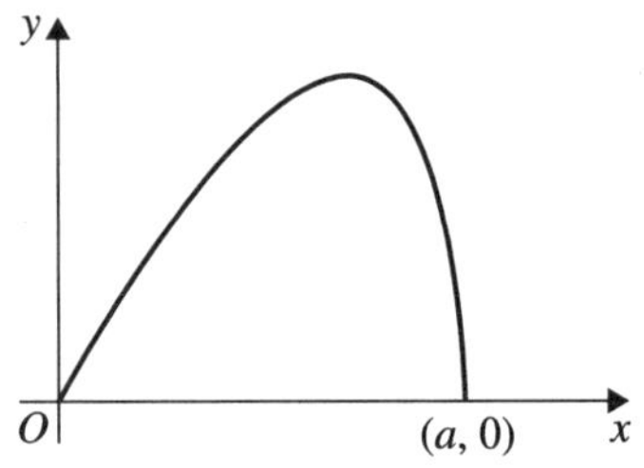

i Find the value of a, giving your answer as a multiple of $\sqrt{2}$.

ii Show that the result of differentiating $\sqrt{9 - 2x^2}$ is $\frac{-2x}{\sqrt{9 - 2x^2}}$.

Hence show that if $y = x\sqrt{9 - 2x^2}$ then

$$\frac{dy}{dx} = \frac{9 - 4x^2}{\sqrt{9 - 2x^2}}.$$

iii Find the x-coordinate of the maximum point on the graph of $y = x\sqrt{9 - 2x^2}$. Write down the gradient of the curve at the origin. What can you say about the gradient at the point $(a, 0)$? [MEI]

14 A curve has equation $y = (x^2 + 5x + 4) \cos 3x$.

a Find $\frac{dy}{dx}$.

b Find the equation of the tangent to the curve at the point where $x = 0$. [AQA]

15 A curve has equation

$$y = \frac{2x}{\sin x}, \quad 0 < x < \pi.$$

a Find $\frac{dy}{dx}$.

b The point P on the curve has coordinates $\left(\frac{\pi}{2}, \pi\right)$.

i Show that the equation of the tangent to the curve at P is $y = 2x$.

ii Find the equation of the normal to the curve at P, giving your answer in the form $y = mx + c$. [AQA]

16 It is given that

$$y = \ln x - 2x^2 + 3, \qquad x > 0$$

a Find $\dfrac{dy}{dx}$.

b Verify that y has a stationary value when $x = \frac{1}{2}$.

c Find the value of $\dfrac{d^2y}{dx^2}$ when $x = \frac{1}{2}$.

d Hence determine whether this stationary value is a maximum or a minimum. [AQA]

17 A curve has equation

$$y = x^2 - 3x + \ln x + 2, \qquad x > 0.$$

a i Find $\dfrac{dy}{dx}$.

ii Hence show that the gradient of the curve at the point where $x = 2$ is $\frac{3}{2}$.

b i Show that the x-coordinates of the stationary points of the curve satisfy the equation

$$2x^2 - 3x + 1 = 0$$

ii Hence find the x-coordinates of each of the stationary points.

iii Find $\dfrac{d^2y}{dx^2}$.

iv Find the value of $\dfrac{d^2y}{dx^2}$ at each of the stationary points.

v Hence show that the y-coordinate of the maximum point is

$$\tfrac{3}{4} - \ln 2.$$

[AQA]

18 A curve has equation

$$y = x^2 - \ln x, \qquad x > 0$$

The curve has one stationary point P.

a Show that the x-coordinate of P is $\dfrac{1}{\sqrt{2}}$.

b Give the y-coordinate of P in the form $a + b \ln 2$, where a and b are rational numbers.

c Find the value of $\dfrac{d^2y}{dx^2}$ at P.

d State, giving a reason, whether P is a maximum or a minimum point on the curve. [AQA]

19 The function f is defined for $x > 0$ by

$$f(x) = e^{-2x} + \frac{3}{x} + 3$$

a i Differentiate f(x) with respect to x to find $f'(x)$.

ii Hence prove that f is a decreasing function.

b Find the range of f. [AQA]

20 A curve has equation $y = e^{3x} - 24x$.

a Show that the curve cuts the x-axis at the point where $x = \alpha$, with $1 < \alpha < \ln 3$.

b Determine $\frac{dy}{dx}$ and $\frac{d^2y}{dx^2}$ as functions of x.

c The curve has a single turning point, P. Find the x-coordinate of P in an exact form, and show that its y-coordinate is $8(1 - 3\ln 2)$. Find the value of $\frac{d^2y}{dx^2}$ at P and hence deduce the nature of the turning point.

21

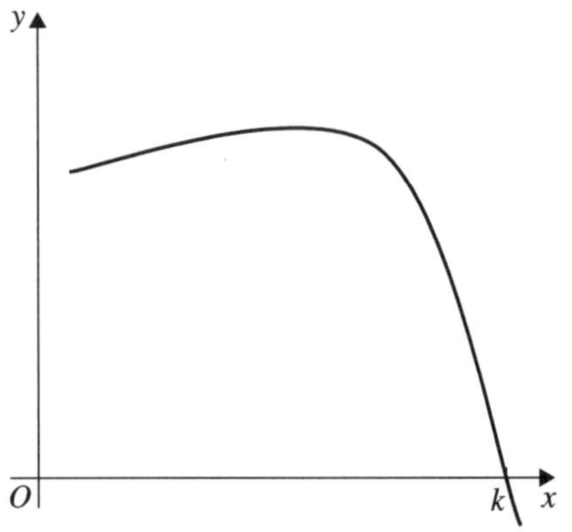

Figure 1 shows a sketch of the curve with equation $y = f(x)$, where

$$f(x) = 10 + \ln(3x) - \tfrac{1}{2}e^x, \qquad 0.1 \leqslant x \leqslant 3.3.$$

Given that $f(k) = 0$,

a Show, by calculation, that $3.1 < k < 3.2$.

b Find $f'(x)$.

The tangent to the graph at $x = 1$ intersects the y-axis at the point P.

c i Find an equation of this tangent.

ii Find the exact y-coordinate of P, giving your answer in the form $a + \ln b$. [EDEXCEL]

PART 5

Numerical methods

5.1 Finding solutions to equations

We know how to find exact solutions to certain types of equations, for example quadratic equations, but there are many types of equations for which we cannot find exact solutions. For example suppose we want to solve $x^3 + x = 20$. We cannot factorise this and so we are left using trial and error to find the solutions to this equation.

We could rewrite the equation $x^3 + x = 20$ as $x^3 + x - 20 = 0$. If we put $f(x) \equiv x^3 + x - 20$ then we see that we are trying to solve $f(x) = 0$.

We can see that $f(1) = -18$, $f(2) = -10$ and $f(3) = 10$.
From this we can deduce that, since $f(2) < 0$ and $f(3) > 0$ and since $f(x)$ is a continuous function there must be a value of x between 2 and 3 for which $f(x) = 0$.

If we had to find x correct to 1 dp we would then need to try more values of x, using a calculator. We would create a table as follows

x	$f(x)$
2	−10
2.1	−8.639
2.2	−7.152
2.3	−5.533
2.4	−3.776
2.5	−1.875
2.6	0.176
2.7	2.383
2.8	4.752
2.9	7.289
3	10

So we see that $f(2.5) < 0$ and $f(2.6) > 0$ and so we see that a solution to $f(x) = 0$ lies between 2.5 and 2.6.
If we want to know x to 1 dp we need to look at the x value of 2.55. We see that $f(2.55) < 0$. So we see that a solution to $f(x) = 0$ lies between 2.55 and 2.6. It is, therefore, 2.6 (to 1 dp).

It is worth noting that it would have been more efficient for us to have looked at the following values:

x	$f(x)$
2.05	−9.335
2.15	−7.912
2.25	−6.359
2.35	−4.672
2.45	−2.844
2.55	−0.869
2.65	1.26
2.75	3.547
2.85	5.999
2.95	8.622

The reason this is more efficient is that we can see that $f(2.55) < 0$ and $f(2.65) > 0$. So we see that a solution to $f(x) = 0$ lies between 2.55 and 2.65. It is, therefore, 2.6 (to 1 dp).

N.B. When we saw that $f(2) = -10$ and $f(3) = 10$ we would have been wrong to have concluded that x must be exactly 2.5.

Example 1

Find the solution to (1 dp) to the equation $x^3 - x - 100 = 0$.

If we let $f(x) = x^3 - x - 100$ we get the following table:

x	$f(x)$
3	−76
4	−40
5	20

So, since $f(4) < 0$ and $f(5) > 0$ there must be a value of x between 4 and 5 for which $f(x) = 0$. When we try more values between 4 and 5 we get the following

x	$f(x)$
4.05	−37.6
4.15	−32.7
4.25	−27.5
4.35	−22
4.45	−16.3
4.55	−10.4
4.65	−4.11
4.75	2.422

From this we see that, since $f(4.65) < 0$ and $f(4.75) > 0$ there must be a value of x between 4.65 and 4.75 and so the solution is 4.7 (to 1 dp).

Exercise 1

1 **a** Show that the equation $x^3 - 3x^2 - 5 = 0$ has a root between 3 and 4.
b Find this root, to 1 dp, by trial and error.

2 **a** Show that the equation $x^3 - 11x - 21 = 0$ has a root between 4 and 5.
b Find this root, to 1 dp, by trial and error.

3 The function $f(x) = x^5 - 3x - 1$ is such that there is a root of the equation $f(x) = 0$ between two consecutive negative integers.
a Find these two integers.
b Calculate the root to 1 dp.
c Calculate also the positive root of the equation to 1 dp.

4 **a** Show by means of a graph that the equation $e^x = 9 - x^2$ has two real roots.
b Show that one of these roots lies in the interval (1, 2).
c Calculate this root to 1 dp.
d Find an interval in the form $(a, a + 1)$ for the other root where a is an integer.

5 **a** Show graphically that the equation $3\cos x = x^2 + 1$ has exactly one positive real root.

b Find an interval for this root in the form $\left(\frac{a}{10}, \frac{a+1}{10}\right)$ where a is an integer.

(N.B.: Make sure that your calculator is in 'radians mode')

c Calculate this root to 1 dp.

6 **a** Given that $f(x) = e^{3x}\cos x - 5$, find (to 3 sf) f(1), f(2) and f(3).

b Hence find an interval for a root of the equation $f(x) = 0$ in the form $(n, n+1)$ where n is an integer.
(N.B.: Make sure that your calculator is in radians mode)

c Calculate this root (to 2 dp).

7 **a** Show graphically that the equation $e^{x-2} = x^3$ has exactly two roots.

b Find both roots to 1 dp.

5.2 Iteration

In example 1 we looked to find a solution to $x^3 - x - 100 = 0$. Our only method was trial and error. If we were asked to find the solution to 5 dp it would take a long time, using many different values for x.
Having got one estimate for the solution, we want to have some method of generating a better estimate. We do this using an iterative method.

If we rearrange the equation $x^3 - x - 100 = 0$ we see that $x^3 = x + 100$ and so $x = \sqrt[3]{x + 100}$. So we see that the solution to $x^3 - x - 100 = 0$ is the x-coordinate of the intersection points of $y = x$ and $y = \sqrt[3]{x + 100}$.

Consider the iteration $x_{n+1} = \sqrt[3]{x_n + 100}$. We see that if this converges to a value of say α then we will get to the stage where $x_{n+1} = x_n = \alpha$. Hence we have $\alpha = \sqrt[3]{\alpha + 100}$ and so $\alpha^3 - \alpha - 100 = 0$.

So we see that if the iteration $x_{n+1} = \sqrt[3]{x_n + 100}$ converges to a value then this value is a solution to the equation $x^3 - x - 100 = 0$

In order to create an iteration for an equation we need to rearrange the equation so that it is written in the form $x = f(x)$. The iteration formula is then $x_{n+1} = f(x_n)$.

Example 2

Show that the equation $3^x = 8x$ has a root lying between $x = 0$ and $x = 1$. Use the iteration formula

$$x_{n+1} = \frac{3^{x_n}}{8}, \quad x_0 = 0.2$$

to find this root, correct to 3 decimal places.

Using the iteration formula $x_{n+1} = \frac{3^{x_n}}{8}$

$x_0 = 0.2$

$x_1 = \frac{3^{0.2}}{8} = 0.155\,7...$

$x_2 = \frac{3^{0.1557}}{8} = 0.148\,32...$

$x_3 = \frac{3^{0.14832}}{8} = 0.147\,12...$

$x_4 = \frac{3^{0.14712}}{8} = 0.146\,92...$

The solution is $x = 0.147$, correct to 3 decimal places.

Fast calculator method

The [ANS] button on calculators can be used to give successive values with an iterative formula.

In the example above ($x_{n+1} = \frac{3^{x_n}}{8}$) proceed as follows:

1. Key [0.2] [=] This makes 'ANS' = 0.2.

2. Key the formula, [3] [∧] [ANS] [÷] [8]

3. Press the [=] button several times. You will obtain the values 0.1557…, 0.14822…, 0.14712… etc.

Example 3

Find the value to which the iterative formula $x_{n+1} = \sqrt[3]{5x_n - 1}$ converges. Use $x_1 = 4$.

1. Key [4] [=]

2. Key the formula, [$\sqrt[3]{\ }$] [(] [5] [×] [ANS] [−] [1] [)]

3. Press [=] several times. You obtain the $x = 2.13$ correct to 2 dp.

Example 4

Find the value to which the iterative formula $x_{n+1} = \frac{x_n}{2} + \frac{3}{(x_n)^2}$ converges. Use $x_1 = 2$.

1. Key [2] [=]

2. Key the formula, [ANS] [÷] [2] [+] [3] [÷] [ANS] [x^2]

3. Press [=] several times to obtain the solution $x = 1.82$ correct to 2 dp.

EXERCISE 2

1 Find the values to which the following iterative formulae converge. Give answers correct to 1 d.p. or to 3 d.p. if you use the fast calculator method described above.

a $x_{n+1} = \sqrt[3]{7x_n - 1}$ $\quad x_1 = 5$

b $x_{n+1} = \sqrt[3]{7x_n - 1}$ $\quad x_1 = -4$

c $x_{n+1} = \dfrac{x_n^3 + 1}{7}$ $\quad x_1 = 1$

d $x_{n+1} = \frac{1}{2}\left(x_n + \dfrac{74}{x_n}\right)$ $\quad x_1 = 8$

e $x_{n+1} = \dfrac{3x_n}{4} + \dfrac{5}{x_n^2}$ $\quad x_1 = 3$

f $x_{n+1} = \sin x_n + 1$ $\quad x_1 = 1.8$

2 Find the equations to which the numbers found in question 1 are solutions.

3 a Draw a sketch of $y = x^2$ for $-3 \leqslant x \leqslant 3$.

b Find the equation of the straight line which must be drawn on this diagram to solve the equation $x^2 + 2x - 1 = 0$.

c Draw this line on your diagram to find the two solutions to $x^2 + 2x - 1 = 0$ (to the nearest integer).

d Use the iteration $x_{n+1} = -\sqrt{1 - 2x_n}$ to find the negative solution to 3 dp.

e What happens if you use the iteration $x_{n+1} = \sqrt{1 - 2x_n}$ with $x_1 = 0$ to find the positive solution?

f Now use the iteration $x_{n+1} = \dfrac{1 - x_n^2}{2}$ with $x_1 = 0$ to find the positive solution to 3 dp.

4 a Find the numbers to (3 dp) to which the following iterations converge:

i $x_{n+1} = \dfrac{2 - x_n^2}{3}$ with $x_1 = 1$

ii $x_{n+1} = -\sqrt{1 - 7x_n}$ with $x_1 = -2$

iii $x_{n+1} = \dfrac{2}{x_n + 5}$ with $x_1 = -2$.

b Find the equations to which these numbers are solutions.

c Hence find the *exact* values to which the values in **4a** are approximations.

5 a Show that there is a solution to the equation $f(x) = 0$, where $f(x) = x^3 - 3x + 4$, in the interval $(-2, -3)$.

b Show that the equation $x = \dfrac{2x^3 - 4}{3x^2 - 3}$ can be rewritten as $x^3 - 3x + 4 = 0$.

c Use the iteration $x_{n+1} = \dfrac{2x_n^3 - 4}{3x_n^2 - 3}$ with $x_1 = -2$ to find the solution to 3 dp.

6 **a** Show that there is a solution to the equation $xe^x = 10$ in the interval (1, 2).

b Show also that $xe^x = 10$ can be rearranged to give $x = \ln\left(\frac{10}{x}\right)$.

c Use the iteration $x_{n+1} = \ln\left(\frac{10}{x_n}\right)$ to find the solution to 1 dp.

7 Suppose the dimensions of cuboid are $x \times y \times z$, where x is the length, y is the width and z is the height. Suppose also that the volume of the cuboid is 200 cm^3, the surface area is 240 cm^3 and that its length is twice its width.

a Write down three equations involving x, y and z.

b Find y and z in terms of x and hence show that $x^3 - 240x + 1200 = 0$.

c Uses the iteration $x_{n+1} = \frac{(x_n)^3 + 1200}{240}$ to find x (to 2 dp) given that x is approximately 5 cm.

8* A man invested £1000 into a high interest bank on 1st January 1994. He put £1000 into this account at the beginning of each year – his final payment was on 1st January 2003. He closed the account on 1st January 2004 when it was worth £15,000.

a If the annual rate of interest is $p\%$ and $r = 1 + \frac{p}{100}$ then show that $r + r^2 + \ldots r^{10} = 15$.

b Hence show that $r^{11} - 16r + 15 = 0$

$\left(\text{You may use } a + ar + ar^2 + \ldots + ar^{n-1} = \frac{a(r^n - 1)}{(r - 1)}\right)$.

c Use the iteration $r_{n+1} = \sqrt[11]{16r_n - 15}$ to find the annual rate of interest (to 2 sf) given that it is about 8% per year.

Review Exercise 5

1 Show the following equations have a root in the given interval.

a $e^x - 3x = 0$ (1, 2)

b $x^3 + x - 8 = 0$ (1, 2)

c $2 + 3x - x^4 = 0$ (1, 2)

d $x^2 = 3x - 1$ (2, 3)

e $2x^3 + x^2 + 6x - 1 = 0$ (0, 1)

f $\sin x - \ln x = 0$ (2, 3) [x is in radians.]

g $e^{2x} + 4x - 5 = 0$ (0, 1)

h $x^4 - 2x^2 - 7 = 0$ (1, 2)

i $x^3 - 3x - 1 = 0$ (−1, 0)

2 a Show that the equation $x^3 - x^2 + 5x - 3 = 0$ has a root between $x = 0$ and $x = 1$.

b Using the iterative formula

$$x_{n+1} = -\tfrac{1}{5}(x_n{}^3 - x_n{}^2 - 3), \text{ with } x_1 = 0,$$

find the root correct to 3 decimal places.

3 a Given $f(x) = 3 + 4x - x^4$, show that the equation $f(x) = 0$ has a root between 1 and 2.

b The iterative formula $x_{n+1} = (3 + 4x_n)^{\frac{1}{4}}$ may be used to obtain an approximate root of the equation $f(x) = 0$. Starting with $x_0 = 1.5$, use the formula to find a root of the equation correct to 2 decimal places.

4 The curve with equation $y = x^3 - 2x^2 - 1 = 0$ intersects the x-axis at the point where $x = \alpha$.

a Show that α lies between 2 and 3.

b Show that the equation $x^3 - 2x^2 - 1 = 0$ can be re-arranged in the form

$$x = 2 + \frac{1}{x^2}$$

c Use the iterative formula $x_{n+1} = 2 + \dfrac{1}{x_n{}^2}$, with $x_1 = 2$, to find x_4, giving your answer to three significant figures.

In the following questions show that the given equations have a root in the given interval.

Using the given iteration formula and starting value, find the root of the equation to 3 significant figures.

5 $x^3 - x + 3 = 0$ $\quad -2 < x < -1$

$x_{n+1} = (x_n - 3)^{\frac{1}{3}}$ $\quad x_0 = -1.5$

6 $x^2 - 5x + 3 = 0$ $\quad 0 < x < 1$

$x_{n+1} = \dfrac{x_n{}^2 + 3}{5}$ $\quad x_0 = 0.5$

7 $x^3 - x^2 - 2 = 0$ $\quad 1 < x < 2$

$x_{n+1} = \sqrt[3]{x_n{}^2 + 2}$ $\quad x_0 = 1.5$

8 $x^4 - 3x + 1 = 0$ $\quad 1 < x < 2$

$x_{n+1} = (3x_n - 1)^{\frac{1}{4}}$ $\quad x_0 = 1.3$

9 $3x - 2\sin x - 1 = 0$ $\quad 0 < x < 1$ (x in radians)

$x_{n+1} = \dfrac{2\sin x_n + 1}{3}$ $\quad x_0 = 0.8$

10 $e^x - x - 3 = 0$ $\quad -2 < x < -1$

$x_{n+1} = e^{x_n} - 2$ $\quad x_0 = -1.5$

11 $e^x - x^2 - 3 = 0$ $\quad 1 < x < 2$

$x_{n+1} = \ln(x_n^2 + 3)$ $\quad x_0 = 1.8$

EXAMINATION EXERCISE 5

1 The curve $y = (x^2 + 4)(2x - 1)$ intersects the line $y = x$ at only one point B.

i Show that the x-coordinate of B satisfies the equation

$$2x^3 - x^2 + 7x - 4 = 0.$$

ii Show that this equation has a root between 0.56 and 0.57. [AQA]

2 a Sketch on one pair of axes the graphs of

$$y = 6 - x \quad \text{and} \quad y = \ln x.$$

b Hence state the number of roots of the equation

$$6 - x = \ln x.$$

c By considering values of the function f, where

$$f(x) = 6 - x - \ln x,$$

i show that the equation in part **b** has a root α such that

$$4 < \alpha < 5,$$

ii determine whether α is closer to 4 or to 5. [AQA]

3 The volume, V cm^3, of liquid in a container when the depth is x cm is given by

$$V = (2x^2 + x^3)^{\frac{1}{2}}.$$

An attempt is made to find the value of x when $V = 10$.

i Show that a possible iterative formula can be used to find x is

$$x_{n+1} = \frac{10}{\sqrt{2 + x_n}}.$$

ii Use the value $x_1 = 4$ in the iterative formula above to find the value of x_3, giving your answer to four significant figures. [AQA]

4 The sequence defined by the iterative formula

$$x_{n+1} = \sqrt[3]{17 - 5x_n},$$

with $x_1 = 2$,converges to α.

i Use the iterative formula to find α correct to 2 decimal places. You should show the result of each iteration.

ii Find a cubic equation of the form

$$x^3 + cx + d = 0$$

which has α as a root. [OCR]

5 The root of the equation $f(x) = 0$, where

$$f(x) = x + \ln 2x - 4$$

is to be estimated using the iterative formula $x_{n+1} = 4 - \ln 2x_n$, with $x_0 = 2.4$.

a Showing your values of $x_1, x_2, x_3, \ldots$, obtain the value, to 3 decimal places, of the root. [EDEXCEL]

6 A curve C has equation

$$y = e^{2x} + x^2 + 4x + 1.$$

i Show that the x-coordinate of any stationary point of C satisfies the equation.

$$x = -2 - e^{2x}.$$

ii

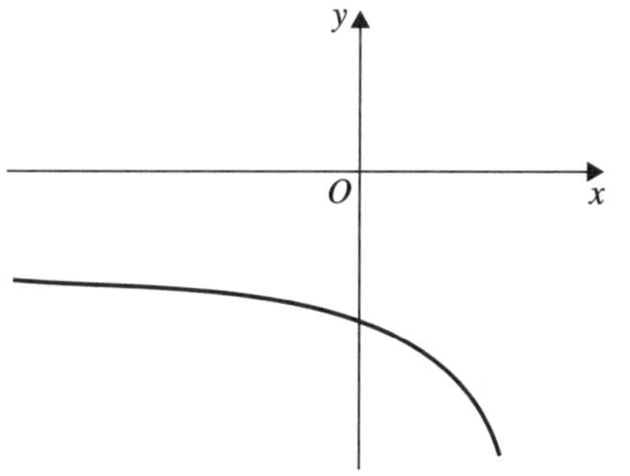

The diagram shows the graph of $y = -2 - e^{2x}$. On a copy of the diagram, draw another graph, the equation of which must be stated, to show that the equation

$$x = -2 - e^{2x}.$$

has exactly one root

iii By using an iteration process based on the equation

$$x = -2 - e^{2x},$$

find, correct to 3 significant figures, the x-coordinate of the stationary point of the curve C. You should show the result of each iteration. [OCR]

7 The diagram shows a sector OAB of a circle, centre O and radius 10 cm. Angle AOB is θ radians. The point C lies on OB and is such that AC is perpendicular to OB. The region R (shaded in the diagram) is bounded by the arc AB and by the lines AC and CB. The area of R is 22 cm^2.

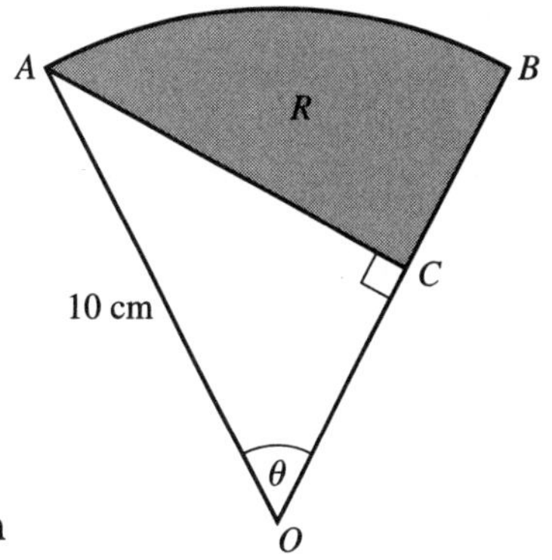

i Show that $\theta = 0.44 + \sin\theta\cos\theta$.

ii Show that θ lies between 0.9 and 1.0.

iii Use an iterative process based on the equation in part (**i**) to find the value of θ correct to 2 decimal places. You should show the result of each iteration. [OCR]

8 **a** Sketch, on the same set of axes, the graphs of

$$y = 2 - e^{-x} \text{ and } y = \sqrt{x}.$$

[It is not necessary to find the coordinates of any points of intersection with the axes.]

Given that $f(x) = e^{-x} + \sqrt{x} - 2,\ x \geqslant 0$,

b explain how your graphs show that the equation $f(x) = 0$ has only one solution,

c show that the solution of $f(x) = 0$ lies between $x = 3$ and $x = 4$.

The iterative formula $x_{n+1} = (2 - e^{-x_n})^2$ is used to solve the equation $f(x) = 0$.

d Taking $x_0 = 4$, write down the values of x_1, x_2, x_3 and x_4, and hence find an approximation to the solution of $f(x) = 0$, giving your answer to 3 decimal places. [EDEXCEL]

9

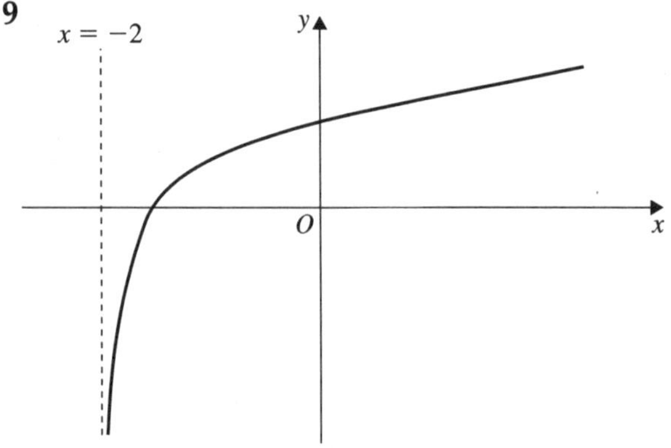

The function f is defined by

$$f : x \mapsto \tfrac{1}{4}\ln(3x + 6) \qquad x > -2,$$

and a sketch of the graph of $y = f(x)$ is shown above.
The equation $x = \frac{1}{4}\ln(3x + 6)$ has two roots α and β, where $\alpha < 0$ and $\beta > 0$.

a Show that $\alpha < -1.9995$, and hence state the value of α correct to 3 decimal places.

b Use an iteration process based on the equation $x = \frac{1}{4}\ln(3x + 6)$ to find the value of β correct to 3 decimal places. You should show the result of each iteration. [OCR]

10 The curve with equation $y = \ln 3x$ crosses the x-axis at the point $P(p, 0)$.

a Sketch the graph of $y = \ln 3x$, showing the exact value of p.

The normal to the curve at the point Q, with x-coordinate q, passes through the origin.

b Show that $x = q$ is a solution of the equation $x^2 + \ln 3x = 0$.

c Show that the equation in part **b** can be rearranged in the form $x = \frac{1}{3}e^{-x^2}$.

d Use the iteration formula $x_{n+1} = \frac{1}{3}e^{-x_n^2}$, with $x_0 = \frac{1}{3}$, to find x_1, x_2, x_3 and x_4. Hence write down, to 3 decimal places, an approximation for q. [EDEXCEL]

11 The diagram shows a circle with centre C and radius r. The chord AB is such that the angle $ACB = \theta$ radians. It is given that the area of the minor segment, shaded in the diagram, is one-fifth of the area of the whole circle.

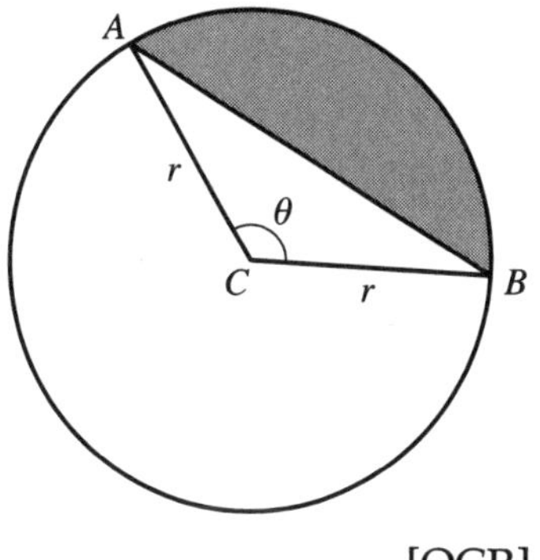

i Show that $\theta = \frac{2}{5}\pi + \sin\theta$.

ii Use an iteration process based on the equation in part **i**, with a starting value of 2, to find the value of θ correct to 1 decimal place. You should state the result of each iteration.

[OCR]

12 The diagram shows the graph of $y = \sqrt{(1 + x^2)}$. The region R, shaded in the diagram, is bounded by the curve and the lines $x = 0$, $x = a$ and $y = 0$. When R is rotated through four right angles about the x-axis, the volume of the solid produced is 25 cubic units.

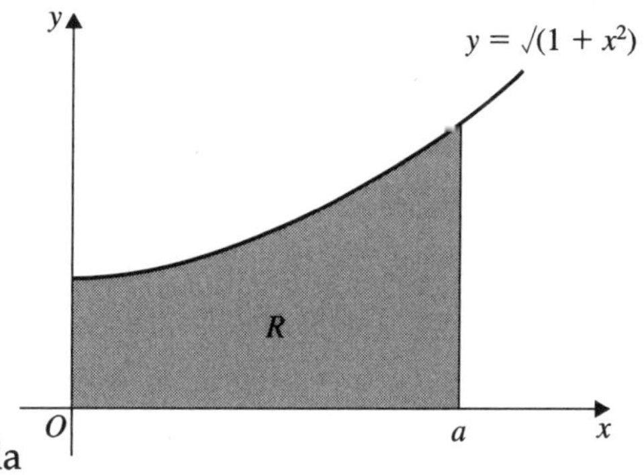

i Find an equation for a and show that a suitable rearrangement of the equation leads to the iterative formula

$$a_{n+1} = \sqrt[3]{\left(\frac{75}{\pi} - 3a_n\right)}.$$

ii Use the iterative formula in part i with $a_1 = 2.5$ to find a_2, a_3 and a_4, giving each value correct to 3 decimal places. Hence state the value of a correct to 2 decimal places.

13

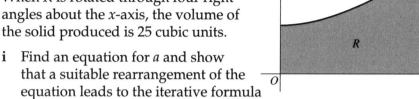

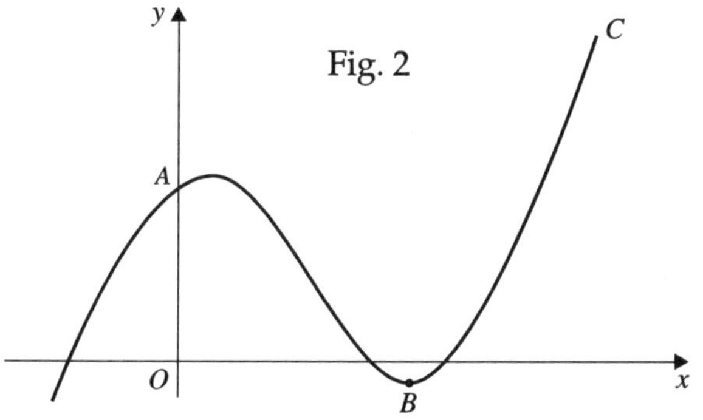

Figure 2 shows part of the curve C with equation $y = f(x)$, where

$$f(x) = 0.5e^x - x^2.$$

The curve C cuts the y-axis at A and there is a minimum at the point B.

a Find an equation of the tangent to C at A.

The x-coordinate of B is approximately 2.15. A more exact estimate is to be made of this coordinate using iterations $x_{n+1} = \ln g(x_n)$.

b Show that a possible form for $g(x)$ is $g(x) = 4x$.

c Using $x_{n+1} = \ln 4x_n$, with $x_0 = 2.15$, calculate x_1, x_2 and x_3. Give the value of x_3 to 4 decimal places.

[EDEXCEL]

PART 6

Partial fractions

We have already looked at how to add two fractions together. We saw in Part 1 that if we want to write $\frac{4}{x} + \frac{5}{x-1}$ as a single fraction we first need to write both fractions over a common denominator.

So in this example our common denominator would be $x(x-1)$.

We see that
$$\frac{4}{x} + \frac{5}{x-1} = \frac{4(x-1)+5x}{x(x-1)}$$
$$= \frac{4x-4+5x}{x(x-1)}$$
$$= \frac{9x-4}{x(x-1)}$$

6.1 Partial fractions

We now want to perform the reverse process, that is we want to go from $\frac{9x-4}{x(x-1)}$ to $\frac{4}{x} + \frac{5}{x-1}$.

When we do this we are said to be expressing $\frac{9x-4}{x(x-1)}$ in partial fractions.

In order to do this we first of all see that we can write $\frac{9x-4}{x(x-1)}$ in the form $\frac{A}{x} + \frac{B}{x-1}$ where A and B are numbers that we need to find.

So we have $\frac{9x-4}{x(x-1)} = \frac{A}{x} + \frac{B}{x-1}$.

We now rewrite the right hand side of the equation, by adding the fractions.

$$\frac{A}{x} + \frac{B}{x-1} = \frac{A(x-1)+Bx}{x(x-1)}$$

and so we see that $\frac{9x-4}{x(x-1)} = \frac{A(x-1)+Bx}{x(x-1)}$

In fact, because this is true for all values of x, we should write

$$\frac{9x-4}{x(x-1)} \equiv \frac{A(x-1)+Bx}{x(x-1)}$$

We see that $9x - 4 \equiv A(x-1) + Bx$.

There are two methods we can use to find A and B.

Method 1

We multiply out the RHS of the equation to get $9x - 4 \equiv Ax - A + Bx$
$\equiv (A + B)x - A.$

It follows from this that $A = 4$ and $A + B = 9$.
Hence we have $A = 4$ and $B = 5$.

Method 2

To solve $9x - 4 \equiv A(x - 1) + Bx$, we substitute certain values of x.

Substituting $x = 0$, we get $9 \times 0 - 4 \equiv A(0 - 1) + B \times 0$,

and so $A = \dfrac{9 \times 0 - 4}{(0 - 1)} = 4$.

Substituting $x = 1$, we get $9 \times 1 - 4 \equiv A(1 - 1) + B \times 1$,

and so $B = \dfrac{9 \times 1 - 4}{1} = 5$.

Example 1

Express $\dfrac{7 + 5x}{(x + 1)(x + 2)}$ in the form $\dfrac{A}{(x + 1)} + \dfrac{B}{(x + 2)}$.

We write $\dfrac{7 + 5x}{(x + 1)(x + 2)} \equiv \dfrac{A}{(x + 1)} + \dfrac{B}{(x + 2)} \equiv \dfrac{A(x + 2) + B(x + 1)}{(x + 1)(x + 2)}$

So we have $7 + 5x \equiv A(x + 2) + B(x + 1)$.

Method 1

We multiply out the RHS of the equation to get $7 + 5x \equiv (A + B)x + (2A + B)$

So we see that $A + B = 5$ and $2A + B = 7$

Solving these gives $A = 2$, $B = 3$.

Method 2

To solve $7 + 5x \equiv A(x + 2) + B(x + 1)$ we substitute certain values of x.

Substituting $x = -1$, we get $7 + 5 \times -1 \equiv A(-1 + 2)$, and so $A = \dfrac{7 + 5 \times -1}{(-1 + 2)} = 2$.

The quickest way to get B is to compare x coefficients, seeing that $A + B = 5$ and so $B = 3$.

We could, however, substitute $x = -2$ and we get

$7 + 5 \times -2 \equiv A(-2 + 2) + B(-2 + 1)$, and so $B = \dfrac{7 + 5 \times -2}{(-2 + 1)} = 3$.

Example 2

Express $\dfrac{3x + 1}{(x + 1)(2x + 1)}$ in the form $\dfrac{A}{(x + 1)} + \dfrac{B}{(2x + 1)}$.

We write $\dfrac{3x + 1}{(x + 1)(2x + 1)} \equiv \dfrac{A}{(x + 1)} + \dfrac{B}{(2x + 1)} \equiv \dfrac{A(2x + 1) + B(x + 1)}{(x + 1)(2x + 1)}$

So we have $3x + 1 \equiv A(2x + 1) + B(x + 1)$.

Method 1

Multiply out the numerator on the RHS of the equation to get
$3x + 1 \equiv (2A + B)x + (A + B)$

So we see that $2A + B = 3$
$A + B = 1$

Solving these gives $A = 2$, $B = -1$.

Method 2

To solve $3x + 1 \equiv A(2x + 1) + B(x + 1)$ we substitute certain values of x.

Substituting $x = -1$, we get $3 \times (-1) + 1 \equiv A(2 \times -1 + 1) + B(-1 + 1)$, and so

$$A = \frac{3 \times -1 + 1}{(2 \times -1 + 1)} = 2.$$

The quickest way to get B is to compare x coefficients, seeing that $2A + B = 3$ and so $B = -1$.

We could, however, substitute $x = -\frac{1}{2}$ and we get

$$3 \times -\tfrac{1}{2} + 1 \equiv A(2 \times -\tfrac{1}{2} + 1) + B(-\tfrac{1}{2} + 1), \text{ and so } B = \frac{3 \times -\frac{1}{2} + 1}{(-\frac{1}{2} + 1)} = -1.$$

NB If we had to express $\dfrac{3x - 2}{x^2 - 4}$ in partial fractions then we would first factorise the denominator as $(x + 2)(x - 2)$ and then write $\dfrac{3x - 2}{x^2 - 4} \equiv \dfrac{A}{(x + 2)} + \dfrac{B}{(x - 2)}$.

The 'cover up rule'

- Consider how we could express $\dfrac{5x + 7}{(x + 1)(x + 2)}$ in partial fractions.

  We need to find A and B where $\dfrac{5x + 7}{(x + 1)(x + 2)} \equiv \dfrac{A}{(x + 1)} + \dfrac{B}{(x + 2)}$.

  Method 2 can be shortened to what is called the 'cover up rule'. The cover up rule says that to find the numerator (e.g. A) above a particular denominator (e.g. $x + 1$) when written as partial fractions, we 'cover up' that factor (i.e. $x + 1$) in the original expression to get $\dfrac{5x + 7}{(x + 2)}$. We then substitute the value of x that makes the 'covered up' factor equal to zero (i.e. $x = -1$) to get

  $$A = \frac{5 \times -1 + 7}{-1 + 2} = 2.$$

To find B we look at the denominator of the fraction B. In this case it is $x + 2$. We then substitute $x = -2$ (the solution to $x + 2 = 0$) into the left hand side with $x + 2$ 'covered up'. That is $B = \frac{5 \times -2 + 7}{-2 + 1} = 3$.

So $\frac{5x + 7}{(x + 1)(x + 2)} \equiv \frac{2}{(x + 1)} + \frac{3}{(x + 2)}$

- If we go right back to the initial question, that is, $\frac{9x - 4}{x(x - 1)} \equiv \frac{A}{x} + \frac{B}{x - 1}$ we can again use the 'cover up rule'.

  To find A we look at what the denominator of the fraction under A is on the right hand side. In this case it is x. We then substitute $x = 0$ into the left hand side with x 'covered up'. That is $A = \frac{9 \times 0 - 4}{(0 - 1)} = 4$.

  To find B we substitute $x = 1$ into the left hand side with $(x - 1)$ covered up. So $B = \frac{9 \times 1 - 4}{1} = 5$.

  We can only use the cover up method to find the numerator of a fraction whose denominator is a single linear expression.

Exercise 1

1 Find A and B in the following:

a $\frac{3x - 5}{(x - 1)(x - 3)} \equiv \frac{A}{(x - 1)} + \frac{B}{(x - 3)}$

b $\frac{5x - 19}{(x - 2)(x - 5)} \equiv \frac{A}{(x - 2)} + \frac{B}{(x - 5)}$

c $\frac{5x + 11}{(x + 1)(x + 4)} \equiv \frac{A}{(x + 1)} + \frac{B}{(x + 4)}$

d $\frac{7x + 17}{(x + 1)(x + 3)} \equiv \frac{A}{(x + 1)} + \frac{B}{(x + 3)}$

e $\frac{5x + 31}{(x + 2)(x + 5)} \equiv \frac{A}{(x + 2)} + \frac{B}{(x + 5)}$

f $\frac{x - 19}{(x + 1)(x - 4)} \equiv \frac{A}{(x + 1)} + \frac{B}{(x - 4)}$

2 Express the following as partial fractions:

a $\frac{9x - 26}{(x - 3)(x - 2)}$

b $\frac{3x - 4}{(x - 1)(x - 2)}$

c $\frac{8x + 11}{(x + 1)(x + 2)}$

d $\frac{7x + 15}{(x + 2)(x + 3)}$

e $\frac{3x}{(x - 1)(x + 2)}$

f $\frac{x + 7}{(x - 1)(x + 3)}$

g $\frac{x + 8}{x(x - 2)}$

h $\frac{5x + 2}{x(x + 1)}$

3 Find A and B in the following:

a $\dfrac{5x+2}{(2x-1)(x+1)} \equiv \dfrac{A}{(2x-1)} + \dfrac{B}{(x+1)}$ **b** $\dfrac{9x+7}{(x-1)(3x+1)} \equiv \dfrac{A}{(x-1)} + \dfrac{B}{(3x+1)}$

c $\dfrac{7x-1}{(2x-1)(x+2)} \equiv \dfrac{A}{(2x-1)} + \dfrac{B}{(x+2)}$ **d** $\dfrac{5x-4}{(3x-2)(x-1)} \equiv \dfrac{A}{(3x-2)} + \dfrac{B}{(x-1)}$

e $\dfrac{x+10}{x(2x-5)} \equiv \dfrac{A}{x} + \dfrac{B}{(2x-5)}$ **f** $\dfrac{4x-23}{(3x+2)(x-3)} \equiv \dfrac{A}{(3x+2)} + \dfrac{B}{(x-3)}$

4 Express the following as partial fractions:

a $\dfrac{11x-21}{(2x-3)(x-3)}$ **b** $\dfrac{5x-2}{(2x+1)(x-1)}$

c $\dfrac{21x+16}{(3x-2)(3x+4)}$ **d** $\dfrac{7x-2}{(2x-1)(5x-1)}$

e $\dfrac{11x+5}{(2x-1)(x+3)}$ **f** $\dfrac{13x+9}{(2x-3)(3x+5)}$

g $\dfrac{12x+8}{4x^2-1}$ **h** $\dfrac{55x-14}{25x^2-4}$

Example 3

Express $\dfrac{27x^2+14x+4}{(3x-1)(x+2)(2x+1)}$ as partial fractions. The denominator is the product of three different linear factors.

Let $\dfrac{27x^2+14x+4}{(3x-1)(x+2)(2x+1)} \equiv \dfrac{A}{(3x-1)} + \dfrac{B}{(x+2)} + \dfrac{C}{(2x+1)}$

$27x^2 + 14x + 4 = A(x+2)(2x+1) + B(3x-1)(2x+1) + C(x+2)(3x-1)$

Substitute $x = -2$ to give $21B = 84$ and so $B = 4$

$x = \dfrac{1}{3}$ gives $\dfrac{28}{9}A = \dfrac{28}{3}$ and so $A = 3$

$x = -\dfrac{1}{2}$ gives $-\dfrac{15}{2}C = \dfrac{15}{2}$ and so $C = -1$

Hence we have $\dfrac{27x^2+14x+4}{(3x-1)(x+2)(2x+1)} \equiv \dfrac{3}{(3x-1)} + \dfrac{4}{(x+2)} - \dfrac{1}{(2x+1)}$

6.2 Repeated factor

We need to be able to deal with a denominator in which there is one linear expression repeated, for example in $\dfrac{12x^2-13x+2}{(2x-1)^2(x-1)}$.

The procedure here is slightly different. We do not try to express $\frac{12x^2 - 13x + 2}{(2x-1)^2(x-1)}$ in the form $\frac{A}{(2x-1)} + \frac{B}{(2x-1)} + \frac{C}{(x-1)}$ but instead we write it in the form $\frac{A}{(2x-1)} + \frac{B}{(2x-1)^2} + \frac{C}{(x-1)}$.

Example 4

Express $\frac{12x^2 - 13x + 2}{(2x-1)^2(x-1)}$ as partial fractions.

Let $\frac{12x^2 - 13x + 2}{(2x-1)^2(x-1)} \equiv \frac{A}{(2x-1)} + \frac{B}{(2x-1)^2} + \frac{C}{(x-1)}$

So we have:

$$12x^2 - 13x + 2 = A(2x-1)(x-1) + B(x-1) + C(2x-1)^2$$

Substitute $x = 1$ to find $C = 1$

Substitute $x = \frac{1}{2}$ to find $B = 3$

Now we could substitute $x = 0$ to give $2 = A - B + C$ and so $A = 4$.
Alternatively we could compare x^2 coefficients to see that $12 = 2A + 4C$ and so $A = 4$

Hence we have $\frac{12x^2 - 13x + 2}{(2x-1)^2(x-1)} \equiv \frac{4}{(2x-1)} + \frac{3}{(2x-1)^2} + \frac{1}{(x-1)}$

EXERCISE 2

1 Find A, B and C in the following:

a $\frac{6x^2 + 23x + 19}{(x+1)(x+2)(x+3)} \equiv \frac{A}{(x+1)} + \frac{B}{(x+2)} + \frac{C}{(x+3)}$

b $\frac{2x^2 + 5x + 5}{(x+2)(x+1)(x-1)} \equiv \frac{A}{(x+2)} + \frac{B}{(x+1)} + \frac{C}{(x-1)}$

c $\frac{11x^2 + 31x + 2}{(x+1)(x+4)(x-2)} \equiv \frac{A}{(x+1)} + \frac{B}{(x+4)} + \frac{C}{(x-2)}$

d $\frac{8x^2 - 31x + 31}{(x-1)(x-2)(x-3)} \equiv \frac{A}{(x-1)} + \frac{B}{(x-2)} + \frac{C}{(x-3)}$

e $\frac{6x^2 - 5x - 34}{(x-4)(x-2)(x+3)} \equiv \frac{A}{(x-4)} + \frac{B}{(x-2)} + \frac{C}{(x+3)}$

2 Express the following as partial fractions:

a $\dfrac{7x^2 + 2x - 12}{(x - 2)(2x + 1)(x + 2)}$

b $\dfrac{11x^2 + 24x + 9}{(x - 3)(x + 1)(2x + 3)}$

c $\dfrac{-5x^2 + 19x - 16}{x(x - 1)(3x - 4)}$

d $\dfrac{91x^2 - 75x + 14}{(5x - 2)(4x - 1)(3x - 2)}$

e $\dfrac{7x^2 + 4x - 21}{(x^2 - 1)(x - 2)}$

f $\dfrac{116x^2 + 63x + 4}{(5x + 1)(4x + 3)(4x - 1)}$

3 Express the following as partial fractions:

a $\dfrac{7x^2 - 2x - 19}{(x + 3)(x - 2)^2}$

b $\dfrac{x^2 + 4x + 13}{(x - 1)(x + 2)^2}$

c $\dfrac{21x^2 + 5x + 11}{(2x - 1)(3x + 1)^2}$

d $\dfrac{32x^2 - 33x + 8}{(3x - 2)(2x - 1)^2}$

4 Express the following as partial fractions:

a $\dfrac{x}{(x - 2)^2}$

b $\dfrac{2x - 3}{(2x - 1)^2}$

c $\dfrac{5x^2 - 6x + 9}{(x + 1)(x - 1)^2}$

d $\dfrac{3x^2 + 17x + 28}{(x - 1)(x + 3)^2}$

e $\dfrac{4x^2 + 55x + 90}{(3x - 2)(x + 5)^2}$

f $\dfrac{x^2 - 10x - 5}{x^3 - x}$

g $\dfrac{8x + 3}{4x^3 - x}$

h $\dfrac{x^2 - 6x + 10}{(x + 4)(x - 1)^2}$

6.3 Improper fractions

If the highest power of x in a polynomial is x^n then n is said to be the degree of that polynomial.

So, for example, the highest power of x in the expression $(3x - 2)(2x - 1)^2$ is x^3 and so the expression has degree 3.

In all the examples we have covered so far the degree of the numerator has been less than the degree of the denominator.

We are now going to deal with examples where the degree of the numerator is equal to, or greater than the degree of the denominator. These are called improper fractions.

If we were asked to write $1 + \dfrac{2}{(x - 1)} + \dfrac{5}{(x - 2)}$ as a single fraction we would look for the common denominator. In this case it is $(x - 1)(x - 2)$ and so we have

$$1 + \frac{2}{(x-1)} + \frac{5}{(x-2)} \equiv \frac{(x-1)(x-2) + 2(x-2) + 5(x-1)}{(x-1)(x-2)}$$

Multiplying this out gives us

$$1 + \frac{2}{(x-1)} + \frac{5}{(x-2)} \equiv \frac{x^2 - 3x + 2 + 2x - 4 + 5x - 5}{(x-1)(x-2)}$$

$$\equiv \frac{x^2 + 4x - 7}{(x-1)(x-2)}.$$

We now need to do the reverse, that is to write $\frac{x^2 + 4x - 7}{(x-2)(x-1)}$ in partial fractions.

Example 5

Express $\frac{x^2 + 4x - 7}{(x-2)(x-1)}$ in partial fractions.

We have seen from the above that the form we are looking for is

$$\frac{x^2 + 4x - 7}{(x-2)(x-1)} \equiv A + \frac{B}{(x-1)} + \frac{C}{(x-2)}$$

So we have $x^2 + 4x - 7 \equiv A(x-1)(x-2) + B(x-2) + C(x-1)$

Substitute $x = 1$ to find $B = 2$

Substitute $x = 2$ to find $C = 5$

Compare x^2 coefficients to see that $A = 1$

Hence we have $\frac{x^2 + 4x - 7}{(x-2)(x-1)} \equiv 1 + \frac{2}{(x-1)} + \frac{5}{(x-2)}$

To write $x + 5 + \frac{3}{(x+3)} + \frac{1}{(x+2)}$ as a single fraction we look for the common denominator. In this case it is $(x+3)(x+2)$ and so we have

$$x + 5 + \frac{3}{(x+3)} + \frac{1}{(x+2)} \equiv \frac{(x+5)(x+3)(x+2) + 3(x+2) + 1(x+3)}{(x+3)(x+2)}$$

$$\equiv \frac{x^3 + 10x^2 + 35x + 39}{(x+3)(x+2)}$$

We now need to do the reverse, that is to write $\frac{x^3 + 10x^2 + 35x + 39}{(x+3)(x+2)}$ in partial fractions.

Example 6

Express $\dfrac{x^3 + 10x^2 + 35x + 39}{(x+3)(x+2)}$ in partial fractions.

We have seen from the above that the form we are looking to write it in is

$$\frac{x^3 + 10x^2 + 35x + 39}{(x+3)(x+2)} \equiv Ax + B + \frac{C}{(x+3)} + \frac{D}{(x+2)}$$

So we have $x^3 + 10x^2 + 35x + 39 \equiv (Ax + B)(x+3)(x+2) + C(x+2) + D(x+3)$

Substitute $x = -2$ to find $(-2)^3 + 10 \times (-2)^2 + 35 \times (-2) + 39 = D(-2+3)$ and so $D = 1$

Substitute $x = -3$ to find $(-3)^3 + 10 \times (-3)^2 + 35 \times (-3) + 39 = C(-3+2)$ and so $C = 3$

Compare x^3 coefficients to see that $A = 1$.

Compare coefficient without any x to see that $39 \equiv 6B + 2C + 3D$ and so $B = 5$

Hence we have $\dfrac{x^3 + 10x^2 + 35x + 39}{(x+3)(x+2)} \equiv x + 5 + \dfrac{3}{(x+3)} + \dfrac{1}{(x+2)}$

Example 7

Express $\dfrac{9x^4 - 27x^3 - 2x^3 + 26x + 35}{(x-2)^2(3x+1)}$ in partial fractions.

As we saw earlier we want to express $\dfrac{9x^4 - 27x^3 - 2x^2 + 26x + 35}{(x-2)^2(3x-1)}$ in the form

$$Ax + B + \frac{C}{(3x+1)} + \frac{D}{(x-2)^2} + \frac{E}{(x-2)}$$

So we have

$$\begin{aligned} 9x^4 - 27x^3 - 2x^2 + 26x + 35 \equiv\ & (Ax + B)(3x+1)(x-2)^2 + C(x-2)^2 + D(3x+1) \\ & + E(3x+1)(x-2) \end{aligned}$$

Substitute $x = 2$ to find $D = 1$

Substitute $x = \dfrac{1}{3}$ to find $C = 5$

Compare x^4 coefficients to see that $9 = 3A$ and so $A = 3$.

Compare x^3 coefficients to see that $-27 = 3B - 12A + A$ and so $B = 2$.

Compare constant coefficients to see that $35 = 4B + 4C + D - 2E$ and so $E = -3$.

$$\frac{9x^4 - 27x^3 - 2x^2 + 26x + 35}{(x-2)^2(3x+1)} \equiv 3x + 2 + \frac{5}{(3x+1)} + \frac{1}{(x-2)^2} - \frac{3}{(x-2)}$$

EXERCISE 3

1 Express the following as partial fractions:

a $\dfrac{x^2 + 3x - 5}{(x + 3)(x - 2)}$ **b** $\dfrac{30x^2 - x}{(5x + 1)(2x - 1)}$

c $\dfrac{x^3 - 4x^2 + 5x - 4}{(x - 3)(x - 1)}$ **d** $\dfrac{2x^3 - 6x^2 + 9x - 8}{(x - 2)(x - 1)}$

e $\dfrac{x^3 + 5x^2 + 14x + 18}{(x + 3)(x + 1)}$ **f** $\dfrac{2x^3 + 5x^2 - 18x - 59}{(x - 3)(x + 4)}$

g $\dfrac{6x^3 + 7x^2 + 2x - 5}{(x + 1)(2x - 1)}$ **h** $\dfrac{12x^3 - 44x^2 + 55x - 26}{(2x - 3)(3x - 5)}$

2 Express the following as partial fractions:

a $\dfrac{3x^3 - 2x^2 - 3x + 6}{(x - 1)^2(x + 1)}$ **b** $\dfrac{36x^3 - 69x^2 + 38x - 7}{(3x - 2)^2(2x - 1)}$

c $\dfrac{x^4 - 4x^3 + 7x^2 - 5x - 1}{(x - 1)^2(x - 2)}$ **d** $\dfrac{x^4 - 3x^3 + 5x^2 - 3x + 6}{(x - 2)^2(x + 1)}$

3 The function f is given by $f(x) = \dfrac{2(x + 1)}{(x + 3)(x - 2)}, x \in \mathbb{R}, x \neq 2, x \neq -3$

a Express f(x) in partial fractions.

b Hence, or otherwise, prove that $f'(x) < 0$ for all values of x in the domain.

4 Use partial fractions to find the following integrals.

a $\displaystyle\int \frac{1}{x(x - 2)}\,dx$ **b** $\displaystyle\int \frac{1}{x^2 - 4x - 5}\,dx$

5 **a** Write $\dfrac{1 - x - x^2}{(1 - x)^2(1 - 2x)}$ in partial fractions.

b Hence expand this expression in ascending power of x up to and including the term in x^3.

EXAMINATION EXERCISE 6

1 The function f is given by

$$f(x) = \frac{3(x + 1)}{(x + 2)(x - 1)}, x \in \mathbb{R}, x \neq -2, x \neq 1.$$

a Express f(x) in partial fractions.

b Hence, or otherwise, prove that $f'(x) < 0$ for all values of x in the domain.

[EDEXCEL]

2 i Express $\dfrac{1+x}{(2-x)^2}$ in the form

$$\frac{A}{2-x}+\frac{B}{(2-x)^2},$$

where A and B are constants.

ii Hence show that

$$\int_0^1 \frac{1+x}{(2-x)^2}\,\mathrm{d}x = \frac{3}{2}-\ln 2.$$ [OCR]

3 a Express $\dfrac{13-2x}{(2x-3)(x+1)}$ in partial fractions.

b Given that $y = 4$ at $x = 2$, use your answer to part **a** to find the solution of the differential equation

$$\frac{\mathrm{d}y}{\mathrm{d}x} = \frac{y(13-2x)}{(2x-3)(x-1)}, \quad x > 1.5$$

Express your answer in the form $y = \mathrm{f}(x)$. [EDEXCEL]

4 i Express $\dfrac{3x+7}{(x+2)(x+3)^2}$ in the form

$$\frac{A}{x+2}+\frac{B}{x+3}+\frac{C}{(x+3)^2},$$

where A, B and C are constants.

ii When x is small, the expansion of

$$\frac{3x+7}{(x+2)(x+3)^2},$$

in ascending powers of x, is

$$p + qx + rx^2 + \dots .$$

Show that $p = \dfrac{7}{18}$, $q = -\dfrac{31}{108}$ and find the value of r. [OCR]

5 The function f is given by

$$\mathrm{f}(x) = \frac{9}{(1+2x)(4-x)}.$$

a Express $\mathrm{f}(x)$ in partial fractions.

b i Show that the first three terms in the expansion of

$$\frac{1}{4-x}$$

in ascending powers of x are

$$\frac{1}{4}+\frac{x}{16}+\frac{x^2}{64}.$$

ii Obtain a similar expansion for

$$\frac{1}{1+2x}.$$

iii Hence, or otherwise, obtain the first three terms in the expression of f(x) in ascending powers of x.

iv Find the range of values of x for which the expansion of f(x) in ascending powers of x is valid.

c i Find $\int f(x)dx$

ii Hence find, to two significant figures, the error in using the expansion of f(x) up to the term in x^2 to evaluate

$$\int_0^{0.25} f(x)dx.$$ [AQA]

6 Figure 1 shows part of the curve with equation $y = f(x)$, where

$$f(x) = \frac{x^2+1}{(1+x)(3-x)}, \quad 0 \leqslant x < 3.$$

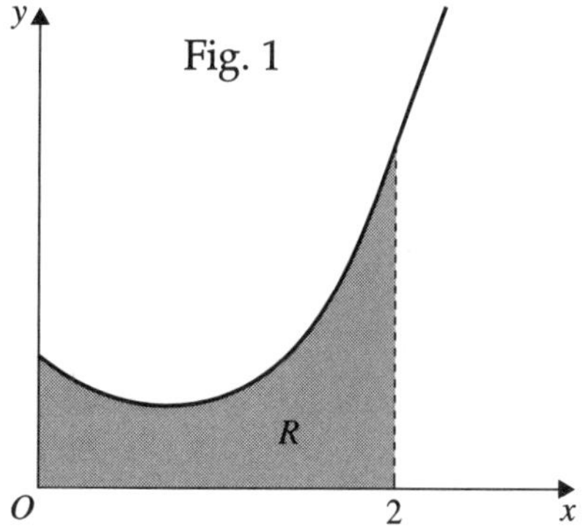

Fig. 1

a Given that $f(x) = A + \frac{B}{1+x} + \frac{C}{3-x}$, find the values of the constants A, B and C.

The finite region R, shown in Fig. 1, is bounded by the curve with equation $y = f(x)$, the x-axis, the y-axis and the line $x = 2$.

b Find the area of R, giving your answer in the form $p + q \ln r$, where p, q and r are rational constants to be found. [EDEXCEL]

7 i Expand $(2-x)(1-x)$. Hence express $\frac{3x}{2-x-x^2}$ in partial fractions.

ii Use the binomial expansion of the partial fractions in part **i** to show that

$$\frac{3x}{2-x-x^2} = \frac{3}{2}x - \frac{3}{4}x^2 + \ldots$$

State the values of x for which this result is valid.

iii Solve the differential equation

$$\frac{dy}{dx} = \frac{3xy}{2-x-x^2}$$

with the condition that $y = 1$ when $x = 0$. Give your answer in the form $y = f(x)$. [MEI]

PART 7

Parametric equations

7.1 Parametric and cartesian form

We are familiar with the idea of writing the equation of a curve in the form $y = f(x)$, where $f(x)$ is a function of x. When we write it in this form we are said to be using the cartesian form of the equation.

We could, however, choose to write the equation in terms of a parameter t.

Consider for example the curve given by $y = (x - 1)^2$.
This curve passes through the points (1, 0), (2, 1) etc.

We could write $y = (x - 1)^2$ by letting $t = x - 1$ and so $y = t^2$. In other words we would say $x = t + 1, y = t^2$.

If $t = 0$ we get $x = 1, y = 0$. so the curve passes through (1, 0).

If $t = 1$ we get $x = 2, y = 1$. So the curve passes through (2, 1).

When we write the curve in the form $x = t + 1, y = t^2$ we are using the parametric form since we are expressing both x and y in terms of the parameter t.

Example 1

A curve has parametric equations $x = t - 2, y = 3t^2$. Find the cartesian equation of this curve.

To find the cartesian equation we eliminate the parameter t.

$x = t - 2$ and so $t = x + 2$. From this we see that $y = 3t^2$
$= 3(x + 2)^2$.

So the cartesian form of the curve is $y = 3(x + 2)^2$.

Example 2

A curve has parametric equations $x = t^3, y = 2t^2 + t$. Find the cartesian equation of this curve.

$x = t^3$ and so $t = \sqrt[3]{x}$. From this we see that $y = 2t^2 + t = 2x^{\frac{2}{3}} + x^{\frac{1}{3}}$.

So the cartesian form of the curve is $y = 2x^{\frac{2}{3}} + x^{\frac{1}{3}}$.

Example 3

A curve has parametric equations $x = 3 \sin \theta, y = 2 \cos \theta$.
Find the cartesian form of this curve.

We will use the identity $\sin^2\theta + \cos^2\theta = 1$.

We have $\sin\theta = \frac{x}{3}$ and $\cos\theta = \frac{y}{2}$

$$\therefore \quad \left(\frac{x}{3}\right)^2 + \left(\frac{y}{2}\right)^2 = 1$$

$$\frac{x^2}{9} + \frac{y^2}{4} = 1$$

The curve is an ellipse.

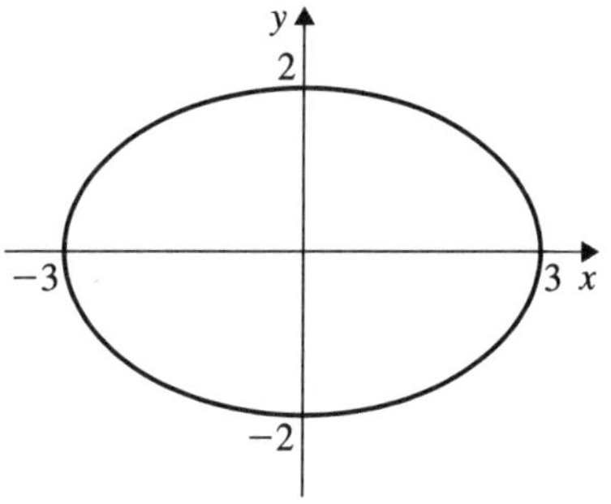

EXERCISE 1

1 Find the cartesian equations in the following curves.

a $x = t + 1, \quad y = t^2$

b $x = t^2, \quad y = t^3$

c $x = 3t, \quad y = \frac{3}{t}$

d $x = 2t, \quad y = 8t^3 - 8t$

e $x = 2t - 1, \quad y = 4t^2 + 1$

f $x = t^2, \quad y = \frac{1}{t}$

g $x = \frac{3}{\sqrt{t}}, \quad y = 2t + 1$

h $x = 2\sqrt{t}, \quad y = 3t^2 + 4$

i $x = \frac{1}{1 + 3t}, \quad y = \frac{1}{1 - 3t}$

j $x = t + 2, \quad y = 3t^2 + 2t + 1$

2 Find y in terms of t in the following curves which have been given in cartesian form:

a $y = x^2 + 1 \qquad x = t$

b $x^2 + y^2 = 1 \qquad x = \cos t$

c $y = \frac{4}{x} \qquad x = 2t$

d $y = \frac{x^3 - x}{3} \qquad x = 3t$

3 A curve is given parametrically by $x = t + 1, y = 2t - 1$. Show that the 'curve' is a straight line and sketch the graph.

4 The parametric equations of a curve are $x = \cos t, y = 2\sin t$. Find the cartesian equation of the curve.

5 a Find the cartesian equation for the curve whose parametric equations are $x = \sin t, y = \cos t$.

b Sketch the curve.

6 a Find the cartesian equation for the curve $x = \cos^2\theta, y = \sin^2\theta$.

b Sketch the curve.

7 Use a formula for $\cos 2\theta$ to obtain the cartesian equation for the curve $x = \cos 2\theta, y = \cos\theta$.

8 a Find the cartesian equation of the curve $x = \sin t, y = \cos 2t$.

b Sketch the curve.

9 The parametric equations of a curve are $x = t^2 + \frac{1}{t}, y = t^2 - \frac{1}{t}$.

Find the cartesian equation of the curve.

10 The diagram shows the curve with parametric equations

$x = \cos t, y = \frac{1}{2}\sin 2t$ for $0 \leqslant t \leqslant 2\pi$.

Show that the cartesian equation of the curve is $y^2 = x^2(1 - x^2)$.

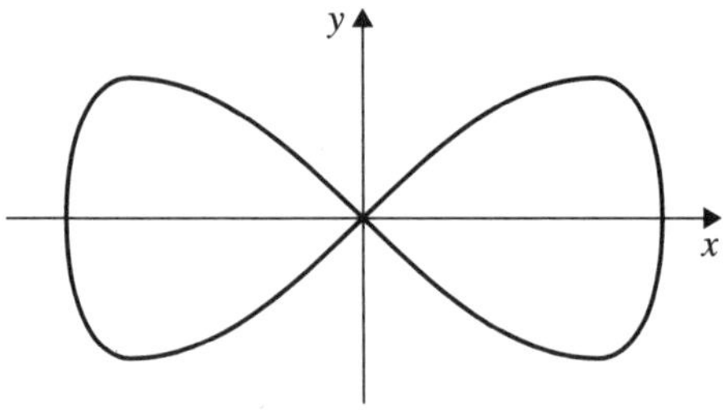

7.2 Parametric differentiation

If x and y are given in terms of t then

$$\frac{dy}{dx} = \frac{dy}{dt} \times \frac{dt}{dx} = \frac{\left(\frac{dy}{dt}\right)}{\left(\frac{dx}{dt}\right)} \qquad \left[\text{remember: } \frac{dt}{dx} = \frac{1}{\left(\frac{dx}{dt}\right)}\right]$$

Example 4

A curve is defined by $x = t^2 + 2, y = 3t^3$. Find $\frac{dy}{dx}$ and hence find the gradient of the curve at the point (3, 3).

$$\frac{dx}{dt} = 2t \text{ and } \frac{dy}{dt} = 9t^2. \qquad \therefore \quad \frac{dy}{dx} = \frac{\left(\frac{dy}{dt}\right)}{\left(\frac{dx}{dt}\right)} = \frac{9t^2}{2t} = \frac{9t}{2}$$

At the point (3, 3) we see that $x = t^2 + 2 = 3$ and $y = 3t^3 = 3$. Solving these equations gives $t = 1$.

So at the point (3, 3) the gradient, $\frac{dy}{dx} = 4\frac{1}{2}$.

Example 5

Find the equation of the tangent to the curve given by

$$x = \frac{2}{t}, y = 2t + 1 \text{ at the point } (2, 5)$$

We find the value of t at the point (2, 5).

$$\frac{2}{t} = 2 \text{ and } 2t + 1 = 5 \text{ so } t = 1.$$

Differentiating, $\frac{dy}{dt} = 2$ and $\frac{dx}{dt} = -\frac{2}{t^2}$

$$\therefore \quad \frac{dy}{dx} = 2 \times \left(-\frac{t^2}{2}\right) = -t^2$$

When $t = 1$, $\frac{dy}{dx} = -1$

The gradient of the tangent is -1.
The tangent passes through the point (2, 5).
The equation of the tangent is $y - 5 = -1(x - 2)$
or $y + x = 7$.

Example 6

Find the coordinates of the stationary point on the curve with equations $x = (t^2 - 3)^2$, $y = t^2 - 4t$

Differentiating, $\frac{dx}{dt} = 2(t^2 - 3) \times 2t$, $\frac{dy}{dt} = 2t - 4$

$$\frac{dy}{dx} = (2t - 4) \times \frac{1}{4t(t^2 - 3)}$$

At a stationary point $\frac{dy}{dx} = 0$

$$\therefore \quad 2t - 4 = 0$$
$$t = 2$$

When $t = 2$, $x = (2^2 - 3)^2 = 1$
$y = 2^2 - 4 \times 2 = -4$

The stationary point is at the point $(1, -4)$.

Exercise 2

1 Find $\frac{dy}{dx}$ in terms of t in the following:

a $x = 4t$ and $y = t^2$

b $x = 2 \sin t$ and $y = 3 \cos t$

2 Find $\frac{dy}{dx}$ in terms of t for the following curves:

a $x = 4t^2$, $y = 8t$

b $x = 5t$, $y = \frac{3}{t}$

c $x = 7 \cos t$, $y = 4 \sin t$

d $x = 2 \sec t$, $y = \tan t$

e $x = e^t$, $y = e^t - e^{-t}$

3 a Show that if $x = \frac{1}{1 + t^3}$ and $y = \frac{t}{1 + t^3}$ then $\frac{dy}{dx} = \frac{2t^3 - 1}{3t^2}$.

b Hence find the equation of the tangent to the curve at the point $(\frac{1}{2}, \frac{1}{2})$.

4 a Find the value of t at the point (2, 2) on the curve $x = 1 - t^3$, $y = 1 + t^2$.

b Hence find the value of $\frac{dy}{dx}$ at (2, 2).

c Use this to find the equation of the normal to the curve at (2, 2).

5 **a** Show that $\frac{dy}{dx} = \left(\frac{1-2t}{1-t}\right)^2$ on the curve $x = \frac{1+t}{1-2t}$, $y = \frac{1+2t}{1-t}$.

b Hence find the equation of the tangent to the curve at the point (1, 1).

c Find also the equation of the normal to the curve at the point $(\frac{1}{4}, 0)$.

6 A curve is given by $x = t^2 - 4t$, $y = t^3 - 4t^2$.

a Find the points at which the gradient is $\frac{3}{2}$.

b Find the point at which the curve is parallel to the y-axis.

7 **a** Find $\frac{dy}{dx}$ for the curve $x = 3t^2 + 5$, $y = 2t^3 - 6t$.

b Hence find the stationary points of the curve.

8 Find the stationary points on the curve $x = \frac{1}{1+t}$, $y = \frac{t^2}{1+t}$.

9 A curve has parametric equations $x = \cos t$, $y = \sin t$.
Find the equation of the tangent to the curve where $t = \frac{\pi}{4}$.

10 A curve has parametric equations $x = t - \cos t$, $y = \sin t$.
Show that the equation of the tangent to the curve, where $t = \pi$, is $x + y = \pi + 1$.

11 The parametric equations of a curve are $x = a\sin\theta$, $y = a\cos^2\theta$, $0 \leqslant \theta \leqslant \frac{\pi}{2}$.

Find the equation of the normal to the curve at the point where $\theta = \frac{\pi}{6}$.

12 The parametric equations of a curve are

$$x = \cos\theta,\ y = 2\sin\theta.$$

Find the equation of the normal to the curve at the point $P(\cos\theta, 2\sin\theta)$.

13 A curve has parametric equations $x = \cos t$, $y = \frac{1}{2}\sin 2t$ for $0 \leqslant t \leqslant 2\pi$.

a Find an expression for $\frac{dy}{dx}$ in terms of the parameter t.

b Find the values of t at the points on the curve where the gradient is -1.

14 A curve has equations $x = t - \sin t$, $y = 1 - \cos t$ for $0 \leqslant t \leqslant 2\pi$.
Find the coordinates of the stationary points on the curve.

15 A curve has parametric equations $x = 2 + 3\cos\theta$
$y = 2 + 3\sin\theta$

a Find the cartesian equation of the curve and describe the curve.

b Find the equation of the tangent at the point with parameter $\theta = \frac{\pi}{4}$.

7.3 The area under a curve

In the same way that we use integration to find the area under a curve in cartesian form so we use integration to find the area under a curve given in parametric form.

We know that the area bounded by the curve $y = f(x)$, the x-axis, the lines $x = a$ and $x = b$ is given by $\int_a^b f(x)dx$.

If the curve is given parametrically we will have $y = f(t)$ and $x = g(t)$. The lines $x = a$ and $x = b$ will correspond to certain t-values, say t_1 and t_2.

In this case the area bounded is given by $\int_{t=t_1}^{t=t_1} y\,dx = \int_{t=t_1}^{t=t_1} f(t)\frac{dx}{dt}dt$.

Example 7

The curve defined by $x = t - 2$, $y = 3t^2$ is shown below. Find the shaded area.

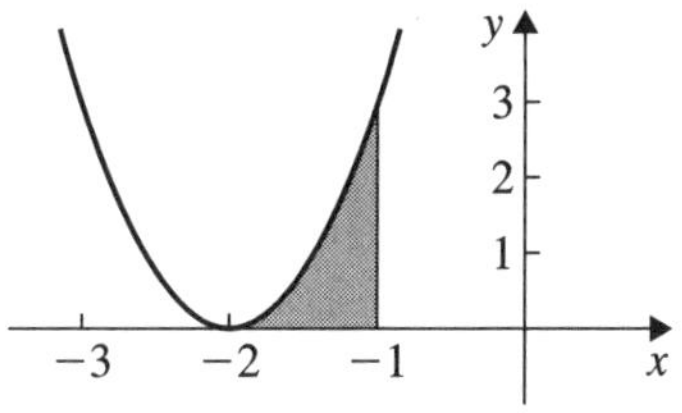

The area is bounded by the values $t = 0$ and $t = 1$. So the area is given by $\int_{t=0}^{t=1} 3t^2\frac{dx}{dt}dt$. Now $x = t - 2$ so $\frac{dx}{dt} = 1$. Therefore the area is $\int_{t=0}^{t=1} 3t^2\,dt = [t^3]_0^1 = 1$ square unit.

[Notice that the curve is $y = 3(x + 2)^2$ and so the area is given by $\int_{-2}^{-1} 3(x + 2)^2\ dx$. We can easily show that the value of this integral is 1 using standard integration.]

Example 8

Find area between the curve defined $x = t^3$ and $y = t^2 - t$, the lines $x = 1$, $x = 8$ and the x-axis.

We can see that when $x = 1, t = 1$, when $x = 8, t = 2$ and also $\frac{dx}{dt} = 3t^2$.

Therefore the area that we are trying to find is given by $\int y.\frac{dx}{dt}dt = \int_1^2 (t^2 - t).3t^2\,dt$

$$\therefore\quad \text{Area} = 3\int_1^2 (t^4 - t^3)\,dt = 3\left[\frac{1}{5}t^5 - \frac{1}{4}t^4\right]_1^2$$

$$= 3\left(\frac{31}{5} - \frac{15}{4}\right) = \frac{147}{20} \text{ square units.}$$

EXERCISE 3

1 Find the area enclosed between the following curves, the x-axis and the vertical lines $x = a$ and $x = b$ which correspond to $t = t_1$ and $t = t_2$.

a $x = t - 1,\quad y = t^2\quad t_1 = 1, t_2 = 3$

b $x = t^2,\quad y = 2t^3\quad t_1 = 1, t_2 = 4$

c $x = 5t,\quad y = \dfrac{5}{t}\quad t_1 = 2, t_2 = 6$

2 Find the area enclosed between the following curves, the x-axis and the vertical lines given below:

a $x = t^2 + 1,\quad y = 2t\ (t > 0)\quad x = 2, x = 5$

b $x = t + 1,\quad y = t^2\quad x = 1, x = 4$

c $x = 2t^2,\quad y = \dfrac{2}{t}\ (t > 0)\quad x = 2, x = 8$

3 Find the area enclosed between the following curves, the x-axis and the vertical lines given below:

a $x = 3t,\quad y = \dfrac{3}{t}\quad x = 3, x = 12$

b $x = 2t - 1,\quad y = 4t^2 + 1\quad x = 1, x = 7$

c $x = \dfrac{3}{\sqrt{t}},\quad y = 2t + 1\quad x = 1.5; x = 3$

d $x = 2\sqrt{t},\quad y = 3t^2 + 4\quad x = 4, x = 6$

4 The graph shows the curve with parametric equations $x = \theta - \sin\theta$, $y = 1 - \cos\theta$ for $0 \leqslant \theta \leqslant 2\pi$. The curve is called a cycloid.

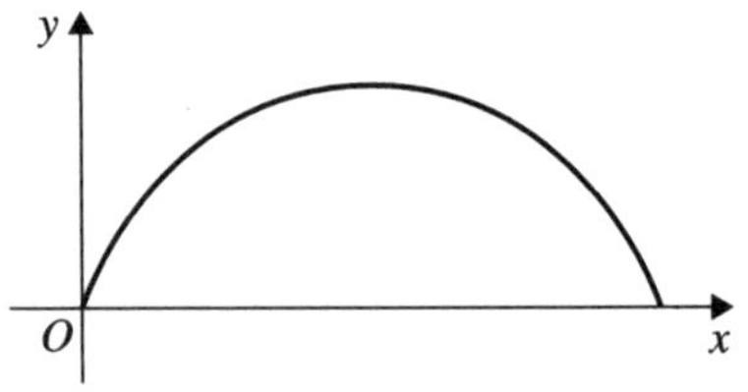

The area A enclosed by the curve and the x-axis is $\displaystyle\int_0^{2\pi} y\frac{dx}{d\theta}\,d\theta$.

Show that $A = \displaystyle\int_0^{2\pi}(1 - \cos\theta)^2 d\theta$ and hence find A, in terms of π.

7.4 Volumes of revolution

We have already seen that when the area bounded between a curve, the x-axis and the lines $x = a$ and $x = b$ is rotated through 360° about the x-axis the volume is given by $\displaystyle\int_a^b \pi y^2\,dx$.

In the same way if the curve is given parametrically we will have $y = f(t)$ and $x = g(t)$. The lines $x = a$ and $x = b$ will correspond to certain t-values, say $t = t_1$ and $t = t_2$.

In this case the volume created is given by $\displaystyle\pi\int_{t=t_1}^{t=t_2} y^2\,dx = \pi\int_{t=t_1}^{t=t_2} [f(t)]^2\frac{dx}{dt}\,dt$.

Example 9

The curve defined by $x = t - 2$, $y = 3t^2$ is shown below. Find the volume created when the shaded area shown below is rotated about the x-axis.

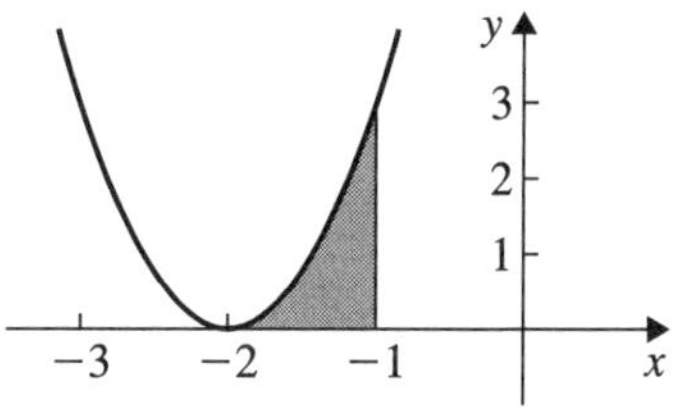

The area is bounded by the values $t = 0$ and $t = 1$. So the volume is given by $\pi\int_{t=0}^{t=1}(3t^2)^2\frac{dx}{dt}\,dt$.

Now $x = t - 2$ so $\frac{dx}{dt} = 1$. Therefore the volume is $\pi\int_{t=0}^{t=1} 9t^4\,dt = \pi\left[\frac{9}{5}t^5\right]_0^1$

$= \frac{9\pi}{5}$ cubic units.

[Notice that the curve is $y = 3(x+2)^2$ and so the volume is given by $\pi\int_{-2}^{-1} 9(x+2)^4\,dx$.

We can show that the value of this integral is $\frac{9\pi}{5}$ using standard integration.]

Exercise 4

1 Find the volume created when the area bounded by the curve $x = t + 1$, $y = t^2$, the vertical lines $x = 1$, $x = 4$ and the x-axis is rotated through 360° about the x-axis.

2 Find the volume created when the area bounded by the curve $x = 3t^2$, $y = 2t^3$, the vertical lines $x = 0$, $x = 3$ and the x-axis is rotated through 360° about the x-axis.

3 Find the volume created when the area bounded by the curve $x = 2t$, $y = \frac{4}{t}$, the vertical lines $x = 2$, $x = 8$ and the x-axis is rotated through 360° about the x-axis.

4 Find the volume created when the area bounded by the curve $x = 2t + 1$, $y = t^2 + 1$, the vertical lines $x = 1$, $x = 9$ and the x-axis is rotated through 360° about the x-axis.

5 Find the volume created when the area bounded by the curve $x = t^3$, $y = \frac{1}{t^2}$, the vertical lines $x = 1$, $x = 8$ and the x-axis is rotated through 360° about the x-axis.

6 **a** Find the area bounded by the curve $x = \frac{2}{\sqrt{t}}$, $y = 3t + 1$, the vertical lines $x = 1$, $x = 2$ and the x-axis.

b Find the volume when this area is rotated through 360° about the x-axis.

REVIEW EXERCISE 7

1 Find the cartesian equation for each curve.

a $x = t + 1, y = t^2$ **b** $x = 3t - 1, y = \frac{1}{t}$

c $x = 2t, y = \frac{4}{t}$ **d** $x = \frac{1}{1+t}, y = \frac{t}{1+t}$

e $x = 5\sin t, y = \cos t$ **f** $x = \cos t, y = 2\cos 2t$

2 Show that the cartesian equation of the curve with parametric equations $x = \frac{1}{2}(1 - t^2),\ \ y = t^3$ is $y^2 = (1 - 2x)^3$.

3 Find the area enclosed between the following curves, the x-axis and the lines given.

a $x = t^2,\ \ y = 2t$ $x = 1, x = 4$

b $x = t + 2,\ \ y = t^3 - 2$ $x = 0, x = 3$

c $x = 2\sqrt{t},\ \ y = 3t^2 + 4$ $x = 4, x = 6$

4 **a** Find the area of the region, R, bounded by the curve $x = t + 1, y = t^2$, the x-axis and the lines $x = 1$ and $x = 3$.

b Find the volume of the solid generated when R is rotated through 360° about the x-axis.

5 The diagram shows the curve with equations $x = t^2, y = t$ for $t > 0$.

Find the volume generated when the shaded region R is rotated 2π about the x-axis.

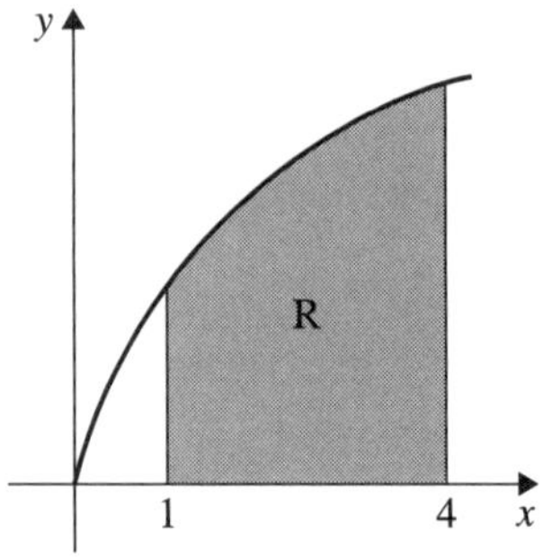

6 Find $\frac{dy}{dx}$, in terms of t, for each of the following curves

a $x = t^2,\ \ y = t$ **b** $x = \sin t,\ \ y = \cos t$ **c** $x = t^3,\ \ y = t^2$

7 Find the turning point of the curve $x = t,\ \ y = t^2 - 1$.

8 Find the turning points of the curve $x = t,\ \ y = t^3 - 3t$.

9 Find the equation of the tangent to the following curves at the given point.

a $x = t,\ \ y = t^2$ at $t = -1$

b $x = \cos\theta,\ \ y = \sin\theta$ at $\theta = \frac{\pi}{4}$

c $x = t,\ \ y = \frac{1}{t}$ at $t = 1$

d $x = 3\cos\theta,\ \ y = 4\sin\theta$ at $\theta = \frac{\pi}{3}$

e $x = 4\sin\theta,\ \ y = 3\cos 2\theta$ at $\theta = \frac{\pi}{6}$

f $x = \cos 2\theta - 2\cos\theta,$
$y = \sin 2\theta - 2\sin\theta$ at $\theta = \frac{\pi}{4}$

Examination Exercise 7

1 The parametric equations of a curve are

$$x = t + \frac{2}{t}, \quad y = t - \frac{2}{t},$$

where $t \neq 0$.

i Show that $t = \frac{1}{2}(x + y)$, and hence write down the value of t corresponding to the point (3, 1) on the curve.

ii Find the gradient of the curve at the point (3, 1). [OCR]

2 A curve is given by the parametric equations

$$x = 1 - t^2, \quad y = 2t.$$

a Find $\frac{dy}{dx}$ in terms of t.

b Hence find the equation of the normal to the curve at the point where $t = 3$. [AQA]

3 A curve is given by the parametric equations

$$x = 2t + 3, \quad y = \frac{2}{t}.$$

Find the equation of the normal at the point on the curve where $t = 2$. [AQA]

4 A curve is given parametrically by the equations

$$x = t(1 + t), \quad y = t^2(1 + t).$$

i Find $\frac{dy}{dx}$ in terms of t.

ii Find the equation of the tangent to the curve at the point where $t = 2$, giving your answer in the form $ax + by + c = 0$.

iii By first simplifying $\frac{y}{x}$, show that the curve has cartesian equation

$$x^3 = xy + y^2.$$

[OCR]

5 A curve is defined by the parametric equations

$$x = 3 \sin t \quad \text{and} \quad y = \cos t.$$

a Show that, at the point P where $t = \frac{\pi}{4}$, the gradient of the curve is $-\frac{1}{3}$.

b Find the equation of the tangent to the curve at the point P, giving your answer in the form $y = mx + c$. [AQA]

6 The curve C is described by the parametric equations

$$x = 3 \cos t, \quad y = \cos 2t, \quad 0 \leqslant t \leqslant \pi.$$

a Find a cartesian equation of the curve C.

b Draw a sketch of the curve C. [EDEXCEL]

7 The parametric equations of a curve are $x = t^2$, $y = 2t$.

i Prove that the equation of the tangent at the point with parameter t is

$$ty = x + t^2.$$

ii The tangent at the point $P(16, 8)$ meets the tangent at the point $Q(9, -6)$ at the point R. Find the coordinates of R.

iii Prove that the line $2y - 4x - 1 = 0$ is a tangent to the curve, and find the parameter at the point of contact. [OCR]

8 The parametric equations of a curve C are

$$x = a\sin\theta, \quad y = 2a\cos\theta,$$

where a is a positive constant and $-\pi < \theta \leqslant \pi$.

i Show that the equation of the tangent to C at the point with parameter θ is

$$2x\sin\theta + y\cos\theta = 2a.$$

ii This tangent passes through the point $(2a, 3a)$.

a Show that θ satisfies an equation of the form $5\sin(\theta + \alpha) = 2$, and state the value of $\tan\alpha$.

b Hence find the two possible values of θ. [OCR]

9 A curve has parametric representation

$$x = 5 - \cos\theta, \quad y = 1 + 4\sin\theta, \quad 0 \leqslant \theta < 2\pi.$$

The region of the plane below the curve and above the x-axis between the points where $\theta = \frac{\pi}{4}$ and $\theta = \frac{3\pi}{4}$ has area A.

a Show that

$$A = \int_{\frac{\pi}{4}}^{\frac{3\pi}{4}} (\sin\theta + 4\sin^2\theta)\,\mathrm{d}\theta.$$

b Hence find the exact value of A. [AQA]

10 The diagram shows a sketch of part of the curve C which is defined parametrically by

$$x = t^2, \ y = \sin t, \ t \geqslant 0.$$

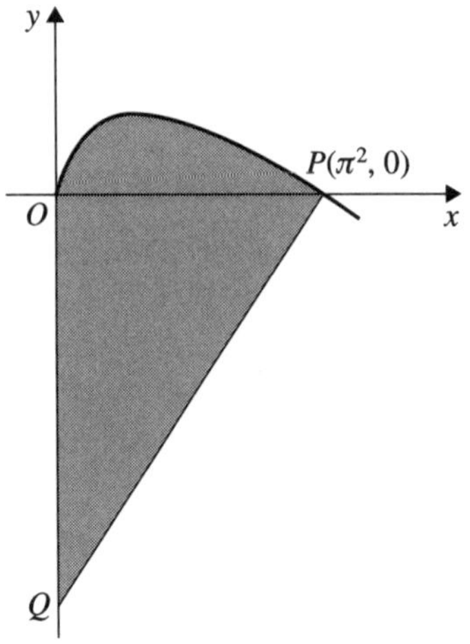

The curve cuts the positive x-axis for the first time at the point $P(\pi^2, 0)$.

The normal to the curve C at P intersects the y-axis at the point Q.

a Show that the equation of the normal PQ is

$$y = 2\pi x - 2\pi^3.$$

b i Find $\int t\sin t\,\mathrm{d}t$.

ii Find, in terms of π, the area of the shaded region bounded by the curve C, the normal PQ and the y-axis. [AQA]

11

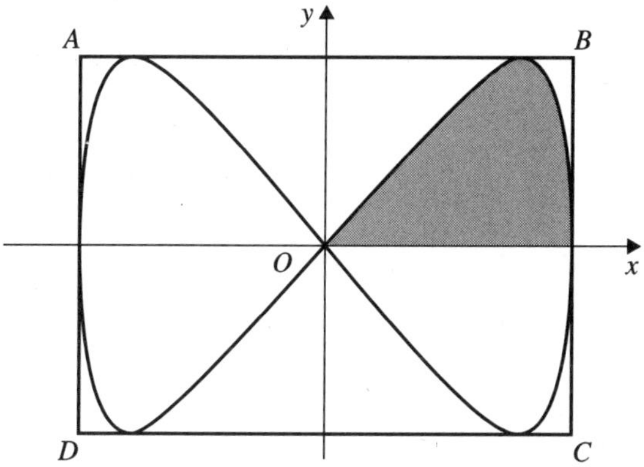

Fig. 2

Part of the design of a stained glass window is shown in Fig. 2. The two loops enclose an area of blue glass. The remaining area within the rectangle *ABCD* is red glass.

The loops are described by the curve with parametric equations

$$x = 3\cos t, \quad y = 9\sin 2t, \; 0 \leqslant t < 2\pi.$$

a Find the cartesian equation of the curve in the form $y^2 = \mathrm{f}(x)$.

b Show that the shaded area in Fig. 2, enclosed by the curve and the x-axis, is given by

$$\int_0^{\frac{\pi}{2}} A \sin 2t \sin t \,\mathrm{d}t, \text{ stating the value of the constant } A.$$

c Find the value of this integral.

The sides of the rectangle *ABCD*, in Fig. 2, are the tangents to the curve that are parallel to the coordinate axes. Given that 1 unit on each axis represents 1 cm,

d Find the total area of the red glass. [EDEXCEL]

Part 8

The binomial series

8.1 The binomial series for any rational index

In the C1 and C2 book we saw that if n is a positive integer

$$(a+b)^n = \binom{n}{0}a^n + \binom{n}{1}a^{n-1}b + \binom{n}{2}a^{n-2}b^2 + \ldots + \binom{n}{r}a^{n-r}b^r + \ldots$$

and

$$(1+x)^n = 1 + nx + \frac{n(n-1)}{1\times 2}x^2 + \ldots + \frac{n(n-1)\ldots(n-r+1)}{1\times 2\times \ldots \times r}x^r + \ldots$$

In both the above cases these expansions were finite, they had $n+1$ terms.

We are now going on to consider such expansions in cases where n is no longer a positive integer but where n is any rational number, that is any fraction, positive or negative.

If we put $n = -1$ into the expression

$$(1+x)^n = 1 + nx + \frac{n(n-1)}{1\times 2}x^2 + \ldots + \frac{n(n-1)\ldots(n-r+1)}{1\times 2\times \ldots \times r}x^r + \ldots$$

We get $(1+x)^{-1} = 1 + (-1)x + \dfrac{(-1)\times(-2)}{1\times 2}x^2 + \dfrac{(-1)\times(-2)\times(-3)}{1\times 2\times 3}x^3 + \ldots$

$$= 1 - x + x^2 - x^3$$

Now we know from work on geometric series that if $|x| < 1$ (that is $-1 < x < 1$) then the sum of the infinite series $1 - x + x^2 - x^3 + \ldots + \ldots$ is $\dfrac{1}{1+x} = (1+x)^{-1}$.

So it follows from the above that

$(1+x)^n = 1 + nx + \dfrac{n(n-1)}{1\times 2}x^2 + \ldots + \dfrac{n(n-1)\ldots(n-r+1)}{1\times 2\times \ldots\times r}x^r + \ldots$ is true not simply for n being a positive integer but also for $n = -1$ provided that $-1 < x < 1$.

In some of the questions in the last exercise in this section we will show that

$(1+x)^n = 1 + nx + \dfrac{n(n-1)}{1\times 2}x^2 + \ldots + \dfrac{n(n-1)\ldots(n-r+1)}{1\times 2\times \ldots \times r}x^r + \ldots$ holds for $n = -2$, $n = -3$, $n = \frac{1}{2}$ and other fractional values, provided that $-1 < x < 1$. It is beyond the scope of this book but it can be proved that

$(1+x)^n = 1 + nx + \dfrac{n(n-1)}{1\times 2}x^2 + \ldots + \dfrac{n(n-1)\ldots(n-r+1)}{1\times 2\times \ldots \times r}x^r + \ldots$ is true for all rational values of n, provided that $-1 < x < 1$.

So we use the following result:

$$(1+x)^n = 1 + nx + \frac{n(n-1)}{1 \times 2}x^2 + \ldots + \frac{n(n-1)\ldots(n-r+1)}{1 \times 2 \times \ldots \times r}x^r + \ldots$$

$-1 < x < 1$, n is rational.

It follows from the above that the expansion of $(1 + kx)^n$ is valid provided $-1 < kx < 1$, that is provided $-\frac{1}{k} < x < \frac{1}{k}$.

Example 1

Find the first four terms in ascending powers of x in the expansion of $(1 + x)^{-5}$, where $-1 < x < 1$.

$$(1+x)^{-5} = 1 + (-5)x + \frac{(-5) \times (-6)}{1 \times 2}x^2 + \frac{(-5) \times (-6) \times (-7)}{1 \times 2 \times 3}x^3 + \ldots$$

So we see that $(1 + x)^{-5} = 1 - 5x + 15x^2 - 35x^3 + \ldots$

Example 2

Find the first four terms in ascending powers of x in the expansion of $(1 + x)^{\frac{1}{3}}$, where $-1 < x < 1$.

$$(1+x)^{\frac{1}{3}} = 1 + \left(\frac{1}{3}\right)x + \frac{\left(\frac{1}{3}\right) \times \left(\frac{1}{3} - 1\right)}{1 \times 2}x^2 + \frac{\left(\frac{1}{3}\right) \times \left(\frac{1}{3} - 1\right) \times \left(\frac{1}{3} - 2\right)}{1 \times 2 \times 3}x^3 + \ldots$$

So we see that $(1 + x)^{\frac{1}{3}} = 1 + \frac{x}{3} - \frac{x^2}{9} + \frac{5x^3}{81} + \ldots$

Example 3

a Find the first four terms in ascending powers of x in the expansion of $(1 + 2x)^{\frac{5}{2}}$, where this expansion is valid.

b For what values of x is the expansion valid?

a $$(1+2x)^{\frac{5}{2}} = 1 + \left(\frac{5}{2}\right)(2x) + \frac{\left(\frac{5}{2}\right) \times \left(\frac{5}{2} - 1\right)}{1 \times 2}(2x)^2 + \frac{\left(\frac{5}{2}\right) \times \left(\frac{5}{2} - 1\right) \times \left(\frac{5}{2} - 2\right)}{1 \times 2 \times 3}(2x)^3 + \ldots$$

So we see that $(1 + 2x)^{\frac{5}{2}} = 1 + 5x + \frac{15x^2}{2} + \frac{5x^3}{2} + \ldots$

b The expansion is valid for $|2x| < 1$. That is $-\frac{1}{2} < x < \frac{1}{2}$.

Example 4

Find $\sqrt[3]{1.03}$ to 4 decimal places.

From example **2**, $(1 + x)^{\frac{1}{3}} = 1 + \frac{x}{3} - \frac{x^2}{9} + \frac{5x^3}{81} + \ldots$ if $-1 < x < 1$.

If we substitute $x = 0.03$ we see that

$$\sqrt[3]{1.03} = (1 + 0.03)^{\frac{1}{3}} = 1 + \frac{0.03}{3} - \frac{(0.03)^2}{9} + \frac{5(0.03)^3}{81} + \dots$$
$$= 1 + 0.01 - 0.001 + \dots \approx 1.0099\dots$$

So we see that $\sqrt[3]{1.03} = 1.0099$ (to 4 dp).

Exercise 1

1 Find the first 3 terms in ascending powers of x in the following expansions, where $-1 < x < 1$:

a $(1 + x)^{-2}$ **b** $(1 + x)^{-3}$ **c** $(1 + x)^{-6}$

d $(1 + x)^{-10}$ **e** $(1 + x)^{\frac{1}{2}}$ **f** $(1 + x)^{\frac{3}{2}}$

g $(1 + x)^{-\frac{1}{2}}$ **h** $(1 + x)^{-\frac{3}{4}}$ **i** $(1 - x)^{\frac{2}{3}}$

j $(1 - x)^{-\frac{2}{5}}$ **k** $\sqrt{(1 - x)}$ **l** $\dfrac{1}{\sqrt[3]{1 - x}}$

2 Use question 1**a** to find the value of $(1.01)^{-2}$ to 3 decimal places.

3 Use the expression in question 1**e** to find the value of $\sqrt{1.02}$, correct to 3 decimal places.

4 Find the limits for x within which the following expressions are valid.

a $(1 + 2x)^{-3}$ **b** $(1 - x)^{-2}$

c $(1 + 5x)^{-4}$ **d** $(1 + 2x)^{-7}$

e $(1 + 2x)^{\frac{1}{2}}$ **f** $(1 + 4x)^{\frac{5}{2}}$

g $(1 - 3x)^{-\frac{1}{2}}$ **h** $(1 - 2x)^{\frac{1}{4}}$

i $\left(1 - \dfrac{x}{3}\right)^{\frac{1}{2}}$ **j** $\left(1 - \dfrac{x}{2}\right)^{-\frac{1}{3}}$

5 Assuming that in each case x lies within the range so that expansion is valid, find the first 4 terms in the following expansions:

a $(1 + 2x)^{-3}$ **b** $(1 - x)^{-2}$

c $(1 + 5x)^{-4}$ **d** $(1 + 2x)^{-7}$

e $(1 + 2x)^{\frac{1}{2}}$ **f** $(1 + 4x)^{\frac{5}{2}}$

g $(1 - 3x)^{-\frac{1}{2}}$ **h** $(1 - 2x)^{\frac{1}{4}}$

i $\left(1 - \dfrac{x}{3}\right)^{\frac{1}{2}}$ **j** $\left(1 - \dfrac{x}{2}\right)^{-\frac{1}{3}}$

6 Find the x^4 term in the expansion of $(1 + 2x)^{\frac{1}{4}}$ where $-\frac{1}{2} < x < \frac{1}{2}$.

7 Find the x^5 term in the expansion of $(1 - 2x)^{-\frac{3}{4}}$ where $-\frac{1}{2} < x < \frac{1}{2}$.

8 Find the x^3 term in the expansion of $\left(1 - \dfrac{x}{2}\right)^{\frac{2}{5}}$ where $-2 < x < 2$.

9 **a** Expand $(x + 2)^3$.

b Evaluate $(\sqrt{3} + 2)^3$ leaving your answer in the form $a\sqrt{3} + b$.

10 **a** Show that $(x + y)^4 - (x - y)^4 = 8xy(x^2 + y^2)$.

b Find the exact value of $(3 + \sqrt{2})^4 - (3 - \sqrt{2})^4$.

11 Expand $(1 + x)^{\frac{1}{2}}$ up to and including the term in x^3. Hence calculate the value of $\sqrt{1.08}$ correct to 4 decimal places.

12 The first four terms in the expansion of $(1 - x)^n$ are $1 - 6x + ax^2 + bx^3$. Show that $a = 15$ and find the value of b.

13 Find the first four terms in the expansion of $\left(1 - \frac{1}{x}\right)^{\frac{1}{2}}$ in *descending* powers of x.

By substituting $x = 100$ find a value of $\sqrt{99}$, giving your answer to 5 decimal places.

14 **a** Find the first three terms in the expansion of $(1 - x)^{\frac{1}{3}}$.

b By letting $x = \frac{1}{1000}$, evaluate $\sqrt[3]{37}$ to six decimal places.

Example 5

Find the first four terms in ascending powers of x in the expansion of $(4 + x)^{\frac{1}{2}}$, stating clearly the values of x for which this is valid.

We first of all need to write $(4 + x)^{\frac{1}{2}} = \left[4\left(1 + \frac{x}{4}\right)\right]^{\frac{1}{2}} = 4^{\frac{1}{2}}\left(1 + \frac{x}{4}\right)^{\frac{1}{2}} = 2\left(1 + \frac{x}{4}\right)^{\frac{1}{2}}$.

So we see that the expansion is valid if $-1 < \frac{x}{4} < 1$, that is provided $-4 < x < 4$.

$$\left(1 + \frac{x}{4}\right)^{\frac{1}{2}} = 1 + \left(\frac{1}{2}\right)\left(\frac{x}{4}\right) + \frac{\left(\frac{1}{2}\right) \times \left(\frac{1}{2} - 1\right)}{2 \times 1}\left(\frac{x}{4}\right)^2 + \frac{\left(\frac{1}{2}\right) \times \left(\frac{1}{2} - 1\right) \times \left(\frac{1}{2} - 2\right)}{3 \times 2 \times 1}\left(\frac{x}{4}\right)^3 + \ldots$$

$$= 1 + \frac{x}{8} - \frac{x^2}{128} + \frac{x^3}{1024} + \ldots$$

Hence we see that $(4 + x)^{\frac{1}{2}} = 2\left(1 + \frac{x}{4}\right)^{\frac{1}{2}} = 2 + \frac{x}{4} - \frac{x^2}{64} + \frac{x^3}{512} + \ldots$

Example 6

Find the first four terms in ascending powers of x in the expansion of $(2x + 3)^{-1}$, stating clearly the values of x for which this is valid.

$$(2x + 3)^{-1} = \left[3\left(\frac{2x}{3} + 1\right)\right]^{-1} = 3^{-1}\left(1 + \frac{2x}{3}\right)^{-1} = \frac{1}{3}\left(1 + \frac{2x}{3}\right)^{-1}$$

So we see that the expansion is valid if $-1 < \frac{2x}{3} < 1$, that is provided $-\frac{3}{2} < x < \frac{3}{2}$.

$$(2x+3)^{-1} = \frac{1}{3}\left(1+\frac{2x}{3}\right)^{-1}$$

$$= \frac{1}{3}\left[1+(-1)\left(\frac{2x}{3}\right)+\frac{(-1)(-2)}{2!}\left(\frac{2x}{3}\right)^2+\frac{(-1)(-2)(-3)}{3!}\left(\frac{2x}{3}\right)^3+\ldots\right]$$

Hence $(2x+3)^{-1} = \frac{1}{3} - \frac{2x}{9} + \frac{4x^2}{27} - \frac{8x^3}{81} + \ldots$

EXERCISE 2

1 Calculate the first three terms (in ascending powers of x) in the expansion of the following, stating also the values of x for which each expansion is valid

a $(2+x)^{-1}$ **b** $(4-x)^{-3}$ **c** $(3+x)^{-2}$

d $(5+x)^{-4}$ **e** $(4-x)^{\frac{1}{2}}$ **f** $(9-x)^{-\frac{1}{2}}$

g $(8-x)^{\frac{1}{3}}$ **h** $(8+x)^{-\frac{2}{3}}$ **i** $(25+x)^{\frac{5}{2}}$

2 Calculate the first three terms (in ascending powers of x) in the expansion of the following, stating also the values of x for which each expansion is valid.

a $(4+3x)^{\frac{1}{2}}$ **b** $(9-2x)^{-\frac{1}{2}}$ **c** $(8-5x)^{\frac{1}{3}}$

d $(25-7x)^{\frac{3}{2}}$

3 Given that, in each case, x lies within the range so that the expansion is valid, find the exact values of a and b in the following

a $(4+3x)^{-\frac{1}{2}} = a + bx + \ldots$ **b** $(8+3x)^{\frac{1}{3}} = a + bx + \ldots$

c $(2-7x)^{-\frac{1}{2}} = a + bx + \ldots$ **d** $(27+2x)^{-\frac{1}{3}} = a + bx + \ldots$

4 Find, in ascending powers of x, up to and including the x^3 term, the expansion of $\sqrt{9+4x}$, stating the values of x for which the expansion is valid.

5 a Calculate the first three terms (in ascending powers of x) in the expansion of $(1-2x)^{\frac{1}{2}}$ and state the values of x for which the expansion is valid.

b Use this to estimate the value of $\sqrt{0.98}$ correct to 3 significant figures.

6 a Expand $(1-3x)^{\frac{1}{5}}$ in ascending powers of up to and including the term in x^2.

b By substituting $x = \frac{1}{32}$ in your series, find an approximation for $\sqrt[5]{29}$, giving your answer correct to 3 decimal places.

8.2 Partial fractions and harder questions

Example 7

Find the first three terms in ascending powers of x in the expansion of $\frac{7+5x}{(x+1)(x+2)}$, stating the values of x for which this is valid.

In example 1 of Part 6 we saw that $\frac{7+5x}{(x+1)(x+2)} = \frac{2}{(x+1)} + \frac{3}{(x+2)}$

So we see that

$$\frac{7+5x}{(x+1)(x+2)} = 2(x+1)^{-1} + 3(x+2)^{-1}$$

$$= 2(1+x)^{-1} + 3\left[2\left(1+\frac{x}{2}\right)\right]^{-1}$$

$$= 2(1+x)^{-1} + \frac{3}{2}\left(1+\frac{x}{2}\right)^{-1}$$

The expansion of $(1+x)^{-1}$ is valid if $-1 < x < 1$, and the expansion of $\left(1+\frac{x}{2}\right)^{-1}$ is valid if $-2 < x < 2$. It follows that the expansion of $\frac{7+5x}{(x+1)(x+2)}$ is valid if $-1 < x < 1$.

Notice that we take the more restricted range of values for x.

So we have $\frac{7+5x}{(x+1)(x+2)} = 2(1 - x + x^2 - x^3) + \frac{3}{2}\left(1 - \frac{x}{2} + \frac{x^2}{4} - \frac{x^3}{8}\right)$

$$= \frac{7}{2} - \frac{11}{4}x + \frac{19}{8}x^2 - \frac{35}{16}x^3 + \ldots$$

EXERCISE 3

1 Express the following in partial fractions and then find the expansion of each one in ascending powers of x, up to and including the term in x^2. State the values of x for which each expansion is valid.

a $\frac{9x-26}{(x-3)(x-2)}$ **b** $\frac{3x-4}{(x-1)(x-2)}$

c $\frac{8x+11}{(x+1)(x+2)}$ **d** $\frac{7x+15}{(x+2)(x+3)}$

e $\frac{3x}{(x-1)(x+2)}$ **f** $\frac{x+7}{(x-1)(x+3)}$

2 **a** Express $\frac{x^2-x-1}{(x-2)(x-1)}$ as partial fractions.

b Hence show that $\frac{x^2-x-1}{(x-2)(x-1)} \equiv 1 - (1-x)^{-1} - \frac{1}{2}\left(1-\frac{x}{2}\right)^{-1}$ and use this to find the first four terms in ascending powers of x in the expansion of $\frac{x^2-x-1}{(x-2)(x-1)}$.

c State the range of values for x for which the expansion is valid.

3 a Find the first four terms in ascending powers of x in the expansion of $(x + 1)\sqrt{1 - 4x}$.

b State the range of values of x for which the expansion is valid.

4 In the series expansion of $(1 + \lambda x)^n$ the x coefficient is -6 and the x^2 coefficient is 24. Find λ and n.

5 Show that the coefficients of x^2 and x^4 in the series expansion of $(1 - 4x^2)^{\frac{1}{2}}$ are equal.

6 Find, in ascending powers of x, up to and including the x^2 term, the expansion of $\dfrac{(x + 1)}{\sqrt{1 - x}}$, stating the values of x for which the expansion is valid.

7 a Find, in ascending powers of x, up to and including the x^2 term, the expansion of $\sqrt{\dfrac{1 + x}{1 - x}}$, stating the values of x for which the expansion is valid.

b By taking $x = \dfrac{1}{9}$, show that $\sqrt{5} \approx \dfrac{181}{81}$.

8 a Find the constants a and b so that the expansions of $(1 + 2x)^{\frac{1}{2}}$ and $\dfrac{1 + ax}{1 + bx}$, up to and including the term in x^2, are the same.

b Use the above result, with $x = -\dfrac{1}{100}$, to obtain an approximate value for $\sqrt{2}$ in the form $\dfrac{m}{n}$, where m and n are integers.

9 In the series expansion of $\sqrt{(1 + px + qx^2)}$ the x and x^2 coefficients are both equal to 1.

a Show that $p = 2$ and find the value of q.

b Find the x^3 term in the series expansion of $\sqrt{(1 + px + qx^2)}$.

c By solving the inequality $-1 < px + qx^2 < 1$, find the values of x for which this expansion is valid.

Questions **10** to **13** are more difficult

10 a Find the sum of the series $1 - x + x^2 - x^3 + \ldots + (-1)^n x^n$.

b For which values of x does the sum to infinity exist?

c If x is restricted in the way described in part **b** then find the sum to infinity of $1 - x + x^2 - x^3 + \ldots$

11 a Deduce from question **10** that $(1 - x)^{-1} = 1 + x + x^2 + x^3 + \ldots$ and state the values of x for which this is true.

b Work out $(1 - x)^{-1}(1 - x)^{-1} = (1 + x + x^2 + \ldots)(1 + x + x^2 + \ldots)$ to express $(1 - x)^2$ as an infinite sum of powers of x.

c Use the fact that $\dfrac{\mathrm{d}}{\mathrm{d}x}((1 - x)^n) = -n(1 - x)^{n-1}$ for all values of n and your answer to **b** to find a formula for $(1 - x)^{-3}$ as an infinite sum of powers of x.

d Use the same method as part **c** to find a formula for $(1-x)^{-4}$ as an infinite sum of powers of x.

12 Show that your answers to 11**a** and **d** both agree with the formula

$$(1+x)^n = 1 + nx + \frac{n(n-1)}{2!}x^2 + \frac{n(n-1)(n-2)}{3!}x^3 + \ldots \text{ for } (-1 < x < 1)$$

13 **a** If $(1+x)^{-\frac{1}{2}} = a_0 + a_1x + a_2x^2 + a_3x^3 + \ldots$ then, by considering $(1+x)^{-\frac{1}{2}}(1+x)^{-\frac{1}{2}} = (1+x)^{-1}$ and using the fact that $(1+x)^{-1} = 1 - x + x^2 - x^3 + \ldots$, find a_0, a_1, a_2 and a_3 (NB a_0 must be positive) stating the values of x for which the expansions are valid.

b Show that your answers to **a** agrees with the formula

$$(1+x)^n = 1 + nx + \frac{n(n-1)}{2!}x^2 + \frac{n(n-1)(n-2)}{3!}x^3 + \ldots \text{ for } (-1 < x < 1)$$

Review Exercise 8

1 State the values of x for which each expansion is valid

a $(2+3x)^{-\frac{1}{2}}$ **b** $(5-3x)^{\frac{1}{3}}$

c $(2-7x)^{-\frac{1}{2}}$ **d** $(9+2x)^{-\frac{1}{3}}$

e $(3-2x)^{-\frac{2}{3}}$ **f** $(5-3x)^{-\frac{3}{2}}$

2 Calculate the first three terms (in ascending powers of x) in the expansion of $(27-3x)^{\frac{1}{3}}$ and state the values of x for which the expansion is valid.

3 Given that, in each cases, x lies within the range so that the expansion is valid, find the values of a and b in the following

a $(27-2x)^{-\frac{2}{3}} = a + bx + \ldots$

b $(4-3x)^{-\frac{3}{2}} = a + bx + \ldots$

4 **a** Express $f(x) = \dfrac{1}{(1+x)(1-2x)}$ in partial fractions.

b Find the first three terms in the expression of $f(x)$ in ascending powers of x.

c State the set of values of x for which the expansion is valid.

5 **a** Express $\dfrac{7x+2}{(2x+1)(x-1)}$ as partial fractions.

b Hence find the first three terms in ascending powers of x in the expansion of $\dfrac{7x+2}{(2x+1)(x-1)}$.

c State the range of values of x for which the expansion in (b) is valid.

6 **a** Express $\dfrac{28x}{(3x-1)(x+2)}$ as partial fractions.

b Hence show that $\dfrac{28x}{(3x-1)(x+2)} \equiv 4\left[\left(1+\dfrac{x}{2}\right)^{-1} - (1-3x)^{-1}\right]$.

c Show that the x^2 coefficient in the expansion of $\dfrac{28x}{(3x-1)(x+2)}$ is -35 and find the x^3 coefficient.

d State the range of values of x for which the expansion is valid.

Examination Exercise 8

1 **i** Expand $(1-2x)^{\frac{1}{2}}$ in ascending powers of x, up to and including the term in x^3, simplifying the coefficients.

ii State the set of values of x for which this expansion is valid. [OCR]

2 **a** Obtain the binomial expansion of $(1+x)^{\frac{1}{2}}$ as far as the term in x^2.

b **i** Hence, or otherwise, find the series expansion of $(4+2x)^{\frac{1}{2}}$ as far as the term x^2

ii Find the range of values of x for which this expansion is valid. [AQA]

3 **i** Expand $(1+4x)^{\frac{1}{2}}$ in ascending powers of x, up to and including the term in x^2, simplifying the coefficients.

ii State the set of values of x for which the expansion is valid.

iii In the expansion of

$$(1+kx)(1+4x)^{\frac{1}{2}},$$

the coefficient of x is 7. Find the value of the constant k and hence the coefficient of x^2. [OCR]

4 **a** Expand $(1+3x)^{-2}$, $|x| < \frac{1}{3}$, in ascending powers of x up to and including the term in x^3, simplifying each term.

b Hence, or otherwise, find the first three terms in the expansion of $\dfrac{x+4}{(1+3x)^2}$ as a series in ascending powers of x. [EDEXCEL]

5 Given that

$$\frac{10(2-3x)}{(1-2x)(2+x)} \equiv \frac{A}{1-2x} + \frac{B}{2+x},$$

a find the values of the constants A and B.

b Hence, or otherwise, find the series expansion in ascending powers of x, up to and including the term in x^3, of $\dfrac{10(2-3x)}{(1-2x)(2+x)}$, for $|x| < \frac{1}{2}$.

[EDEXCEL]

6 Let $f(x) = \dfrac{1+x}{2+x} - \dfrac{1-x}{2-x}$.

i Show that $f(x)$ may be expressed as $\dfrac{2x}{4-x^2}$.

ii Hence or otherwise show that, for small x,

$$f(x) = \frac{1}{2}x + \frac{1}{8}x^3 + \frac{1}{32}x^5 + \ldots$$ [OCR]

7 **a** Prove that, when $x = \frac{1}{15}$, the value of $(1+5x)^{-\frac{1}{2}}$ is exactly equal to $\sin 60°$.

b Expand $(1+5x)^{-\frac{1}{2}}$, $|x| < 0.2$, in ascending powers of x up to and including the term in x^3, simplifying each term.

c Use your answer to part **b** to find an approximation for $\sin 60°$.

d Find the difference between the exact value of $\sin 60°$ and the approximation in part **c**. [EDEXCEL]

8 $$f(x) = (1+3x)^{-1}, \ |x| < \tfrac{1}{3}.$$

a Expand $f(x)$ in ascending powers of x up to and including the term in x^3.

b Hence show that, for small x,

$$\frac{1+x}{1+3x} \approx 1 - 2x + 6x^2 - 18x^3.$$

c Taking a suitable value for x, which should be stated, use the series expansion in part **b** to find an approximate value for $\frac{101}{103}$, giving your answer to 5 decimal places. [EDEXCEL]

9 **i** Obtain the expansion of $(8+y)^{\frac{1}{3}}$ in ascending powers of y, up to and including the term in y^2.

ii State the set of values of y for which the expansion is valid.

iii Show that, if k^3 and higher powers of k are neglected,

$$(8 + 2k + k^2)^{\frac{1}{3}} = 2 + \frac{1}{6}k + \frac{5}{72}k^2.$$

iv Write down the expansion of $(8 - 2k + k^2)^{\frac{1}{3}}$ in ascending powers of k, up to and including the term in k^2. [OCR]

10 When $(1+ax)^n$ is expanded as a series in ascending powers of x, the coefficients of x and x^2 are -6 and 27 respectively.

a Find the value of a and the value of n.

b Find the coefficient of x^3.

c State the set of values of x for which the expansion is valid. [EDEXCEL]

11 **a** Express $\dfrac{4-x}{(1-x)(2+x)}$ in the form $\dfrac{A}{1-x} + \dfrac{B}{2+x}$.

b i Show that the first **three** terms in the expansion of

$$\frac{1}{2+x}$$

in ascending powers of x are $\frac{1}{2} - \frac{x}{4} + \frac{x^2}{8}$.

ii Obtain also the first **three** terms in the expansion of

$$\frac{1}{1-x}$$

in ascending powers of x.

c Hence, or otherwise, obtain the first **three** terms in the expansion of

$$\frac{4-x}{(1-x)(2+x)}$$

in ascending powers of x. [AQA]

12 Given that

$$\frac{3+5x}{(1+3x)(1-x)} \equiv \frac{A}{1+3x} + \frac{B}{1-x},$$

a Find the values of the constants A and B.

b Hence, or otherwise, find the series expansion in ascending powers of x, up to and including the term in x^2, of

$$\frac{3+5x}{(1+3x)(1-x)}.$$

c State, with a reason, whether your series expansion in part **b** is valid for $x = \frac{1}{2}$. [EDEXCEL]

13 a Find the binomial expression of $\frac{1}{\sqrt{1-4x^2}}$, up and including the term in x^4.

State the range of values of x for which the expansion is valid.

b Show that $\sqrt{\frac{1+2x}{1-2x}} = \frac{1+2x}{\sqrt{1-4x^2}}$.

Deduce the binomial expansion of $\sqrt{\frac{1+2x}{1-2x}}$, up and including the term in x^5.

[MEI]

PART 9

Differentiation 2

9.1 Implicit functions

In an implicit function the relationship between, say, two variables x and y cannot easily be expressed in the form $y = \mathrm{f}(x)$.
Thus $y = x^3 - 3x + 1$ is an explicit function of x but $y^2 + 4xy + 1 = x^3$ is an implicit function. The method of differentiating an implicit function is shown in the examples below.

Example 1

Given $y^2 + 5x = x^2$, find $\frac{\mathrm{d}y}{\mathrm{d}x}$ in terms of x and y.

Differentiate each term with respect to x.

$$\frac{\mathrm{d}}{\mathrm{d}x}(y^2) + \frac{\mathrm{d}}{\mathrm{d}x}(5x) = \frac{\mathrm{d}}{\mathrm{d}x}(x^2)$$

$$\frac{\mathrm{d}}{\mathrm{d}y}(y^2)\frac{\mathrm{d}y}{\mathrm{d}x} + \frac{\mathrm{d}}{\mathrm{d}x}(5x) = \frac{\mathrm{d}}{\mathrm{d}x}(x^2)$$

$$2y\frac{\mathrm{d}y}{\mathrm{d}x} + 5 = 2x$$

$$2y\frac{\mathrm{d}y}{\mathrm{d}x} = 2x - 5$$

$$\frac{\mathrm{d}y}{\mathrm{d}x} = \frac{2x - 5}{2y}$$

Notice that we obtained $\frac{\mathrm{d}}{\mathrm{d}x}(y^2)$ using the chain rule.

Example 2

Find $\frac{\mathrm{d}y}{\mathrm{d}x}$ in terms of x and y: **a** $y^3 + x^2y - x^2 = 0$ **b** $\ln y + x^2 = 3x$

a $$\frac{\mathrm{d}}{\mathrm{d}x}(y^3) + \frac{\mathrm{d}}{\mathrm{d}x}(x^2y) - \frac{\mathrm{d}}{\mathrm{d}x}(x^2) = 0$$

$$\frac{\mathrm{d}}{\mathrm{d}y}(y^3)\frac{\mathrm{d}y}{\mathrm{d}x} + \left(x^2\frac{\mathrm{d}y}{\mathrm{d}x} + y \times 2x\right) - 2x = 0$$ [Use the product rule for the x^2y term.]

$$3y^2\frac{\mathrm{d}y}{\mathrm{d}x} + x^2\frac{\mathrm{d}y}{\mathrm{d}x} + 2xy - 2x = 0$$

$$\frac{\mathrm{d}y}{\mathrm{d}x}(3y^2 + x^2) = 2x - 2xy$$

$$\frac{\mathrm{d}y}{\mathrm{d}x} = \frac{2x(1 - y)}{3y^2 + x^2}$$

b $\frac{d}{dy}(\ln y)\frac{dy}{dx} + \frac{d}{dx}(x^2) = \frac{d}{dx}(3x)$

$$\frac{1}{y}\frac{dy}{dx} + 2x = 3$$

$$\frac{1}{y}\frac{dy}{dx} = 3 - 2x$$

$$\frac{dy}{dx} = y(3 - 2x)$$

Example 3

Find the equation of the tangent to the curve $2x^2 - 2xy + y^2 = 5$ at the point (1, 3).

Differentiate with respect to x.

$$4x - \left[2x\frac{dy}{dx} + y \times 2\right] + 2y\frac{dy}{dx} = 0$$

$$4x - 2y + \frac{dy}{dx}(-2x + 2y) = 0$$

$$\frac{dy}{dx} = \frac{2y - 4x}{-2x + 2y}$$

At the point (1, 3), $\frac{dy}{dx} = \frac{2 \times 3 - 4 \times 1}{-2 \times 1 + 2 \times 3} = \frac{2}{4} = \frac{1}{2}$

The tangent passes through the point (1, 3) and has gradient $\frac{1}{2}$.

The equation of the tangent is $\frac{y - 3}{x - 1} = \frac{1}{2}$

$$2y - 6 = x - 1$$

$$2y = x + 5.$$

Example 4

Find the coordinates of the stationary points on the curve $x^2 + xy + y^2 = 3$.

Differentiating, $2x + \left(x\frac{dy}{dx} + y \times 1\right) + 2y\frac{dy}{dx} = 0$

$$\frac{dy}{dx}(x + 2y) + 2x + y = 0$$

$$\frac{dy}{dx} = -\frac{(2x + y)}{(x + 2y)}$$

At stationary points $\frac{dy}{dx} = 0$.

$\therefore \quad 2x + y = 0$

$\therefore \quad y = -2x$ [1]

Substitute $y = -2x$ in $x^2 + xy + y^2 = 3$.

$$x^2 + x(-2x) + (-2x)^2 = 3$$
$$x^2 - 2x^2 + 4x^2 = 3$$
$$3x^2 = 3$$
$$x^2 = 1$$
$$x = 1 \text{ or } -1$$

From [1], when $x = 1$, $y = -2$
when $x = -1$, $y = 2$

The curve has stationary points at $(1, -2)$ and $(-1, 2)$.

[The second derivative will not be used here as the second derivative for implicit functions is not on the syllabus for C4.]

EXERCISE 1

1 For each curve find $\frac{dy}{dx}$ in terms of x and y.

a $x^2 + y^2 = 9$
b $xy = 4$
c $x^2 + xy + y^2 = 0$
d $x^2 + y^2 - 6x + 8y = 0$
e $x^2 + 3y^2 - 4y = 0$
f $y^3 = x^2 + 10$
g $3x^3 + 4xy^2 + y^3 = 0$

2 Find $\frac{dy}{dx}$ for the following curves at the points indicated:

a $x^2 + y^2 = 10$ $(3, 1)$
b $xy = 9$ $(3, 3)$
c $x^2 + y^2 + 4x = 9$ $(1, 2)$
d $y^2 = x^3$ $(4, 8)$

3 The diagram shows a circle with equation $x^2 + y^2 = 8$.

By finding $\frac{dy}{dx}$, calculate the gradient of the tangent to the circle at the point (2, 2).

Using the equation $y - y_1 = m(x - x_1)$, find the equation of the tangent at the point (2, 2).

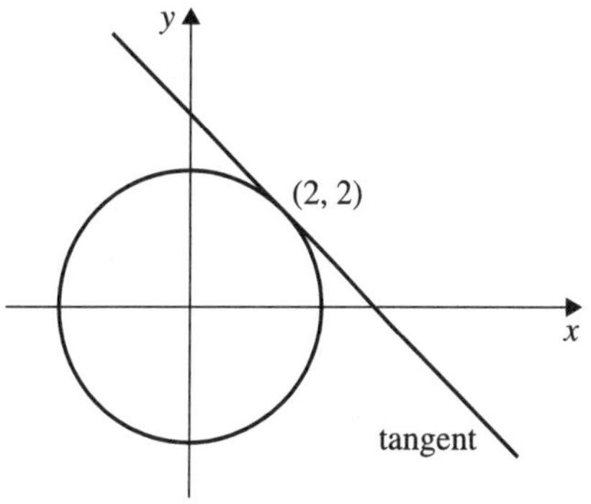

4 Find the equation of the normal to $x^2 + y^2 = 8$ at the point (2, 2).
[Hint: Remember that gradient of tangent × gradient of normal = −1]

5 Find the equation of the tangent to the curve $xy = 4$ at the point (2, 2).

6 Find the equation of the normal to the curve $xy = 4$ at the point $(-2, -2)$.

7 Find the equation of the tangent to the circle $x^2 + y^2 - 2x + 4y = 20$ at the point (4, 2).

8 Find the equation of the normal to the curve $x^2 + xy + y^2 = 3$ at the point (1, 1).

9 For the curve $\ln y + x^2 - x = 0$, show that $\frac{dy}{dx} = y(1 - 2x)$.

10 Find the gradient of each curve at the point given.

a $x^3 + xy + y^3 = 20$ $(2, 2)$

b $2x^3 - 3x^2y - y^2 = 0$ $(-1, -2)$

c $\ln y + x^2 = 4$ $(2, 1)$

d $3 \ln y + 2x^3 = 16$ $(2, 1)$

11 The diagram shows the graph of $xy - x^2 = 4$.

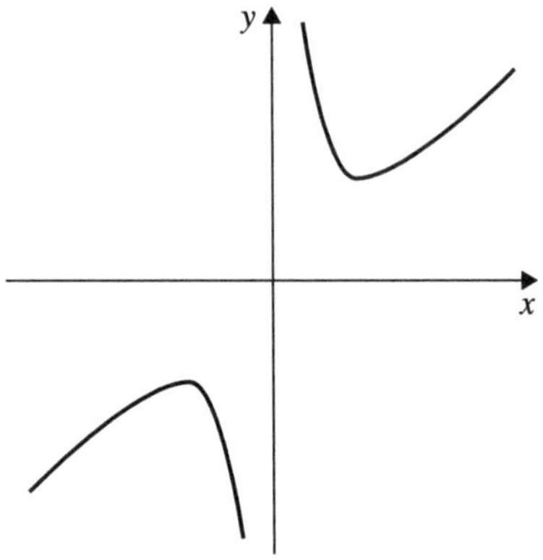

a Show that when $\frac{dy}{dx} = 0$, $y = 2x$.

b Substitute $y = 2x$ into the equation $xy - x^2 = 4$ to find the x coordinates of the two points where the gradient is zero.

c Find the y coordinates of the two turning points.

12 Show that the stationary points on the curve $x^2 + 4xy - 2y^2 + 24 = 0$ occur when $x = -2y$. Hence find the stationary points.

13 A curve is defined by the equation $x^2 - 4x - 2xy + 3y^2 - 2 = 0$.

a Find $\frac{dy}{dx}$ in terms of x and y.

b Show that the equation on which the stationary points must lie is $y = x - 2$.

c Show that one of the stationary points is $(1, -1)$ and find the co-ordinates of the other one.

14 **a** Show that the tangents to the curve $y^2 - 16x - 2y = 47$, at the points $(-2, -3)$ and $(13, 17)$, are perpendicular.

b Find the co-ordinates of their points of intersection.

15 Calculate $\frac{dy}{dx}$ for the following curves at the given points:

a $x^2 - 7xy + y^2 = -29$ $(2, 3)$

b $xy^2 - 7 \sin y + x = 5$ $(5, 0)$

c $y + x \ln y + x^2 = 5$ $(2, 1)$

d $e^x y + y^2 = 6$ $(0, 2)$

e $(x - y)^2 = 4xy - 7$ $(2, 1)$

16* **a** Given that $x = \cos y$,

i find $\frac{dy}{dx}$ in terms of y (by differentiating both sides with respect to x)

ii find $\frac{dy}{dx}$ in terms of x (using the fact that $\cos^2 y + \sin^2 y = 1$)

b Use part **a** to find $\frac{dy}{dx}$ in terms of x when $y = \cos^{-1} x$.

17* **a** Given that $x = \tan y$,

i find $\frac{dy}{dx}$ in terms of y (by differentiating both sides with respect to x)

ii find $\frac{dy}{dx}$ in terms of x (using the fact that $1 + \tan^2 y = \sec^2 y$)

b Use part **a** to find $\frac{dy}{dx}$ in terms of x when $y = \tan^{-1} x$.

18* Given $y = x^{x^2}$, find $\frac{dy}{dx}$ in terms of x.

19* Find $\frac{dy}{dx}$ in terms of x and y.

a $e^y = x$

b $e^{xy} = 2$

c $\frac{1}{x} + \frac{1}{y} = \frac{1}{4}$

d $\cos x \sin y = \frac{1}{2}$

e $\sin y + \cos y = x$

9.2 Exponential growth and decay

Under certain conditions, the number of bacteria in a particular culture doubles every 10 minutes.

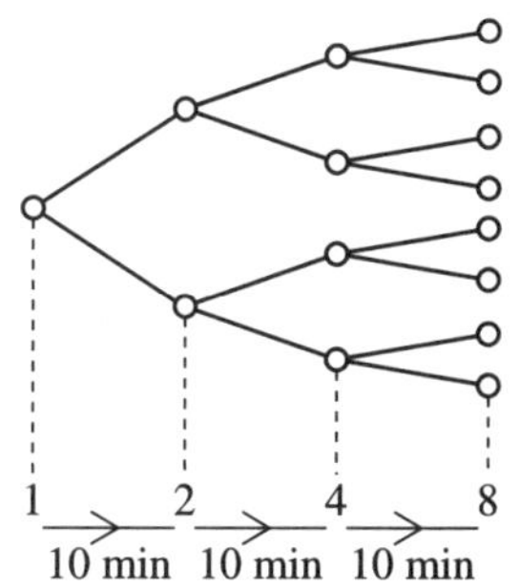

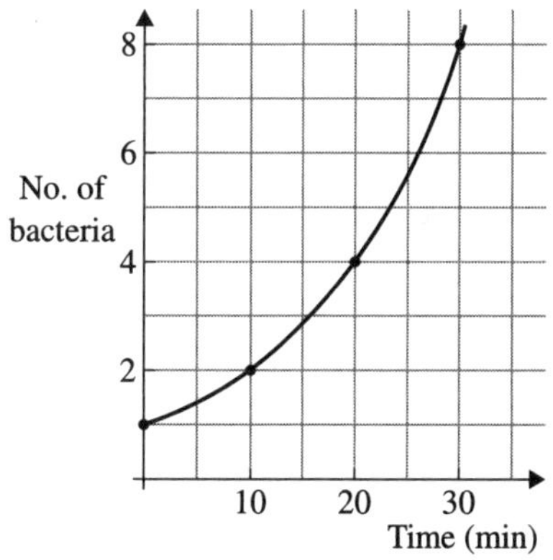

This is an example of *exponential growth*. The number of bacteria is multiplied by the same number in equal periods of time.

After t minutes the number of bacteria N is given by $N = 2^{\frac{t}{10}}$. You can check this formula by substituting $t = 0, 10, 20$ and so on.

Radioactive materials like uranium or plutonium decay very, very slowly. The *half life* is the length of time for half the material to decay. Suppose uranium has a half life of 40 years and there is 100 g of it now. In 40 years there would be 50 g. After another 40 years there would still be 25 g and so on.

After t years the mass of uranium m is given by $m = 100\,e^{-0.0173t}$. [Check by substituting $t = 40, 80$.] This is an example of exponential decay.

The derivative of a^x (a is a constant)

Let $y = a^x$.
Take logarithms to base e.

$$\ln y = \ln a^x$$
$$\ln y = x \ln a$$

Differentiate both sides with respect to x

$$\frac{\mathrm{d}}{\mathrm{d}x}(\ln y) = \ln a \quad \text{[Remember: } \ln a \text{ is a constant.]}$$

$$\Rightarrow \quad \frac{1}{y}\frac{\mathrm{d}y}{\mathrm{d}x} = \ln a$$

$$\frac{\mathrm{d}y}{\mathrm{d}x} = y \ln a = a^x \ln a$$

So $\boldsymbol{\frac{\mathrm{d}}{\mathrm{d}x}(a^x) = a^x \ln a}$

Progress in this topic is easier if you are confident about applying the basic laws of logarithms and indices. The first examples are intended to remind you of work covered previously.

Example 1

If $18 = 5 \times 2^t$, find the value of t.

Write $2^t = \frac{18}{5} = 3.6$.

Take logarithms,

$$\log_{10} 2^t = \log_{10} 3.6 \quad [\log_{10} \text{ is written 'log'}]$$
$$t \log 2 = \log 3.6$$
$$t = \frac{\log 3.6}{\log 2} = 1.85 \text{ (correct to 2 decimal places)}$$

Example 2

If $P = 4000\, e^{-\frac{1}{2}t}$, find

a the initial value of P

b the value of P when $t = 5$.

a The initial value is the value of P when $t = 0$.
When $t = 0$, $P = 4000 \times e^0$
$P = 4000$

b When $t = 5$, $P = 4000 \times e^{-\frac{5}{2}}$
$= 4000 \times 0.082\,085 \ldots$
$= 328$ (correct to 3 significant figures)

Example 3

A substance is decaying exponentially. Its mass, m grams, after t years is given by

$$m = 200\,\mathrm{e}^{-0.04t}.$$

a Find the value of t when $m = 50$.

b Find the rate at which the mass is decreasing when $t = 10$.

a When $m = 50$,

$$\begin{aligned} 50 &= 200\,\mathrm{e}^{-0.04t} \\ \mathrm{e}^{-0.04t} &= 0.25 \\ -0.04t &= \ln 0.25 = -1.386\,29\ldots \\ t &= \frac{-1.38629}{-0.04} \\ t &= 34.7 \text{ years (to 3 significant figures)} \end{aligned}$$

b The rate of decrease of the mass is given by $\dfrac{\mathrm{d}m}{\mathrm{d}t}$.

$$\frac{\mathrm{d}m}{\mathrm{d}t} = 200 \times (-0.04) \times \mathrm{e}^{-0.04t}$$

When $t = 10$, $\dfrac{\mathrm{d}m}{\mathrm{d}t} = 200 \times (-0.04) \times \mathrm{e}^{-0.4}$

$= -5.36$ grams/year (to 3 significant figures)

The negative sign indicates that the mass is decreasing.

Example 4

The number of microbes N in a culture, after t hours, is given by $N = 50 \times 3^t$. Find the rate of increase of the number of microbes when $t = 5$ hours.

$$N = 50 \times 3^t \qquad [1]$$

The rate of increase is given by $\dfrac{\mathrm{d}N}{\mathrm{d}t}$.

Take logarithms to base e.

$$\begin{aligned} \ln N &= \ln(50 \times 3^t) = \ln 50 + \ln 3^t \\ \ln N &= t \ln 3 + \ln 50 \end{aligned}$$

Differentiate both sides with respect to t.

$$\begin{aligned} \frac{\mathrm{d}}{\mathrm{d}t}(\ln N) &= \ln 3 \\ \frac{1}{N}\frac{\mathrm{d}N}{\mathrm{d}t} &= \ln 3 \\ \frac{\mathrm{d}N}{\mathrm{d}t} &= N \ln 3 \end{aligned}$$

We require the value of N when $t = 5$.

From equation [1], $N = 50 \times 3^5$

$= 12\,150$

$\therefore \quad \dfrac{\mathrm{d}N}{\mathrm{d}t} = 12\,150 \ln 3 = 13\,300$ microbes/hour (3 significant figures)

The rate of increase of the number of microbes is 13 300 per hour.

Example 5

Solve the inequality $0.3^x < 0.02$.

Take logarithms (base 10).

$$\log 0.3^x < \log 0.02$$
$$x \log 0.3 < \log 0.02$$

Be careful here. Since log 0.3 is a negative number, the inequality must be reversed when dividing both sides by log 0.3.

$$x > \frac{\log 0.02}{\log 0.3}$$

$x > 3.25$, correct to three significant figures.

EXERCISE 2

1 Find x, correct to 3 significant figures.

a $4^x = 11$ b $2^x = 19$ c $5^{x-1} = 100$

d $e^x = 19$ e $5e^{2x} = 32$ f $3e^{-x} = 11$

g $3(7^{\frac{x}{2}}) = 20$ h $9^{3x-1} = 1$ i $\ln x = 3$

j $\ln 2x = 5$ k $2 \ln x = -1$ l $\ln x^2 - \ln 2x = 4$

2 Given $m = 40e^{-4t}$, find the value of m when $t = 0.1$.

3 Given $P = 6000e^{\frac{t}{100}}$, find the value of P when $t = 20$.

4 Given $x = 200 \times 4^{2t-1}$, find the value of x when $t = 1.2$.

5 Find $\frac{dy}{dx}$ in the following.

a $y = 3^x$ b $y = 5^{x-1}$ c $y = 3(4^x)$

d $y = 2^{x^2}$ e $y = e^{-\frac{x^2}{3}}$

6 a Find $\frac{d}{dt}(7^t)$ b Find $\frac{d}{dt}(4^t)$ c Find $\frac{d}{dt}\left(\frac{1}{3^t}\right)$

7 Find $\frac{d\theta}{dt}$ in the following.

a $\theta = 150e^{-\frac{t}{10}}$ b $\theta = 100 - e^{-2t}$ c $\theta = 10 \times 6^t$

8 A substance is decaying exponentially. Its mass, m grams, after t years is given by $m = 1000e^{-0.02t}$.

a Write down the initial value of m.

b Find the value of t when $m = 400$.

9 Find t, correct to 3 significant figures.

a $5e^{-0.2t} = 11$ b $1000 = 2500 - 1800e^{-0.1t}$

c $450 = 600 - 500e^{-0.01t}$

10 At time t hours the temperature θ°C of a cooling body is given by $\theta = 80\,e^{kt}$, where k is a constant.

a Given that $\theta = 40$°C when $t = 10$ hours, find the value of k.

b Find the value of θ when $t = 40$.

11 At time t minutes the temperature θ°C of a cooling furnace is given by $\theta = 2500\,e^{-0.9t}$.
Find the rate of decrease of the temperature when $t = 2$.

12 After t years the mass m grams of a piece of radioactive material is given by $m = 1000 \times 2^{-0.002t}$.

a State the value of m when $t = 0$.

b Find the value of t when the material has decayed to one tenth of its original mass.

13 A population of flies P is given by the formula $P = A\,e^{-kt}$, where t is the time in days measured from a time when $P = 2000$.

a Write down the value of A.

b Given that $P = 500$ when $t = 5$, show that $k = \frac{1}{5}\ln 4$.

c Find the value of P when $t = 8$ days.

14 A radioactive substance decays so that after t days the amount remaining is 0.75^t units.

a How many units remain after 10 days?

b After how many days will the amount of the substance remaining be less than 0.1 units?

15 A radioactive substance is decaying exponentially. Its mass, m grams, after t years is given by $m = 300\,e^{-0.005t}$.

a Find the value of t when $m = 100$.

b Find the rate at which the mass is decreasing when $t = 20$.

16 Make t the subject in the following formulae.

a $m = a\,e^{3t}$ **b** $v = \dfrac{b}{10}e^{-t}$ **c** $P = 150\,e^{t^2}$

17 At time t minutes after being switched on, the temperature of a furnace θ°C is given by $\theta = 2000 - 1800\,e^{-0.1t}$.

a State the value which θ approaches after a long time.

b Find the time taken to reach a temperature of 1500°C.

c Find the rate at which the temperature is increasing when $t = 10$.

18 At time t minutes after being switched on, the temperature of an oven θ°C is given by $\theta = 190 - 175\,e^{-0.08t}$.

a State the value which θ approaches after a long time.

b Find the time taken to reach a temperature of 160°C.

c Find the rate at which the temperature is increasing when $\theta = 50$°C.

9.3 Connected rates of change

Suppose we are told that the area A of a circle is increasing at a rate of 10 cm^2 per second. Can we find the rate at which the radius is increasing at the instant when the radius is 4 cm?

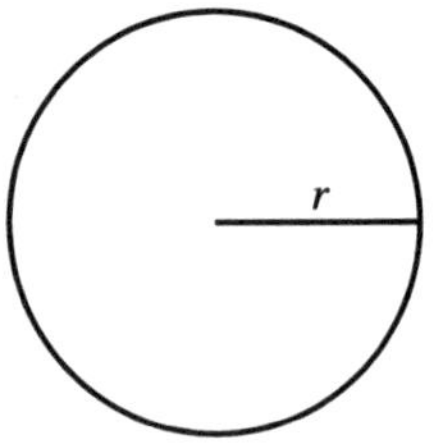

$\frac{dA}{dt} = 10\,cm^2\,s^{-1}$

We can write $\frac{dA}{dt} = 10$ because $\frac{dA}{dt}$ is the rate of change of the area with respect to time. The question asks us to find $\frac{dr}{dt}$, the rate of change of the radius with respect to time.
The formula connecting A and r is $A = \pi r^2$.
Differentiate with respect to r.

$$\frac{dA}{dr} = 2\pi r, \text{ and when } r = 4, \frac{dA}{dr} = 8\pi.$$

Now $\frac{dA}{dt} = \frac{dA}{dr} \times \frac{dr}{dt}$ [1] [Chain rule]

$$\therefore \quad 10 = 8\pi \times \frac{dr}{dt}$$

$$\frac{dr}{dt} = \frac{10}{8\pi} = \frac{5}{4\pi}\text{ cm/sec}$$

When the radius is 4 cm, the radius is increasing at a rate of $\frac{5}{4\pi}$ cm per second.

In questions on connected rates of change you will always write an equation similar to equation [1] above and then use it to find the required rate of change.

Example 1

The side of a square is increasing at a rate of 30 cm/second. Find the rate at which the area of the square is increasing when the side of the square is 11 cm.

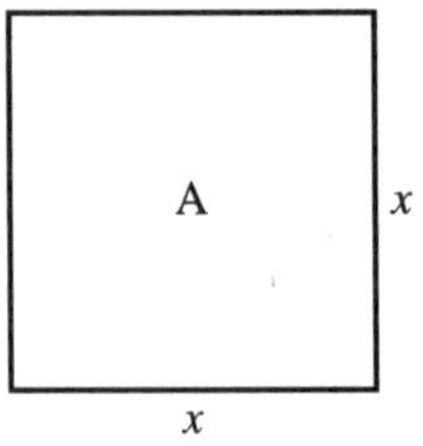

The area A of a square of side x is given by $A = x^2$.

$$\frac{dA}{dx} = 2x, \text{ and when } x = 11, \frac{dA}{dx} = 22.$$

We are given $\frac{dx}{dt} = 30\text{ cm s}^{-1}$, we have $\frac{dA}{dx} = 22$ and we want to find $\frac{dA}{dt}$.

By the chain rule, $\frac{dA}{dt} = \frac{dA}{dx} \times \frac{dx}{dt}$

$$\frac{dA}{dt} = 22 \times 30 = 660\text{ cm}^2\text{ s}^{-1}.$$

The area of the square is increasing at a rate of 660 $cm^2\,s^{-1}$.

Example 2

The volume of a sphere is increasing at a rate of $100\ \text{cm}^3\,\text{s}^{-1}$. Find the rate of increase of the radius of the sphere at the instant when the radius is 3 cm.

For a sphere, $V = \frac{4}{3}\pi r^3$

$\frac{dV}{dr} = 4\pi r^2$, and when $r = 3$, $\frac{dV}{dr} = 36\pi$.

We are given $\frac{dV}{dt} = 100$ and we want to find $\frac{dr}{dt}$.

By the chain rule, $\frac{dV}{dt} = \frac{dV}{dr} \times \frac{dr}{dt}$

$$100 = 36\pi \times \frac{dr}{dt}$$

$$\frac{dr}{dt} = \frac{100}{36\pi} = \frac{25}{9\pi}\ \text{cm}\,\text{s}^{-1}.$$

Example 3

A hollow right circular cone, whose height of the cone is twice the radius, is held with its vertex downwards and water is poured in at a rate of $5\pi\ \text{cm}^3\,\text{s}^{-1}$. Find the rate of rise of the water level when the depth is 2 cm.

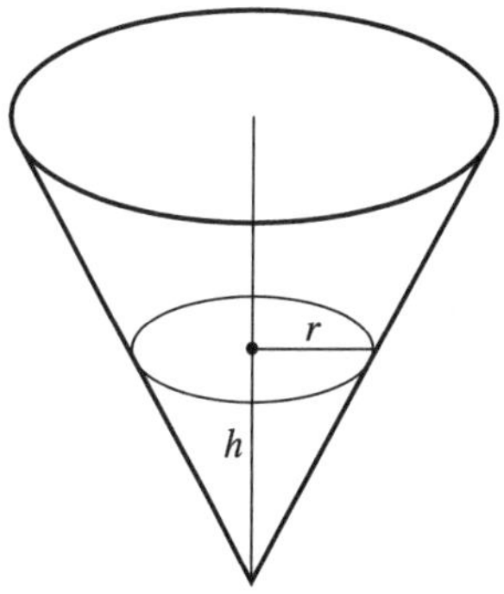

The volume of the water in the cone V when the height is h is given by $V = \frac{1}{3}\pi r^2 h$. We are given that $h = 2r$.

$$\therefore \quad V = \tfrac{1}{3}\pi\left(\frac{h}{2}\right)^2 h = \tfrac{1}{12}\pi h^3$$

$\frac{dV}{dh} = \frac{1}{4}\pi h^2$, and when $h = 2$, $\frac{dV}{dh} = \frac{1}{4} \times \pi \times 2^2 = \pi$

We are given $\frac{dV}{dt} = 5\pi\ \text{cm}^3\,\text{s}^{-1}$ and we want to find $\frac{dh}{dt}$.

By the chain rule, $\frac{dV}{dt} = \frac{dv}{dh} \times \frac{dh}{dt}$

$$5\pi = \pi \times \frac{dh}{dt} \quad \Rightarrow \quad \frac{dh}{dt} = 5\ \text{cm}\,\text{s}^{-1}$$

The water level is rising at $5\ \text{cm}\,\text{s}^{-1}$.

Exercise 3

Give answers to three significant figures where necessary.

1 Given that $y = x^2$ and that x is increasing at a rate $0.1\ \text{cm}\,\text{s}^{-1}$, find the rate at which y is increasing when $x = 3$ cm.

2 If the radius of a circle increases at a rate of $0.5\ \text{cm}\,\text{s}^{-1}$, find the rate at which the area of the circle increases when the radius is 10 cm.

3 The area of a square is increasing at a rate of $4\text{ cm}^2\text{ s}^{-1}$. Find the rate of increase of each side when each side is of length 10 cm.

4 A cube is expanding. Each edge of the cube increases at a rate of 0.01 cm s^{-1}. When the edges of the cube are 8 cm find the rate of increase of

a the volume

b the surface area of the cube.

5 If the radius of a sphere is increasing at 3 cm s^{-1}, find the rate at which the volume is increasing when the radius is 5 cm.

6 The area of a circular ink blot is increasing at a rate of $0.8\text{ cm}^2\text{ s}^{-1}$. Find the rate at which the radius is increasing when the radius of the blot is 2 cm.

7 The radius of a sphere is increasing at the rate of 2 cm s^{-1}. Find, in terms of π, the rate of increase of the volume when the radius is 10 cm.

8 The side length of a cube is increasing at the rate of 3 cm s^{-1}. Find the rate of increase of the volume when the length of a side is 5 cm.

9 The volume of a cube is increasing at the rate of $96\text{ cm}^3\text{ s}^{-1}$. Find the rate of change of the side of the base when its length is 4 cm.

10 The surface area of a sphere is increasing at the rate of $40\pi\text{ cm}^2\text{ s}^{-1}$. Find the rate of change of the radius when its length is 1 cm. [The surface area of a sphere, $S = 4\pi r^2$]

11 A hollow right circular cone, whose height of the cone is twice the radius, is held with its vertex downwards and water is poured in at a rate of $4\text{ cm}^3\text{ s}^{-1}$.

a Show that when the depth of the water is h the volume in the cone is
$$V = \frac{\pi h^3}{12}.$$

b Find, in terms of π, the rate of rise of the water level when the depth is 2 cm.

12 The empty water container shown is a right prism with AD = 3 m and AC = AB = BC = 1 m. It lies with CF on horizontal ground and with ABDE horizontal.

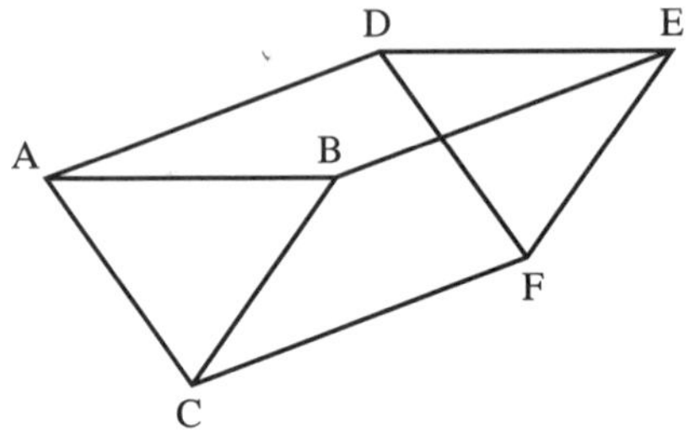

a Show that the volume (in m^3) in the container when the water has height h is $V = \sqrt{3}h^2$.

Water is poured into the container in such a way that the height is rising at $5\sqrt{3}\text{ cm s}^{-1}$ when the height of the water is 50 cm.

b Find the rate (in $\text{m}^3\text{ s}^{-1}$) at which the volume of water is increasing at this time.

13 Given that $y^2 = x^3$ and that x is increasing at a rate of 0.5 units per second, find the rate at which y is increasing when $x = 4$ units.

14 A liquid drips onto a surface and forms a circular film of uniform depth 0.1 cm. The liquid drips at a rate of $0.2\ \text{cm}^3\,\text{s}^{-1}$. When the radius of the film is 3 cm find the rate at which the radius is increasing.

15* **a** Show that the volume obtained when the part of the curve $y = \sqrt{9 - x^2}$ that is bounded by the x-axis and the line $x = 3 - h$ is rotated about the x-axis is $\dfrac{\pi h^2}{3}(9 - h)$.

Water is poured at a rate of $18\ \text{cm}^3\,\text{s}^{-1}$ into a hemisphere whose radius is 3 cm.

b Find, in terms of π, the rate of rise of the water level when the depth is just about to fill up.

REVIEW EXERCISE 9

1 Find $\dfrac{dy}{dx}$ (using implicit differentiation), in terms of x and y.

a $y^2 + y = x$ **b** $y + xy + y^2 = 2$ **c** $x^2 + y^2 = 2$

2 For the following curves express $\dfrac{dy}{dx}$ in terms of x and y.

a $x^2 + 3y^2 = 7$ **b** $4x^2 - y^3 + 2x + 3y = 0$

c $3x^2 + 4xy - 5y^2 = 20$ **d** $xy^3 + x^3y = x - y$

e $\sin x + 2\cos y = 1$ **f** $x^2y^2 - y = 2$

3 Find the gradient of the curve $x^3 - 2y^3 = 3xy$ at the point (2, 1).

4 A curve is defined by the equation $x^2 + 7xy + 3y^2 = 27$.

a Find $\dfrac{dy}{dx}$ in terms of x and y.

b Hence find the value of $\dfrac{dy}{dx}$ at the point (1, 2) on the curve.

5 Find, in the form $ax + by + c = 0$, the equation of the tangent to the curve $x^2 + xy + 4y^2 = 16$ at the point (3, 1).

6 Find, in the form $ax + by + c = 0$, the equation of the normal to the curve $x^2y^2 = x^2 + 5y^2$ at the point $(3, \frac{3}{2})$.

7 **a** For the curve $x^2 + 4y^2 - 4x - 8y + 4 = 0$, express $\dfrac{dy}{dx}$ in terms of x and y.

b What can you say about the gradient of the curve at the points (2, 2) and (0, 1) on the curve?

8 Show that $(-1, 3)$ and $(0, 0)$ are stationary points on the curve $3x^2 + 2xy - 5y^2 + 16y = 0$.

9 For the curve $\ln y = 3x^2 + 4$, show that $\dfrac{dy}{dx} = 6xy$.

10 **a** For the curve $3\sin x + 2\cos y = \sqrt{3}$ show that $\dfrac{dy}{dx} = \dfrac{3\cos x}{2\sin y}$.

b Find the gradient of the curve at the point $\left(0, \dfrac{\pi}{6}\right)$.

11 Find the gradient of each curve at the point given.

a $y + x\ln y + x^2 = 10$ at $(3, 1)$

b $e^x y + 2y = 12$ at $(0, 4)$

12 Find $\dfrac{dy}{dx}$ in terms of x.

a $y = 8^x$ **b** $y = 2 \times 3^x$ **c** $y = 4^{x^2}$

d $y = 4e^{-3x}$ **e** $y = x^x$

13 Show that the gradient of the curve $y = 3^x$ at the point $(1, 3)$ is $\ln 27$.

14 A radioactive substance is decaying exponentially. Its mass, m grams, after t years is given by $m = 300\,e^{-0.05t}$.

a Write down the initial value of m.

b Find the value of t when $m = 100$.

15 At time t minutes the temperature θ°C of a cooling oven is given by $\theta = 450\,e^{-0.8t}$.
Find the rate of decrease of the temperature when $t = 10$.

16 At time t hours the temperature θ°C of a cooling body is given by $\theta = 100\,e^{kt}$, where k is a constant.

a Given that $\theta = 60$ when $t = 2$, find the value of k.

b Find the value of θ when $t = 6$.

17 Make t the subject in the following formulae.

a $m = Ae^{2t}$ **b** $s = \dfrac{a}{4}e^{-t}$ **c** $v = 20\,e^{t^2}$

18 A substance is decaying exponentially. After t years, its mass m grams is given by $m = 1000\,e^{-0.01t}$.

a Find the value of t when $m = 800$.

b Find the rate at which the mass is decreasing when $t = 10$.

19 For a body falling vertically, its velocity v m s^{-1} after t seconds is given by

$$v = 50(1 - e^{-0.2t})$$

i State the value which the velocity approaches after a long time.

ii Find the time for the velocity to reach 30 m s^{-1}.

iii Find the rate at which the velocity is increasing when the velocity reaches 30 m s^{-1}.

20 The population P of a particular species is modelled by the formula

$$P = A\,e^{kt},$$

where t is the time in years measured from a date when $P = 1000$.

a Write down the value of A.

b Given that $P = 2000$ when $t = 5$, show that $k \simeq 0.139$.

c Find the value of the population 10 years after the initial date.

21 The radius of a sphere is increasing at the rate of 1 cm s^{-1}. Find, in terms of π, the rate of increase of the volume when the radius is 5 cm.

22 The side length of a cube is increasing at the rate of 2 cm s^{-1}. Find the rate of increase of the volume when the length of a side is 6 cm.

23 The surface area S of a sphere of radius r is given by the formula $S = 4\pi r^2$. Find the rate of increase of the surface area when $r = 4$ cm, given that the radius increases at 2 cm s^{-1}.

24 The volume of a cube is increasing at the rate of $81\text{ cm}^3\text{ s}^{-1}$. Find the rate of change of the side of the base when its length is 3 cm.

25 The surface area of a sphere is increasing at the rate of $2\pi\text{ cm}^2\text{ s}^{-1}$. Find the rate of change of the radius when its length is 1 cm.

26 A hollow right circular cone, whose height is three times the radius, is held with its vertex downwards and water is poured in at a rate of $10\text{ cm}^3\text{ s}^{-1}$.

a Show that when the depth of the water is h the volume in the cone is $V = \dfrac{\pi h^3}{27}$.

b Find, in terms of π, the rate of rise of the water level when the depth is 2 cm.

27 When a metal cube is heated, its volume, $V\text{ cm}^3$, increases at a constant rate of $3.7 \times 10^{-6}\text{ cm}^3\text{ s}^{-1}$. Each edge of the cube has length x cm at time t seconds.

a Find $\dfrac{dx}{dt}$ when $x = 2$.

b Find the rate of increase of the total surface area of the cube in $\text{cm}^2\text{ s}^{-1}$, when $x = 2$.

28 The area of a circle is increasing at a rate of $6\text{ cm}^2\text{ s}^{-1}$. Find the rate of increase of the circumference when the radius is 3 cm.

29 The area and perimeter of the rectangle shown are denoted by A and P respectively. Find $\dfrac{dP}{dt}$, given $\dfrac{dA}{dt} = 16\text{ cm}^2\text{ s}^{-1}$ at the instant when $x = 2$ cm.

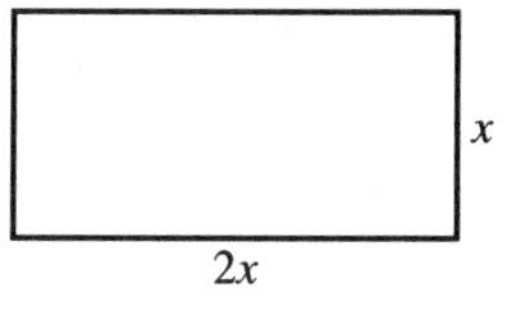

Examination Exercise 9

1 Find the gradient of the curve $x^3 + y^3 = 19$ at the point where $x = -2$. [OCR]

2 i For the curve $2x^2 + xy + y^2 = 14$, find $\dfrac{dy}{dx}$ in terms of x and y.

ii Deduce that there are two points on the curve $2x^2 + xy + y^2 = 14$ at which the tangents are parallel to the x-axis, and find their coordinates. [OCR]

3 A curve has equation $\dfrac{x^2}{9} + \dfrac{y^2}{25} = 1$.

a Find the y-coordinates of the two points on the curve at which the x-coordinate is 2.

b Find the values of the gradient of the curve at these two points, giving your answers to two significant figures. [AQA]

4 A curve has equation $7x^2 + 48xy - 7y^2 + 75 = 0$.
A and B are two distinct points on the curve. At each of these points the gradient of the curve is equal to $\frac{2}{11}$.

a Use implicit differentiation to show that $x + 2y = 0$ at the points A and B.

b Find the coordinates of the points A and B. [EDEXCEL]

5 The curve C has equation $5x^2 + 2xy - 3y^2 + 3 = 0$. The point P on the curve C has coordinates $(1, 2)$.

a Find the gradient of the curve at P.

b Find the equation of the normal to the curve C at P, in the form $y = ax + b$, where a and b are constants. [EDEXCEL]

6 A curve is given by the equation

$$3(x + 1)^2 - 9(y - 1)^2 = 32.$$

a Find the coordinates of the two points on the curve at which $x = 3$.

b Find the gradient of the curve at each of these two points. [AQA]

7 a Given that $a^x \equiv e^{kx}$, where a and k are constants, $a > 0$ and $x \in \mathbb{R}$, prove that $k = \ln a$,

b Hence, using the derivative of e^{kx}, prove that when $y = 2^x$,

$$\frac{dy}{dx} = 2^x \ln 2$$

c Hence deduce that the gradient of the curve with equation $y = 2^x$ at the point $(2, 4)$ is $\ln 16$. [EDEXCEL]

8 a A quantity N is growing exponentially. The table shows values of N at different times t. Determine the values of P and q.

t	20	30	40	q
N	50	100	P	800

b A radioactive substance is decaying exponentially. Its mass, m grams, after t years is given by

$$m = 600\,e^{-0.0035t}.$$

i Find the value of t for which $m = 300$.

ii Find, correct to 2 significant figures, the rate at which the mass is decreasing when $t = 80$. [OCR]

9 A radioactive substance is decaying exponentially. Initially its mass is 480 g. Its mass, M grams, at a time t years after the initial observation is given by

$$M = 480\,\mathrm{e}^{kt},$$

where k is a constant. When $t = 330$, the mass of the substance will be 240 g.

i State the value of the mass when $t = 660$.

ii Determine the value of k.

iii Find the rate at which the mass will be decreasing when $t = 200$. [OCR]

10 The decay of a radioactive substance can be modelled by the equation

$$m = m_0\,\mathrm{e}^{-kt},$$

where m grams is the mass at time t years, m_0 grams is the initial mass, and k is a constant.

a The time taken for a sample of the radioactive substance strontium 90 to decay to half of its initial mass is 28 years. Show that the value of k is approximately 0.024 755.

b A sample of strontium 90 has a mass of 1 gram. Assuming this mass has resulted from radioactive decay, use the model to find the mass this sample would have had 100 years ago. Give your answer to three significant figures. [AQA]

11 The mean annual temperature of the water in a certain large lake is expected to increase due to climatic change. A model giving the mean annual temperature, θ°C, at a time t years after the first observation is

$$\theta = 0.0006t^2 + 0.05t + 4.3.$$

The number of crustaceans in the lake depends on the water temperature. A model giving the number, N, of crustaceans is

$$N = 320\,\mathrm{e}^{0.4\theta}.$$

i According to the models, how many crustaceans were present in the lake when the first observation was made?

ii By first writing down expressions for $\dfrac{\mathrm{d}\theta}{\mathrm{d}t}$ and $\dfrac{\mathrm{d}N}{\mathrm{d}\theta}$, find the rate at which the models predict that the number of crustaceans will be increasing when $t = 30$. [OCR]

12 Observations were made of the number of bacteria in a certain specimen. The number N present after t minutes is modelled by the formula

$$N = Ac^t,$$

where A and c are constants.

Initially there are 1000 bacteria in the specimen.

a Write down the value of A.

b Given that there are 12 000 bacteria after 60 minutes, show that the value of c is 1.0423 to four decimal places.

c **i** Express t in terms of N.

ii Calculate, to the nearest minute, the time taken for the number of bacteria to increase from one thousand to one million. [AQA]

13 The volume of a sphere is increasing at a rate of $75\text{ cm}^3\text{ s}^{-1}$. Find the rate at which the radius is increasing at the instant when the radius of the sphere is 15 cm. Give your answer correct to 2 significant figures.

[The volume, V, of a sphere of radius r is given by $V = \frac{4}{3}\pi r^3$.] [OCR]

14 At a time t minutes, a circular puddle has radius r cm and area $A\text{ cm}^2$.

a Find $\frac{\mathrm{d}A}{\mathrm{d}r}$ in terms of r.

b The radius is increasing at a rate of 3 cm per minute.

Find the rate at which the area is increasing at the instant when the radius is 50 cm. Give your answer to the nearest cm^2 per minute. [AQA]

15 **a** Given that $y = 5\,\mathrm{e}^{3x} + \ln(2 - 5x)$, find an expression for $\frac{\mathrm{d}y}{\mathrm{d}x}$.

b A water tank is such that, when the depth of water is h cm, the volume, $V\text{ cm}^3$, of water in the tank is given by $V = 28h^3$. Water is flowing into the tank at a constant rate of 400 cm^3 per minute. Find the rate at which the depth of water is increasing at the instant when $h = 10$. [OCR]

16 As part of a laser light show, a rectangular shape continuously changes its dimensions whilst its perimeter remains constant at 20 metres.

a At time t seconds, the length of one side of the rectangle is x metres.

i Show that the area $A\text{ m}^2$ of the rectangle is given by

$$A = x(10 - x).$$

ii Find $\frac{\mathrm{d}A}{\mathrm{d}x}$.

b Find the rate at which the area of the rectangle is increasing when the length of one of the sides is 4 metres and is increasing at a rate of 0.5 m s^{-1}. [AQA]

17 A scientific study involves monitoring the growth of a particular species of weed. The area of ground affected by the weed after t years is $A\text{ m}^2$. Two models giving A in terms of t are as follows.

Model 1: $A = 10 + 8\ln(t + 1)$.
Model 2: $A = 5(4 + 3t)^{\frac{1}{2}}$.

i For Model 1:

a find the value of A when $t = 15$, giving your answer correct to 3 significant figures;

b find by differentiation the rate at which the area affected by the weed is increasing when $t = 15$.

ii For Model 2:

a find the value of A when $t = 15$;

b find by differentiation the rate at which the area affected by the weed is increasing when $t = 15$.

18 The volume of oil, V m^3, in a tank changes with time t hours ($1 \leqslant t \leqslant T$) according to the formula

$$V = 32 - 10 \ln t,$$

where T represents the time when the tank is empty of oil.

a State the volume of oil in the tank when $t = 1$.

b Find the rate of change, in m^3 per hour, of the volume of oil in the tank at the time when $t = 8$, interpreting the sign of your answer.

c Determine the value of T. [AQA]

19 Given that $y = \log_a x$, $x > 0$, where a is a positive constant,

a **i** express x in terms of a and y,

ii deduce that $\ln x = y \ln a$.

b Show that $\dfrac{dy}{dx} = \dfrac{1}{x \ln a}$.

The curve C has equation $y = \log_{10} x$, $x > 0$. The point A on C has x-coordinate 10. Using the result in part **b**,

c find an equation for the tangent to C at A.

The tangent to C at A crosses the x-axis at the point B.

d Find the exact x-coordinate of B. [EDEXCEL]

20 Let $y = \log_{10} x$, so that $x = 10^y$.

i Show that x may be expressed in the form $e^{y \ln 10}$.

ii Write down an expression for $\dfrac{dx}{dy}$ in terms of y.

iii Hence show that the derivative of $\log_{10} x$ with respect to x is $\dfrac{1}{x \ln 10}$. [OCR]

21 A curve is defined by the equation $e^x + e^y = 2$.

i By differentiating implicitly, or otherwise, show that $\dfrac{dy}{dx} = -e^{x-y}$.

ii Verify that

$$x = \ln(1 + t), \quad y = \ln(1 - t)$$

are parametric equations for the curve.

iii Find $\dfrac{dy}{dx}$ in terms of t, and hence or otherwise find the exact coordinates (in terms of logarithms) of the point on the curve where the gradient is -2. [MEI]

PART 10

Integration

10.1 Standard functions

In 'Pure Mathematics C1 C2' we integrated functions of the form x^n ($n \neq -1$).

We found that $\int x^n\,dx = \dfrac{x^{n+1}}{n+1} + c$

We can now integrate several more functions:

$$\int e^x\,dx = e^x + c \qquad \int \frac{1}{x}\,dx = \ln x + c$$

$$\int \sin x\,dx = -\cos x + c \qquad \int \cos x = \sin x + c$$

Earlier in this book we differentiated functions such as $(x + 1)^5$ which is a function of a function. We can now integrate functions such as this.

In general, differentiation is a straight forward procedure but integration can be more difficult if you cannot differentiate standard functions with confidence. The approach to integration is illustrated in the following examples.

Example 1

a $\int (x+1)^5 dx$. We have $\dfrac{d}{dx}(x+1)^6 = 6(x+1)^5$.

$\therefore \quad \int (x+1)^5\,dx = \frac{1}{6}(x+1)^6 + c$

b $\int (3x-1)^4 dx$. We have $\dfrac{d}{dx}(3x-1)^5 = 5(3x-1)^4 \times 3$.

$\therefore \quad \int (3x-1)^4\,dx = \frac{1}{15}(3x-1)^5 + c$

c $\int \dfrac{1}{(1+4x)^2}\,dx$. We have $\dfrac{d}{dx}(1+4x)^{-1} = -1(1+4x)^{-2} \times 4$.

$\therefore \quad \int \dfrac{1}{(1+4x)^2}\,dx = -\frac{1}{4}(1+4x)^{-1} = -\dfrac{1}{4(1+4x)} + c.$

Example 2

a $\int 3e^x dx = 3e^x + c$

b $\int e^{4x} dx$. We have $\dfrac{d}{dx}(e^{4x}) = 4e^{4x}$

$\therefore \quad \int e^{4x} dx = \frac{1}{4}e^{4x} + c$

c $\int e^{-2x} dx$. We have $\dfrac{d}{dx}(e^{-2x}) = -2e^{-2x}$

$\therefore \quad \int e^{-2x} dx = -\frac{1}{2}e^{-2x} + c$

Example 3

a $\int (\sin x + 2\cos x)\,dx = -\cos x + 2\sin x + c$

b $\int \cos 5x\,dx.$ We have $\frac{d}{dx}(\sin 5x) = 5\cos 5x$

$\therefore \int \cos 5x\,dx = \frac{1}{5}\sin 5x + c$

c $\int 2\sin(7x-2).$ We have $\frac{d}{dx}[\cos(7x-2)] = -7\sin(7x-2)$

$\therefore \int 2\sin(7x-2)\,dx = -\frac{2}{7}\cos(7x-2) + c$

Example 4

a $\int 3 \times \frac{1}{x}\,dx = 3\ln x + c$

b $\int \frac{1}{2x}\,dx = \frac{1}{2}\int \frac{1}{x}\,dx = \frac{1}{2}\ln x + c$

c $\int \frac{1}{5x-1} = \frac{1}{5}\ln(5x-1) + c$

Example 5

Evaluate the definite integrals.

a $\int_0^{\frac{\pi}{4}} \cos 2x\,dx = \left[\frac{1}{2}\sin 2x\right]_0^{\frac{\pi}{4}} = \frac{1}{2}\left[\sin\left(2 \times \frac{\pi}{4}\right) - \sin 0\right]$

$= \frac{1}{2}\sin\frac{\pi}{2} - 0 = \frac{1}{2}.$

b $\int_0^1 (1 + e^x + e^{2x})\,dx = \left[x + e^x + \frac{1}{2}e^{2x}\right]_0^1$

$= 1 + e + \frac{1}{2}e^2 - \left(0 + 1 + \frac{1}{2}\right)$

$= e + \frac{1}{2}e^2 - \frac{1}{2}.$

c $\int_2^3 \left(1 + \frac{1}{x}\right)^2 dx = \int_2^3 \left(1 + \frac{2}{x} + \frac{1}{x^2}\right) dx = \int_2^3 \left(1 + \frac{2}{x} + x^{-2}\right) dx$

$= \left[x + 2\ln x + \frac{x^{-1}}{(-1)}\right]_2^3 = \left[x + 2\ln x - \frac{1}{x}\right]_2^3$

$= 3 + 2\ln 3 - \frac{1}{3} - \left(2 + 2\ln 2 - \frac{1}{2}\right)$

$= 1\frac{1}{6} + 2\ln 3 - 2\ln 2$

$= 1\frac{1}{6} + 2\ln\frac{3}{2}.$

EXERCISE 1

1 Integrate with respect to x.

a $(1+x)^3$ **b** $(x+4)^6$ **c** $(2x+1)^4$ **d** $(1-x)^4$

e $\dfrac{1}{(x+3)^2}$ **f** $2(3x-1)^2$ **g** $\dfrac{4}{(x+1)^3}$ **h** $(5x+3)^{-2}$

i $7e^x$ **j** $x+5e^x$ **k** e^{5x} **l** $e^{6x}-1$

m $3e^{2x}$ **n** $9e^{-x}$ **o** $\dfrac{4}{e^x}$ **p** $\dfrac{1}{e^x}+2x$

2 Integrate with respect to x.

a $\int \sin x \, dx$ **b** $\int \cos 4x \, dx$ **c** $\int \sin 10x \, dx$

d $2\int \cos 5x \, dx$ **e** $\int \frac{1}{2}\sin 6x \, dx$ **f** $\int (\cos 2x + 2x) \, dx$

g $\int \sin(4x-1) \, dx$ **h** $\int [x - \cos(x+1)] \, dx$ **i** $\int \left(2x + \dfrac{1}{x}\right) dx$

j $\int \left(\cos x + \dfrac{2}{x}\right) dx$ **k** $\int \dfrac{1}{4x} \, dx$ **l** $\int \dfrac{1}{x+3} \, dx$

m $\int \dfrac{6}{1+x} \, dx$ **n** $\int \dfrac{1}{3x+2} \, dx$ **o** $\int \left(e^{2x} + \dfrac{2}{x}\right) dx$

3 **a** Work out $\dfrac{d}{dx}(4x+1)^{\frac{3}{2}}$.

b Hence find $\int (4x+1)^{\frac{1}{2}} \, dx$.

4 Work out,

a $\dfrac{d}{dx}\sqrt{5x-2}$ **b** $\int (5x-2)^{-\frac{1}{2}} \, dx$

5 Work out the following

a $\int (6x+1)^{\frac{1}{3}} \, dx$ **b** $\int \dfrac{1}{\sqrt{x+2}} \, dx$ **c** $\int (1+2x)^{\frac{3}{2}} \, dx$

6 Find the shaded area under the curve $y = e^x$.

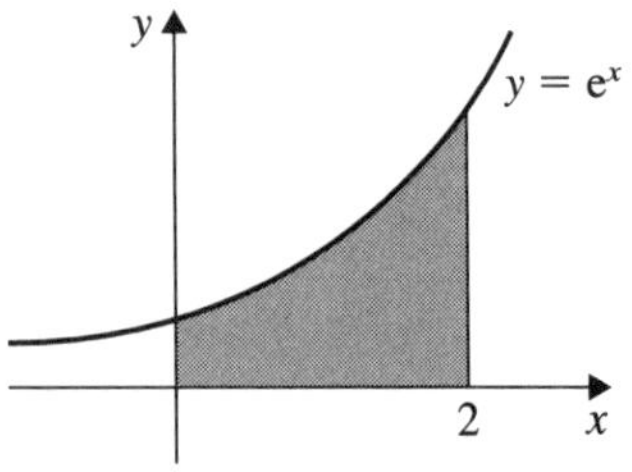

7 Calculate the following indefinite integrals:

a $\int e^{3x+1} \, dx$ **b** $\int 4e^{2x-3} \, dx$ **c** $\int \dfrac{5}{e^x} \, dx$

d $\int (2x+1)^3\,dx$

e $\int \frac{1}{(4x-3)^2}\,dx$

f $\int \frac{e^{2x}+1}{e^x}\,dx$

g $\int \sqrt{2x+1}\,dx$

h $\int \frac{1}{\sqrt{4x+5}}\,dx$

i $\int (e^x+1)^2\,dx$

8 Find $\int y\,dx$ for the following:

a $y = \frac{4}{2x+7}$

b $y = (e^x+1)(e^{-x}+1)$

c $y = \frac{2}{(3x-2)^2}$

d $y = \sqrt{4x+3}$

e $y = \frac{1}{\sqrt{(8x-1)}}$

f $y = \sqrt[3]{6x+5}$

9 Show that $\frac{x^3+5x+1}{x^2} = x + \frac{5}{x} + \frac{1}{x^2}$.

Hence find $\int \frac{x^3+5x+1}{x^2}\,dx$.

10 Find the following indefinite integrals:

a $\int \left(1-\frac{2}{x}\right)^2 dx$

b $\int \frac{x^2+5x+3}{x^2}\,dx$

c $\int \frac{e^x+1}{2}\,dx$

d $\int \frac{(x-2)(x-3)}{x^2}\,dx$

e $\int \frac{(x-5)(x+5)}{x}\,dx$

f $\int \frac{e^{3x+2}+1}{e^{2x}}\,dx$

11 The curve $y = \sin 4x$ meets the x-axis at $x = 0$ and $x = a$.

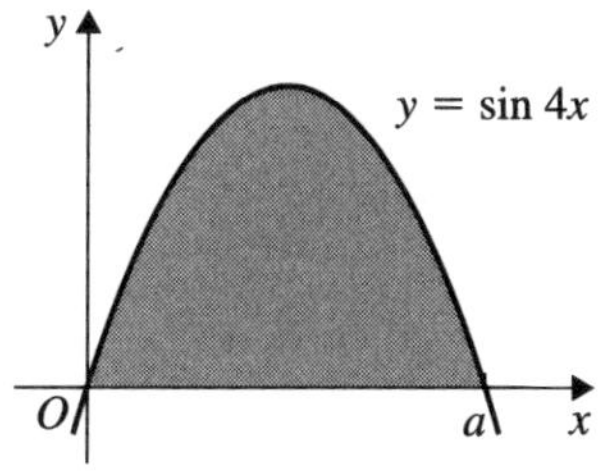

a Write down the value of a in radians.

b Find the area enclosed by the curve $y = \sin 4x$ and the x-axis between $x = 0$ and $x = a$.

Evaluate:

12 $\int_{\frac{\pi}{6}}^{\frac{\pi}{2}} \cos x\,dx$

13 $\int_0^1 e^{3x}\,dx$

14 $\int_2^5 \frac{1}{x}\,dx$

15 $\int_0^{\frac{1}{3}} (3x-1)^3\,dx$

16 $\int_{\frac{\pi}{6}}^{\frac{\pi}{4}} \sin 2x\,dx$

17 $\int_{-1}^{1} e^{2x+1}\,dx$

18 $\int_2^4 \frac{10}{x}\,dx$

19 $\int_1^3 \frac{1}{x+3}\,dx$

20 $\int_0^5 \sqrt{x+4}\,dx$

21 $\int_1^{1.5} \cos(2x-1)\,dx$

22 $\int_1^3 2x + \frac{1}{2x}\,dx$

23 $\int_0^{\frac{\pi}{4}} \sec^2 x\,dx$

24 $\int_{\frac{\pi}{6}}^{\frac{\pi}{4}} \sec^2 x\,dx$

25 $\int_1^2 \left(x+\frac{1}{x}\right)^2$

26 $\int_0^2 \frac{1}{2x+1}\,dx$

27 Show that $5\cos^2 x + 3\sin^2 x - 1 \equiv \cos 2x + 3$

Hence evaluate $\int_0^{\frac{\pi}{4}} (5\cos^2 x + 3\sin^2 - 1)\,dx$.

28 Evaluate $\int_0^{\frac{\pi}{6}} (\cos 3x + \sin 2x)\,dx$.

29 The sketch shows the curve $y = e^x$ and the lines $x = 1$ and $x = 3$.

Show that the area of the shaded region is $e(e^2 - 2)$.

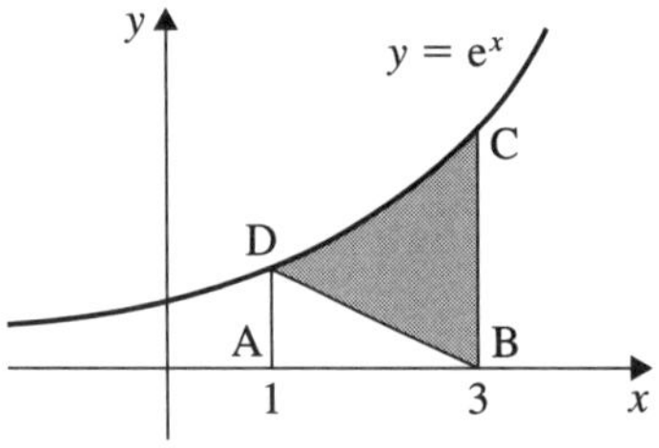

30 **a** Find the x coordinates of the points where the line $y = x - 4$ cuts the curve $y = 1 - \dfrac{6}{x}$.

b Find the exact area enclosed between the line $y = x - 4$ and the curve $y = 1 - \dfrac{6}{x}$.

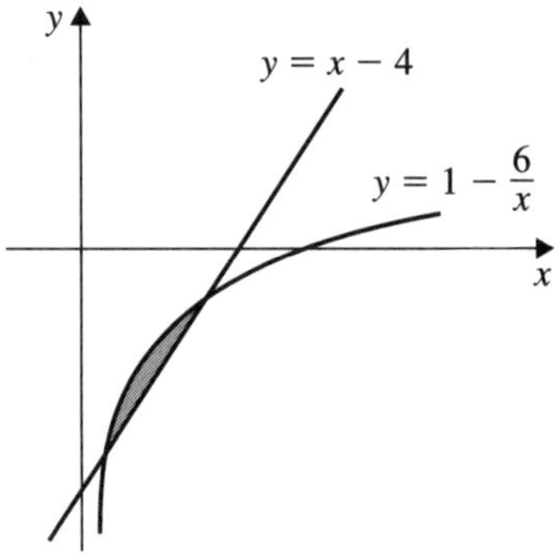

31 Find the exact area enclosed between the curve $y = 4 - \dfrac{3}{x}$ and the line $y = x$.

32 **a** Draw a sketch of the curve $y = \sqrt{9 - x}$ and the line $y = -\frac{1}{5}x + 3$ for x between 0 and 9.

b Show that where the curve and the line meet, $25(9 - x) = (-x + 15)^2$ and solve this to find the x coordinates of the points of intersection.

c Hence find the exact area enclosed between the curve and the line.

33* **a** Show that the area enclosed between $y = e^{2x} + 1$ and $y = 3e^x - 1$ is $\dfrac{3 - 4\ln 2}{2}$.

b Show that the area enclosed between $y = 2e^{2x}$ and $y = 18(e^x - 2)$ is $27 - 36\ln 2$.

34* **a** Find the exact x coordinates (in terms of logarithms) of the points of intersection of the curves $y = (e^x + 1)^2$ and $y = 10e^x - 11$.

b Hence find the exact area enclosed between the two curves.

10.2 Integrals of the form $\int \frac{f'(x)}{f(x)}\,dx$

In Part 4 of this book we found that $\frac{d}{dx}[\ln f(x)] = \frac{1}{f(x)} \times f'(x)$.

It follows that $\int \frac{f'(x)}{f(x)}\,dx = \ln f(x) + c$

The function $\ln x$ is only valid provided $x > 0$ so we need to be able to deal with integrals involving negative values of x.

The area under the curve $y = \frac{1}{x}$ between $x = -1$ and $x = -2$

is equal to the area between $x = 1$ and $x = 2$.

We can write $\int \frac{1}{x}\,dx = \ln |x| + c$ and further that $\int \frac{f'(x)}{f(x)} = \ln |f(x)| + c$

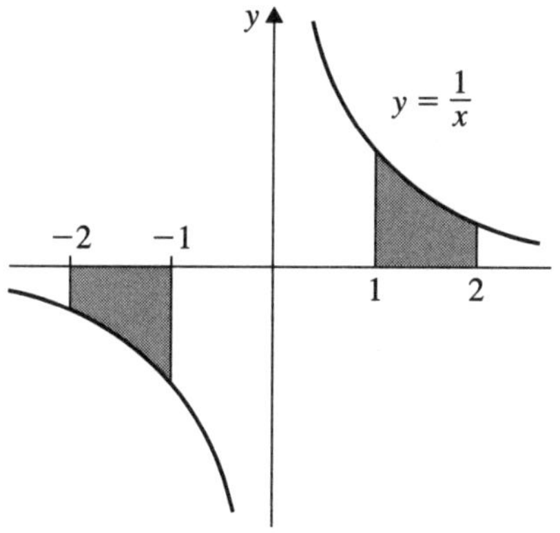

Normally we write the modulus sign only when dealing with definite integrals.

Note that if the interval between the limits a and b of the integral $\int_a^b \frac{1}{x}\,dx$ includes $x = 0$, then the integration is not valid. We cannot integrate across a discontinuity.

Example 1

Integrate the indefinite integrals.

a $\int \frac{2x}{x^2 + 1}\,dx = \ln(x^2 + 1) + c$

b $\int \frac{3x^2 - 1}{x^3 - x}\,dx = \ln(x^3 - x) + c$

c $\int \frac{\cos x}{\sin x}\,dx = \ln \sin x + c$

d $\int \frac{1}{4x + 3} = \frac{1}{4}\int \frac{4}{4x + 3} = \frac{1}{4}\ln(4x + 3) + c$

e $\int \frac{x}{x^2 - 7}\,dx = \frac{1}{2}\int \frac{2x}{x^2 - 7}\,dx$

$= \frac{1}{2}\ln(x^2 - 7) + c$

f $\int \frac{e^x}{3e^x + 11} = \frac{1}{3}\int \frac{3e^x}{3e^x + 11}$

$= \frac{1}{3}\ln(3e^x + 11) + c$

Example 2

Evaluate the definite integral $\int_1^2 \frac{1}{1-2x}\,dx$.

$$\int_1^2 \frac{1}{1-2x}\,dx = -\frac{1}{2}\int_1^2 \frac{-2}{1-2x}\,dx$$

$$= -\frac{1}{2}\Big[\ln|1-2x|\Big]_1^2$$

We use the modulus function because when $x = 1$ and $x = 2$, $1 - 2x < 0$.

$$= -\frac{1}{2}[\ln 3 - \ln 1]$$

[Remember $|-3| = 3$ and $|-1| = 1$.]

$$= -\frac{1}{2}\ln 3$$

Example 3

Find $\int \tan x\,dx$.

$$\int \tan x\,dx = \int \frac{\sin x}{\cos x}\,dx$$

[Remember this method.]

$$= -\int \frac{-\sin x}{\cos x}\,dx$$

$$= -\ln\cos x + c$$

$$= \ln(\cos x)^{-1} + c$$

$$= \ln\sec x + c$$

EXERCISE 2

Find the following indefinite integrals.

1 $\int \frac{1}{x+3}\,dx$ **2** $\int \frac{2}{2x+1}\,dx$ **3** $\int \frac{2x}{x^2+5}\,dx$

4 $\int \frac{3x^2}{x^3+2}\,dx$ **5** $\int \frac{1}{4x+1}\,dx$ **6** $\int \frac{1}{7x-1}\,dx$

7 $\int \frac{e^x}{e^x+3}\,dx$ **8** $\int \frac{x}{x^2+3}\,dx$ **9** $\int \cot x\,dx$

10 $\int \frac{1}{x+1} + \frac{1}{x-2}\,dx$ **11** $\int \frac{2}{2x+1} + \frac{5}{5x+2}\,dx$ **12** $\int \frac{6}{2x+1} - \frac{1}{x+1}\,dx$

13 Find the area under the curve $y = \frac{1}{3x-1}$ between $x = 1$ and $x = 2$.

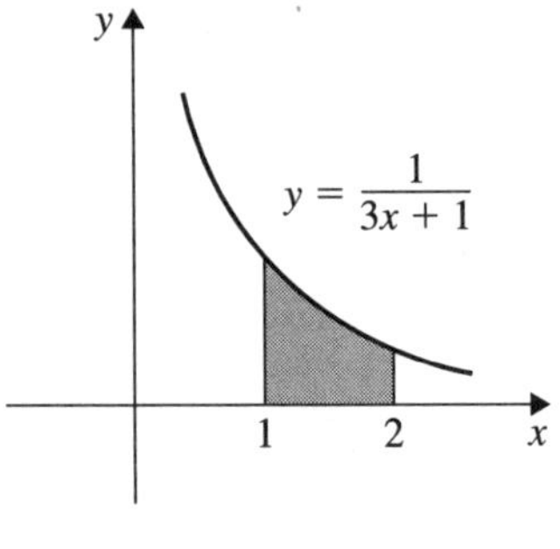

14 Find the exact values of the following definite integrals, leaving your answers in terms of logarithms:

a $\displaystyle\int_1^4 \frac{2}{x}\,dx$ **b** $\displaystyle\int_1^8 \frac{1}{3x}\,dx$ **c** $\displaystyle\int_3^6 \frac{2x+1}{x}\,dx$

d $\displaystyle\int_2^8 \frac{4x-1}{2x}\,dx$ **e** $\displaystyle\int_2^3 \frac{3}{3x-5}\,dx$ **f** $\displaystyle\int_2^4 \frac{2}{x-1}\,dx$

g $\displaystyle\int_1^4 \frac{6}{2x+1} - \frac{3}{3x-2}\,dx$ **h** $\displaystyle\int_2^5 \frac{1}{x+1} - \frac{1}{x-1}\,dx$ **i** $\displaystyle\int_5^{15} \left(\frac{x+1}{x}\right)dx$

j $\displaystyle\int_0^3 \frac{3}{3x+1} + \frac{2}{2x+3}\,dx$ **k** $\displaystyle\int_2^4 \left(\frac{x^2+2x+3}{x}\right)dx$ **l** $\displaystyle\int_1^3 \left(\frac{x^2-4}{x}\right)dx$

15 Obtain the area enclosed by the curve $y = \frac{1}{x}$, the x-axis and the lines $x = -2$ and $x = -3$.

16 Evaluate the following, giving your answers correct to two decimal places.

a $\displaystyle\int_{-2}^0 \frac{1}{x-1}\,dx$ **b** $\displaystyle\int_1^2 \frac{x}{x^2-9}\,dx$ **c** $\displaystyle\int_1^3 \frac{1}{1-4x}\,dx$

17 Find the following

a $\displaystyle\int \tan x\,dx$ **b** $\displaystyle\int \frac{\cos 2x}{\sin 2x}\,dx$ **c** $\displaystyle\int \frac{\sec^2 x}{\tan x}\,dx$

d $\displaystyle\int \frac{\sin x}{\cos x + 3}\,dx$ **e** $\displaystyle\int \frac{2\sin x\cos x}{\cos 2x}\,dx$ **f** $\displaystyle\int \frac{1}{x\ln x}\,dx$

10.3 Integration by substitution

Some integrals are easier to perform if we change the variable making a substitution.

Let f(x) be a function of x and let $y = \int f(x)\,dx$

So $\frac{dy}{dx} = f(x)$.

If u is a function of x then y is a function of u.

$$\frac{dy}{du} = \frac{dy}{dx}\cdot\frac{dx}{du}$$
$$= f(x)\frac{dx}{du}$$

Integrating with respect to u, $y = \int f(x)\frac{dx}{du}\,du$

$$\text{or } \int f(x)\,dx = \int f(x)\frac{dx}{du}\,du$$

So an integral with respect to x is transformed into an integral with respect to u. We substitute for f(x) and $\frac{dx}{du}$ in terms of u.

Example 1

Find $\int x(2x+1)^2\,dx$

Let $u = 2x+1 \Rightarrow 2x = u-1$

$$x = \frac{1}{2}u - \frac{1}{2}$$

$$\frac{dx}{du} = \frac{1}{2}$$

$$\int x(2x+1)^2 \frac{dx}{du}\,du = \int\left(\frac{u-1}{2}\right)u^2\frac{1}{2}\,du$$

$$= \frac{1}{4}\int u^3 - u^2\,du = \frac{1}{4}\left(\frac{u^4}{4} - \frac{u^3}{3}\right) + c$$

$$= \frac{u^3}{48}(3u-4) + c = \frac{(2x+1)^3}{48}[3(2x+1)-4] + c$$

$$= \left(\frac{2x+1}{48}\right)^3(6x-1) + c$$

Example 2

Find $I = \int x\sqrt{x-1}\,dx$

Let $u = \sqrt{x-1} \Rightarrow x - 1 = u^2$

$x = u^2 + 1$

$\frac{dx}{dx} = 2u$

$$\therefore\quad I = \int (u^2+1)u\,.\,2u\,du$$

↑ ↑ ↑

x $\sqrt{x-1}$ $\frac{dx}{du}$

$$= 2\int u^4 + u^2\,du$$

$$= 2\left(\frac{u^5}{5} + \frac{u^3}{3}\right) + c$$

$$= 2\left(\frac{3u^5 + 5u^3}{15}\right) + c$$

$$= \frac{2}{15}(x-1)^{\frac{3}{2}}(3x+2) + c$$

Example 3

Find $I = \int 2x(x^2+1)^2\,dx$

Let $u = x^2 + 1$

$\frac{du}{dx} = 2x$

$$\therefore\quad I = \int 2x(x^2+1)^2\frac{dx}{du}dx$$

$$= \int 2x(x^2+1)^2\frac{1}{2x}\,du$$

$$= \int u^2\,du$$

$$= \frac{u^3}{3} + c$$

$$= \frac{1}{3}(x^2+1)^3 + c$$

Example 4

Evaluate the definite integral $\int_1^2 x(2x-3)^3\,dx$

This is a definite integral and we must remember to change the limits to the corresponding values of the new variable u.

Let $u = 2x - 3 \Rightarrow x = \dfrac{u+3}{2} \Rightarrow \dfrac{dx}{du} = \dfrac{1}{2}$

When $x = 1$, $u = -1$ and when $x = 2$, $u = 1$

$$\therefore \quad \int_1^2 x(2x-3)^3\,dx = \int_{u=-1}^{u=1} x(2x-3)^3 \frac{dx}{du}\,du$$

$$= \int_{-1}^{1} \left(\frac{u+3}{2}\right)u^3 \frac{1}{2}\,du = \frac{1}{4}\int_{-1}^{1} u^4 + 3u^3\,du$$

$$= \frac{1}{4}\left[\frac{u^5}{5} + \frac{3}{4}u^4\right]_{-1}^{1}$$

$$= \frac{1}{4}\left[\frac{1}{5} + \frac{3}{4} - \left(-\frac{1}{5} + \frac{3}{4}\right)\right] = \frac{1}{4}\left[\frac{2}{5}\right] = \frac{1}{10}$$

Example 5

Find $\displaystyle\int \frac{1}{\sqrt{1-x^2}}\,dx$

In this case we use a non linear substitution.

Let $x = \sin u$

$\dfrac{dx}{du} = \cos u$

$$\therefore \quad \int \frac{1}{\sqrt{1-x^2}}\,dx = \int \frac{1}{\sqrt{1-\sin^2 u}} \frac{dx}{du}\,.\,du$$

$$= \int \frac{1}{\sqrt{\cos^2 u}} \cos u\,du$$

$$= \int 1\,du = u + c$$

$$= \sin^{-1} x + c$$

EXERCISE 3

1 Find the following integrals, using the given substitutions.

a $\int x(x+2)^2\,dx \quad u = x+2$

b $\int x(x-3)^2\,dx \quad u = x-3$

c $\int 2x(x+4)^3\,dx \quad u = x+4$

d $\int x^2(x-1)^2\,dx \quad u = x-1$

2 Find the following indefinite integrals using the given substitutions:

a $\int x(x+1)^3\,dx \quad u = x+1$

b $\int x(x-1)^5\,dx \quad u = x-1$

c $\int x\sqrt{x+3}\,dx \quad u = \sqrt{x+3}$

d $\int \dfrac{x^2}{\sqrt{x-2}}\,dx \quad u = x-2$

e $\int x(2x+1)^3\,dx \quad u = 2x+1$

f $\int x\sqrt{(4x-1)}\,dx \quad u = 4x-1$

g $\int \dfrac{x}{\sqrt{(5x+1)}}\,dx \quad u = 5x+1$

h $\int (x+1)(3x-2)^4\,dx \quad u = 3x-2$

3 Find the following indefinite integrals using the given substitutions:

a $\int 12x(x^2+1)^3\,dx \quad u = x^2+1$

b $\int 18x^2(x^3-3)^2\,dx \quad u = x^3-3$

c $\int e^x(e^x-1)^3\,dx \quad u = e^x-1$

d $\int \frac{e^x}{\sqrt{e^x+2}}\,dx \quad u = e^x+2$

e $\int \sin^2 x \cos x\,dx \quad u = \sin x$

f $\int \cos^3 x \sin x\,dx \quad u = \cos x$

g $\int \tan^3 x \sec^2 x\,dx \quad u = \tan x$

h $\int \frac{1}{\sqrt{1-x^2}}\,dx \quad x = \sin\theta$

4 Evaluate the following definite integrals using the given substitutions:

a $\int_0^1 x(x+1)^3\,dx \quad u = x+1$

b $\int_3^4 x(x-3)^4\,dx \quad u = x-3$

c $\int_3^6 x\sqrt{x-2}\,dx \quad u = x-2$

d $\int_0^4 \frac{x}{\sqrt{2x+1}}\,dx \quad u = 2x+1$

5 Find the following definite integrals using the given substitutions:

a $\int_0^{\pi} e^{\cos x}\sin x\,dx \quad u = \cos x$

b $\int_1^3 \frac{x}{\sqrt{x+2}}\,dx \quad u = x+2$

c $\int_0^{\frac{\pi}{4}} \frac{\sin x}{\cos x}\,dx \quad u = \cos x$

d $\int_0^{\frac{\pi}{2}} \sin x \sin 2x\,dx \quad u = \sin x$

6 Use the substitution $x = \sin\theta$ to show that, for $|x| \leqslant 1$,

$$\int \frac{1}{(1-x^2)^{\frac{3}{2}}}\,dx = \frac{x}{(1-x^2)^{\frac{1}{2}}} + c,$$ where c is an arbitrary constant.

7 By making the substitution $x = 2\sin t$, evaluate $\int_0^1 \frac{x^2}{\sqrt{(4-x^2)}}\,dx$.

Give your answer correct to 2 decimal places. [Hint: see section 10.5]

10.4 Integration by parts

In Part 4 we obtained a formula for differentiating the product of two functions u and v.

$$\frac{d}{dx}(u\,v) = u\frac{dv}{dx} + v\frac{du}{dx}$$

Integrate both sides of the equation with respect to x.

$$\int \frac{d}{dx}(u\,v)\,dx = \int u\frac{dv}{dx}\,dx + \int v\frac{du}{dx}\,dx$$

$$u\,v = \int u\frac{dv}{dx}\,dx + \int v\frac{du}{dx}\,dx$$

or $$\int u\frac{dv}{dx}\,dx = u\,v - \int v\frac{du}{dx}\,dx$$

This is the formula for integrating by parts.

Example 1

Find $\int x\,e^{2x}\,dx$

Let $u = x$ and $\dfrac{dv}{dx} = e^{2x}$

Then $\dfrac{du}{dx} = 1$ and $v = \dfrac{1}{2}e^{2x}$

By the formula: $\displaystyle\int u\frac{dv}{dx}\,dx = u\,v - \int v\frac{du}{dx}\,dx$

$$\int x\,e^{2x}\,dx = x\,.\,\frac{1}{2}e^{2x} - \int\frac{1}{2}e^{2x}\,.\,1\,dx$$

$$= \frac{x}{2}e^{2x} - \frac{1}{4}e^{2x} + c$$

Notice that we had to make a choice in the original integral. We could have chosen $x = u$ or $x = \dfrac{dv}{dx}$. We chose to let $x = u$ because $\dfrac{du}{dx} = 1$.

You will learn by experience which substitution to make in different questions. If you find that you have made the wrong choice abandon the working and start again.

Example 2

Find $I = \int x \ln x\,dx$

Let $u = \ln x$ and $\dfrac{dv}{dx} = x$. Notice that we could not choose $\dfrac{dv}{dx} = \ln x$ because it is not easy to integrate $\ln x$.

So $\dfrac{du}{dx} = \dfrac{1}{x}$ and $v = \dfrac{1}{2}x^2$

$$\therefore\quad I = \underset{\underset{u}{\uparrow}}{\ln x} \times \underset{\underset{v}{\uparrow}}{\frac{1}{2}x^2} - \int \underset{\underset{v}{\uparrow}}{\frac{1}{2}x^2} \times \underset{\underset{\frac{du}{dx}}{\uparrow}}{\frac{1}{x}}\,dx$$

$$I = \frac{1}{2}x^2 \ln x - \frac{1}{2}\int x\,dx$$

$$= \frac{1}{2}x^2 \ln x - \frac{1}{4}x^2 + c$$

Example 3

Find $I = \int \ln x\,dx$

There is a neat 'trick' involved here.

Write $I = \int 1 \times \ln x\,dx$ and let $u = \ln x$ and $\dfrac{dv}{dx} = 1$

Then $\frac{du}{dx} = \frac{1}{x}$ and $v = x$

$$\therefore \quad I = (\ln x) \times x - \int x \times \frac{1}{x} dx$$

$$\uparrow_{u} \quad \uparrow_{v} \quad \uparrow_{v} \quad \uparrow_{\frac{du}{dx}}$$

$$= x \ln x - \int 1\, dx$$

$$= x \ln x - x + c$$

Example 4

Find $I = \int x^2 e^x dx$

Let $u = x^2$, $\frac{dv}{dx} = e^x$ so that $\frac{du}{dx} = 2x$ and $v = e^x$

$$\therefore \quad I = x^2 e^x - \int e^x . 2x\, dx$$

Now consider $\int 2x e^x dx$

Let $u = 2x$, $\frac{dv}{dx} = e^x$ so that $\frac{du}{dx} = 2$ and $v = e^x$

$$\therefore \quad \int 2x e^x dx = 2x e^x - \int e^x . 2\, dx$$

$$= 2x e^x - 2e^x + c$$

$$\therefore \quad I = x^2 e^x - (2x e^x - 2e^x) + c$$

$$= e^x (x^2 - 2x + 2) + c$$

This example shows that in some questions you may have to integrate by parts more than once.

EXERCISE 4

1 Use integration by parts to find the following.

a $\int x(1 + x)^2 dx$ $\left[\text{Let } u = x \text{ and } \frac{dv}{dx} = (1 + x)^2\right]$

b $\int x \cos x\, dx$ $\left[\text{Let } u = x \text{ and } \frac{dv}{dx} = \cos x\right]$

2 Integrate the following:

a $\int x e^x dx$ b $\int x(1 + x)^3 dx$ c $\int x e^{-x} dx$

d $\int x e^{3x} dx$ e $\int x \sin x\, dx$ f $\int 3x \cos 2x\, dx$

g $\int 2x(x-1)^3\,dx$ **h** $\int x\sqrt{x+1}\,dx$ **i** $\int x\ln 2x\,dx$

j $\int x^2\ln x\,dx$ **k** $\int \frac{\ln x}{x^3}\,dx$ **l** $\int x^2 e^x\,dx$

3 The sketch shows the curve $y = x\,e^{-x}$ and the line $x = 1$. Find the shaded area.

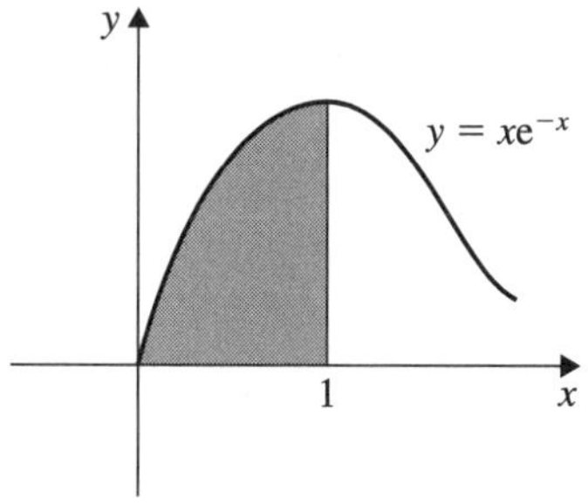

4 Calculate the exact values of following definite integrals.

a $\int_0^1 xe^{-3x}\,dx$ **b** $\int_0^{\frac{\pi}{4}} x\sin 2x\,dx$

c $\int_0^1 x(2x+1)^2\,dx$ **d** $\int_3^8 x\sqrt{(x+1)}\,dx$

e $\int_1^e x^2\ln x\,dx$ **f** $\int_0^{\frac{\pi}{2}} x^2\sin x\,dx$

5 **a** Find the exact value of $\int_0^1 x(2x+1)^3\,dx$ by using integration by parts.

b Show that you get the same value for $\int_0^1 x(2x+1)^3\,dx$ by using the method of integration by substitution, with $u = 2x + 1$.

6 **a** Show that $\int 3x(1+2x)^4\,dx = \frac{(1+2x)^5(10x-1)}{40} + c$.

b Show that $\int \frac{x}{e^{2x}}\,dx = -\left(\frac{2x+1}{4e^{2x}}\right) + c$.

c Show that $\int \frac{4x}{\sqrt{x+1}}\,dx = \frac{8(x-2)\sqrt{x+1}}{3} + c$.

7 The curve $y = e^x(1-x)$ is shown.

a Write down the coordinates of point A and point B where the curve cuts the axes.

b Calculate the area enclosed by the curve and the x and y axes.

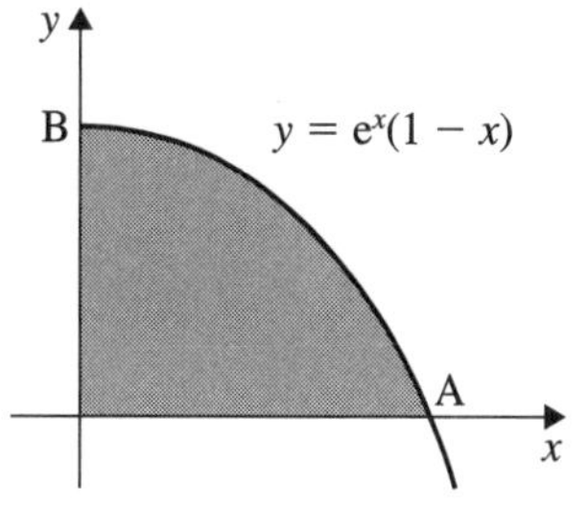

8 **a** Find $\frac{d}{dx}(e^{x^2})$.

b By writing $x^3e^{x^2} = \frac{x^2}{2}(2xe^{x^2})$, find $\int x^3e^{x^2}\,dx$.

9* a Show that $\int e^x \cos x\, dx = e^x \sin x - \int e^x \sin x\, dx$.

b Find a similar expression for $\int e^x \sin x\, dx$.

c Hence find $\int e^x \cos x\, dx$.

10* Find, a $\int e^{5x} \sin 3x\, dx$.

b $\int e^{ax} \sin bx\, dx$.

10.5 Using trigonometric identities

Sometimes we use trigonometric identities to transform an integral into a function which can be integrated. The most frequently used formulae are below:

$$\cos 2x = 2\cos^2 x - 1$$
$$\cos 2x = 1 - 2\sin^2 x$$
$$1 + \tan^2 x = \sec^2 x$$

Remember also that $\cos 4x = 2\cos^2 2x - 1$ and so on.

Example 1

Find $I = \int \sin^2 x\, dx$

Rearrange the formula $\cos 2x = 1 - 2\sin^2 x$

$$2\sin^2 x = 1 - \cos 2x$$
$$\sin^2 x = \tfrac{1}{2}(1 - \cos 2x)$$

$$\therefore \quad I = \tfrac{1}{2}\int (1 - \cos 2x)\, dx = \tfrac{1}{2}(x - \tfrac{1}{2}\sin 2x) + c$$
$$= \tfrac{1}{2}x - \tfrac{1}{4}\sin 2x + c$$

Example 2

Find $I = \int \cos^2 4x\, dx$

Rearrange the formula $\cos 8x = 2\cos^2 4x - 1$

$$\cos^2 4x = \tfrac{1}{2}(\cos 8x + 1)$$

$$\therefore \quad I = \tfrac{1}{2}\int (\cos 8x + 1)\, dx = \tfrac{1}{16}\sin 8x + \tfrac{1}{2}x + c$$

Example 3

Find $I = \int \tan^2 x\, dx$

From above $\tan^2 x = \sec^2 x - 1$

$$\therefore \quad I = \int (\sec^2 x - 1)\, dx$$
$$= \tan x - x + c$$

Exercise 5

1 Integrate with respect to x:

a $\int \sin^2 x \, dx$ b $\int \cos^2 x \, dx$ c $\int \tan^2 x \, dx$

d $\int \sin^2 3x \, dx$ e $\int \cos^2 2x \, dx$ f $\int (1 - 2\sin x)^2 \, dx$

g $\int (\cos x + \sec x)^2 \, dx$ h $\int \tan^2 2x \, dx$ i $\int (1 + \tan x)^2 \, dx$

2 Evaluate the following

a $\int_0^{\frac{\pi}{4}} \sin^2 x \, dx$ b $\int_0^{\frac{\pi}{2}} \cos^2 \tfrac{1}{2} x$ c $\int_0^{\frac{\pi}{3}} \tan^2 x \, dx$

d $\int_0^{\frac{\pi}{1}} (1 + 2\cos 2x)^2 \, dx$ e $\int_{\frac{\pi}{6}}^{\frac{\pi}{4}} (\sin^2 x + \cos^2 x) \, dx$

10.6 Volume of revolution

When the area between a curve and the x-axis is rotated through one revolution (360°) about the x-axis a *volume of revolution* is formed.

Consider the area below the line $y = x$ between $x = 0$ and $x = 2$. When this area is rotated through one revolution about the x-axis we obtain a solid cone of radius and height 2 units.

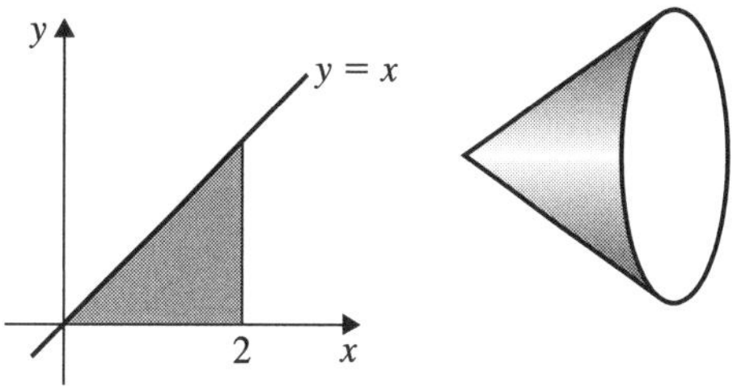

The diagram shows the curve $y = f(x)$ with an elementary strip, width δx and length y. When this strip is rotated through one revolution about the x-axis it produces an elementary disc of radius y and thickness δx. If the volume of this elementary disc is δV, then

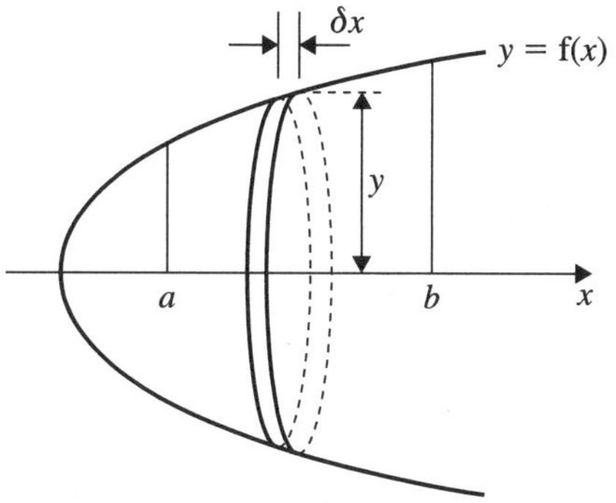

$$\delta V = \pi y^2 \, \delta x$$

As $\delta x \to 0$, the sum of all these elementary discs tends towards V, the volume of the solid of revolution.

$$\therefore \quad V = \lim_{\delta x \to 0} \sum_{x=a}^{x=b} \pi y^2 \, \delta x$$

$$\mathbf{V = \int_b^a \pi y^2 \, dx}$$

If an area is rotated about the y-axis in a similar way the volume, V, of the solid is given by $V = \int \pi x^2 \, dy$.

Example 1

Find the volume of revolution formed by rotating the area enclosed by the curve $y = x^2$, the x-axis and the lines $x = 1$ and $x = 2$ through one revolution about the x-axis.

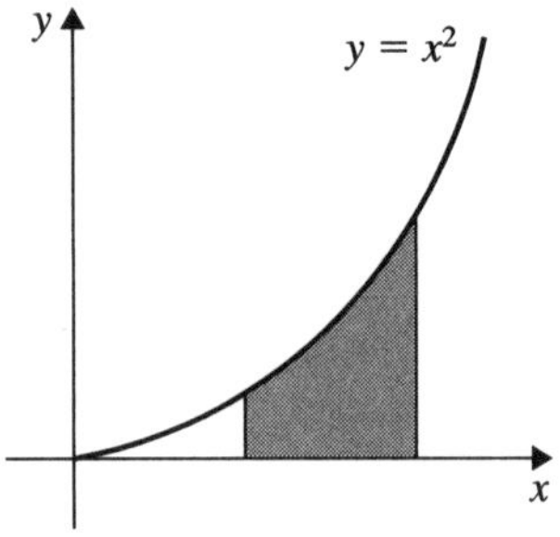

The volume formed, $V = \int_1^2 \pi y^2 \, dx$

$$= \pi \int_1^2 x^4 \, dx$$

$$= \pi \left[\frac{x^5}{5} \right]_1^2$$

$$= \pi \left[\frac{32}{5} - \frac{1}{5} \right] = \frac{31}{5} \pi \text{ cubic units.}$$

EXERCISE 6

1 Find the volume of revolution when the area enclosed between the curve $y = \frac{1}{x}$, the x-axis and the lines $x = 1$ and $x = 3$ is rotated through 360° about the x-axis.

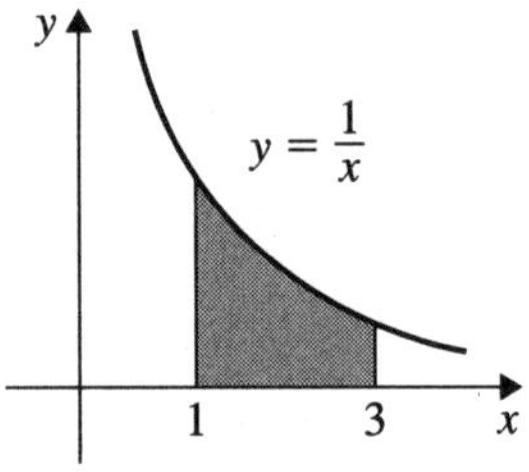

2 Find the volume of revolution when the area enclosed between the curve $y = \frac{1}{\sqrt{x}}$, the x-axis and the lines $x = 1$ and $x = 4$ is rotated through 360° about the x-axis.

3 Find the volume of revolution when the area enclosed between the curve $y = \sqrt{4 - x^2}$, the x-axis and the lines $x = -2$ and $x = 2$ is rotated through 360° about the x-axis.

In Questions **4** to **11** find the volume of revolution formed when the area enclosed between each curve, the x-axis and the values of x given is rotated through one revolution about the x-axis.

4 $y = x + 1$, between $x = 0$ and $x = 4$.

5 $y = \frac{x}{2}$, between $x = 2$ and $x = 4$.

6 $y = x^2 - 4$, between $x = -2$ and $x = 2$.

7 $y = 2x^{\frac{1}{4}}$, between $x = 1$ and $x = 16$.

8 $y = e^{2x}$, between $x = 0$ and $x = 1$.

9 $x^2 + y^2 = 9$, between $x = 0$ and $x = 3$.

10 $y = x^{\frac{1}{2}} e^x$, between $x = 1$ and $x = 2$.

11 $y = \frac{x}{\sqrt{x^3 + 4}}$, between $x = 0$ and $x = 2$.

12 The diagram shows part of the graph of $y = \cos x$.

Find the volume obtained when the region R is revolved through 2π radians about the x-axis.

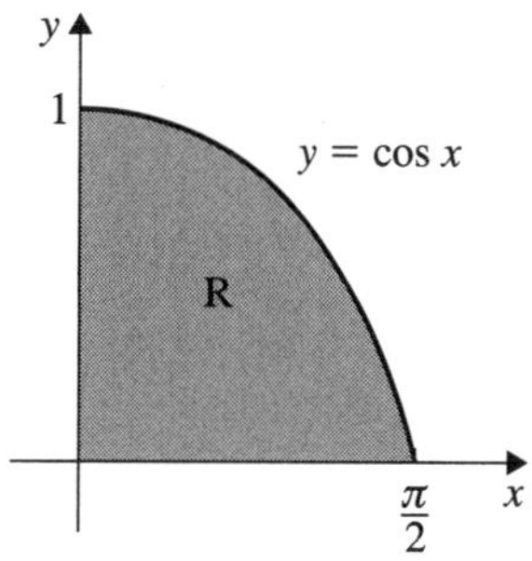

13 The shaded region is bounded by the curve, $y = x^{\frac{1}{2}} \sin x$, the x-axis and the line $x = \dfrac{\pi}{2}$. This shaded region is rotated through 2π radians about the x-axis to form a solid of revolution. Calculate the volume of the solid of revolution formed, giving your answer in terms of π.

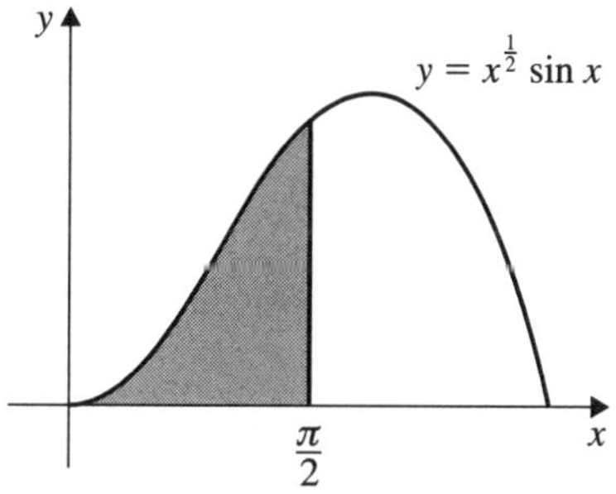

14 The region bounded by the curve $y = x + 2\sin x$, the x-axis and the line $x = \pi$, is rotated through 2π radians about the x-axis. Find the volume of revolution formed, giving your answer in terms of π.

15 The region R is bounded by the curve $y = x^{-\frac{1}{2}}$ and the lines $x = 1$ and $x = 4$. Find the volume generated when this area is rotated 360° about the x-axis.

16 The area in the first quadrant bounded by the curve $y = x^2$, the y-axis and the line $y = 3$ is rotated through 360° about the y-axis (not the x-axis!).

Calculate the volume of the solid formed.

$\left[\text{Hint: Use } V = \int \pi x^2 \, dy\right]$

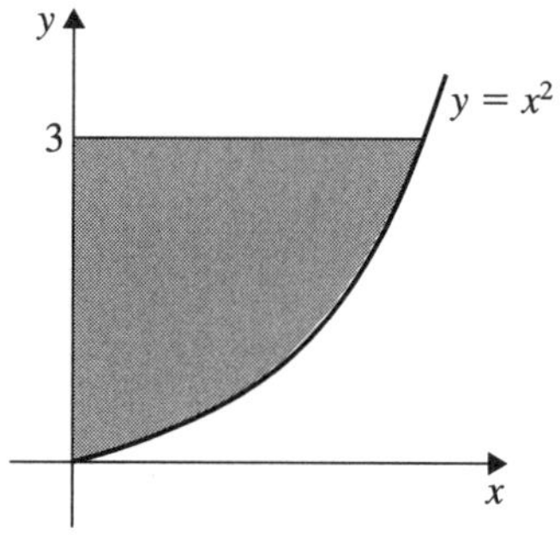

17 The area in the first quadrant bounded by the curve $y = \sqrt{x}$, the y-axis and the line $y = 2$ is rotated through 360° about the y-axis. Calculate the volume of the solid formed.

10.7 Integration using partial fractions

Example 1

Evaluate $I = \displaystyle\int_2^3 \frac{12x^2 - 13x + 2}{(2x-1)^2(x-1)} \, dx$

Using the methods from section 6 we obtain

$$\frac{12x^2 - 13x + 2}{(2x-1)^2(x-1)} \equiv \frac{4}{2x-1} + \frac{3}{(2x-1)^2} + \frac{1}{x-1}$$

$$\therefore\ \mathrm{I} = \int_2^3 \frac{4}{2x-1} + \frac{3}{(2x-1)^2} + \frac{1}{x-1}\,\mathrm{d}x$$

$$= \left[2\ln|2x-1| - \frac{3}{2}(2x-1)^{-1} + \ln|x-1|\right]_2^3$$

$$= 2\ln 5 - \frac{3}{2}\times\frac{1}{5} + \ln 2 - \left(2\ln 3 - \frac{3}{2}\times\frac{1}{3} + \ln 1\right)$$

$$= \ln 5^2 + \ln 2 - \ln 3^2 - \frac{3}{10} + \frac{1}{2}$$

$$= \ln\left(\frac{50}{9}\right) + \frac{1}{5}$$

EXERCISE 7

1 a Express $\dfrac{x}{(1+x)(1+2x)}$ in partial fractions.

b Find $\displaystyle\int \frac{x}{(1+x)(1+2x)}\,\mathrm{d}x$.

2 Using partial fractions, calculate the following indefinite integrals:

a $\displaystyle\int \frac{2}{x^2-1}\,\mathrm{d}x$

b $\displaystyle\int \frac{x+6}{x^2+9x+20}\,\mathrm{d}x$

c $\displaystyle\int \frac{x+18}{x^2-9x+14}\,\mathrm{d}x$

3 Use partial fractions to show that $\displaystyle\int_0^1 \frac{18-4x-x^2}{(4-3x)(1+x)^2} = \frac{7}{3}\ln 2 + \frac{3}{2}$.

4 The diagram shows the graph of

$$y = \frac{1}{(x-1)(x-3)}$$

The line $y = -\frac{4}{3}$ intersects the curve at points P and Q.
Calculate the area of the shaded region, giving your answer correct to three decimal places.

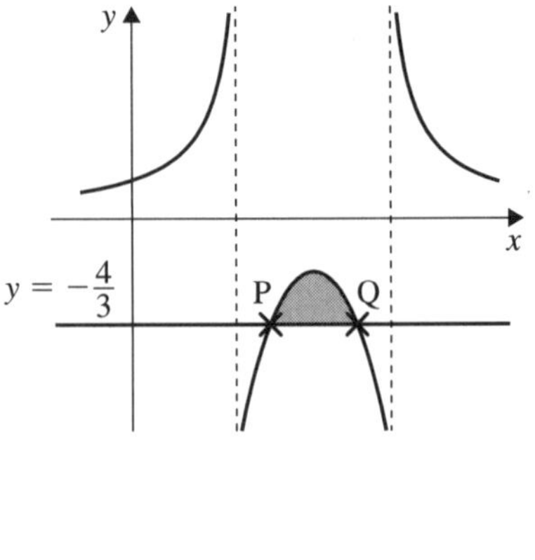

5 a Express $\dfrac{8x^2+16x+7}{(x+1)^2(2x+1)}$ in partial fractions.

b Hence show that $\displaystyle\int_0^1 \frac{8x^2+16x+7}{(x+1)^2(2x+1)}\,\mathrm{d}x = \frac{1}{2} + 2\ln 6$.

6 a Express $\dfrac{6x^3-7x^2+14x-7}{(2x-1)(3x-2)}$ in partial fractions.

b Hence show that $\displaystyle\int_1^2 \frac{6x^3-7x^2+14x-7}{(2x-1)(3x-2)}\,\mathrm{d}x = \frac{3}{2} + \ln 12$.

7 Show that $\int_2^3 \frac{2x^3 - 9x^2 + 11x - 9}{(2x-7)(x-1)}\,dx = \frac{5}{2} + \ln\left(\frac{2}{3}\right)$.

8 Show that $\int_2^4 \frac{4x^4 - 28x^3 + 51x^2 - 39x + 1}{(2x-1)^2(x-5)}\,dx = \frac{88}{21} - \frac{3}{2}\ln 3 + \frac{1}{2}\ln 7$

9 $f(x) \equiv x^3 + 6x^2 + 11x + 6$

A root of the equation $f(x) = 0$ is -2.

a Using algebra and showing all your working, factorise $f(x)$ completely.

b Express $\frac{2}{f(x)}$ in partial fractions.

c Evaluate $\int_0^1 \frac{2}{f(x)}\,dx$, giving your answer in the form $\ln\left(\frac{a}{b}\right)$ where a and b are integers.

10 Using the substitution $u = e^x$ and partial fractions, show that

$$\int_0^{\ln 2} \frac{1}{1+e^x}\,dx = \ln\left(\frac{4}{3}\right)$$

10.8 Integration methods, summary

One of the main problems in integration is in deciding which of several different methods to use. In this section we will briefly review six main types of integral and provide a large number of mixed questions so that you can practise making decisions about the best method to adopt. Use the exercise sensibly, it is not necessary to do *all* the questions.

Type 1 Standard functions

$\int \cos ax\,dx = \frac{1}{a}\sin ax,$ $\int \sin ax\,dx = -\frac{1}{a}\cos ax,$

$\int \sec^2 x\,dx = \tan x,$ $\int e^{ax}\,dx = \frac{1}{a}e^{ax},$ $\int \frac{1}{x}\,dx = \ln x$

$\int (ax+b)^n\,dn = \frac{1}{a(n+1)}(ax+b)^{n+1}$

Examples

a $\int \sin 3x\,dx = -\frac{1}{3}\cos 3x + c$

b $\int \sec^2 5x = \frac{1}{5}\tan 5x + c$

c $\int e^{7x-1}\,dx = \frac{1}{7}e^{7x-1} + c$

d $\int \frac{1}{3x+1}\,dx = \frac{1}{3}\ln(3x+1) + c$

e $\int (2x-3)^4\,dx = \frac{1}{10}(2x-3)^5 + c$

f $\int (5x+3)^{\frac{1}{2}} = \frac{2}{15}(5x+3)^{\frac{3}{2}} + c$

Type 2 The form $\int \frac{f'(x)}{f(x)}\,dx = \ln f(x) + c$

Examples

a $\int \frac{2x}{x^2+7}\,dx = \ln(x^2+7) + c$

b $\int \frac{x^2}{x^3-5} = \frac{1}{3}\int \frac{3x^2}{x^3-5} = \frac{1}{3}\ln(x^3-5) + c$

c $\int \frac{\cos x}{\sin x}\,dx = \ln \sin x + c$

d $\int \frac{e^x}{e^x+11}\,dx = \ln(e^x+11) + c$

Type 3 Using trigonometric identities

For $\int \sin^2 x\,dx$ or $\int \cos^2 x\,dx$ use $\cos 2x = 2\cos^2 - 1$

or $\cos 2x = 1 - 2\sin^2 x$

For $\int \tan^2 x\,dx$ use $1 + \tan^2 x = \sec^2 x$

Examples

a $\int \cos^2 x\,dx = \frac{1}{2}\int(\cos 2x + 1)\,dx = \frac{1}{2}\left(\frac{1}{2}\sin 2x + x\right) + c$

b $\int \tan^2 x\,dx = \int(\sec^2 x - 1)\,dx = \tan x - x + c$

Type 4 By substitution

Example

$\int x(1+x)^3\,dx$ Let $u = 1 + x$

$\therefore \quad x = u - 1$

$\frac{dx}{du} = 1$

$$\int x(1+x)^3 \frac{dx}{du}\,du = \int (u-1)u^3\,du = \frac{u^5}{5} - \frac{u^4}{4} + c$$

$$= \frac{4u^5 - 5u^4}{20} + c$$

$$= \frac{(1+x)^4}{20}\,[4(1+x) - 5] + c$$

$$= \frac{(1+x)^4}{20}\,(4x - 1) + c$$

Type 5 Integration by parts

$$\int u\frac{dv}{dx}\,dx = uv - \int v\frac{du}{dx}\,dx$$

Example

$\int x\,e^x\,dx$ Let $u = x$ and $\frac{dv}{dx} = e^x$

Then $\frac{du}{dx} = 1$ and $v = e^x$

$$\therefore \int \underset{u}{x}\, \underset{\frac{dv}{dx}}{e^x}\,dx = \underset{u}{x}\, \underset{v}{e^x} - \int \underset{v}{e^x} \times \underset{\frac{du}{dx}}{1}\,dx$$

$$= x\,e^x - e^x + c$$

Type 6 Using partial fractions

Example

$$\int \frac{2+3x}{(1+x)(1+2x)}\,dx = \int \frac{1}{1+x} + \frac{1}{1+2x}\,dx \qquad \text{(working omitted)}$$

$$= \ln(1+x) + \frac{1}{2}\ln(1+2x) + c$$

EXERCISE 8

Section A Find the following.

1 $\int (x+2)^3\,dx$

2 $\int \cos 3x\,dx$

3 $\int 4e^x\,dx$

4 $\int \frac{1}{x}\,dx$

5 $\int \frac{1}{x^2}\,dx$

6 $\int \frac{2}{x^2-1}\,dx$

7 $\int x\cos x\,dx$

8 $\int \frac{5}{5x-1}\,dx$

9 $\int \sec^2 2x\,dx$

10 $\int \frac{2x}{x^2+a}\,dx$

11 $\int x(x+4)^2\,dx$

12 $\int x\,e^{5x}\,dx$

13 $\int \frac{\cos 3x}{\sin 3x}\,dx$

14 $\int e^{3x+2}\,dx$

15 $\int x^3\sqrt{x^4-1}\,dx$

16 $\int \frac{5}{x-7}\,dx$

17 $\int \frac{5x+7}{(x^2+3x+2)}\,dx$

18 $\int \cos^2 x\,dx$

19 $\int x\ln x\,dx$

20 $\int \frac{x+1}{x}\,dx$

21 $\int \frac{1}{1+x}\,dx$

22 $\int x^2 e^x\,dx$

23 $\int \sin(3-4x)\,dx$

24 $\int \frac{4}{\cos^2 x}\,dx$

25 $\int (e^x - e^{-x})^2$

26 $\int \frac{x^2+3x+1}{x}\,dx$

27 $\int \tan x\,dx$

28 $\int \tan^2 x\,dx$

29 $\int x(x^2+1)^3\,dx$

30 $\int x\sin 2x\,dx$

31 Integrate by using the given substitution.

a $\int 3x^2 e^{x^3}\,dx \qquad u = x^3$

b $\int \cos x\, e^{\sin x}\,dx \qquad u = \sin x$

c Show that $\int_0^1 \frac{1}{1+x^2}\,dx = \frac{\pi}{8} + \frac{1}{4} \qquad$ [Let $x = \tan u$]

d Evaluate $\int_2^{10} \frac{x}{\sqrt{2x+5}}\,dx \qquad u^2 = 2x + 5$

32 Find the exact values of the following:

a $\int_0^1 e^x\,dx$

b $\int_0^2 \frac{e^x}{2}\,dx$

c $\int_1^3 \frac{1}{x}\,dx$

d $\int_1^4 \frac{1}{2x}\,dx$

33 Find the exact values of the following:

a $\int_1^3 \frac{x+1}{x}\,dx$

b $\int_1^2 \left(\frac{x+2}{x}\right)^2 dx$

c $\int_3^6 \left(\frac{1}{x} + 3\right)^2 dx$

d $\int_1^{e^2} \left(\frac{e}{x} + 1\right)^2 dx$

34 Given that $\frac{dy}{dx} = \frac{1}{2x}$ and that $y = 3$ when $x = 1$, find y in terms of x.

35 Given that $\frac{dy}{dx} = \frac{5}{3x}$ and that $y = 2$ when $x = e$, find y in terms of x.

36 Given that $\frac{dy}{dx} = \frac{x^2+1}{3x}$ and that $y = 1$ when $x = 1$, find y in terms of x.

37 Given that $\frac{dy}{dx} = e^x + 1$ and that $y = e$ when $x = 1$, find y in terms of x.

38 a Express $\cos 2x$ in terms of $\sin x$.

b Hence find the exact value of $\int_{\frac{\pi}{4}}^{\frac{\pi}{2}} \sin^2 x\,dx$.

39 a Show that $\tan^2 x + 1 = \sec^2 x$.

b Use this to find $\int_{\frac{\pi}{4}}^{\frac{\pi}{3}} \tan^2 x\,dx$.

40 Find $\int x^2 \sin x\,dx$ by integrating by parts twice.

Section B

1 Calculate the exact values of following definite integrals:

a $\int_0^{\pi} \sin x \, dx$ **b** $\int_0^{\frac{\pi}{2}} \cos 2x \, dx$ **c** $\int_0^{\frac{\pi}{12}} \sin 3x \, dx$

d $\int_0^{\frac{\pi}{3}} \sec^2 x \, dx$ **e** $\int_0^{1} e^{2x+1} \, dx$ **f** $\int_0^{\infty} e^{-3x} \, dx$

g $\int_1^{2} (5x-1)^2 \, dx$ **h** $\int_4^{12} \sqrt{2x+1} \, dx$ **i** $\int_2^{10} \frac{1}{\sqrt{x-1}} \, dx$

Find the following indefinite integrals.

2 $\int \frac{3}{2x+5} \, dx$ **3** $\int 2x \, e^{x^2} \, dx$ **4** $\int \frac{\cos x}{4+\sin x} \, dx$

5 $\int \frac{3x-6}{x(x-3)} \, dx$ **6** $\int 3x \ln x \, dx$ **7** $x(1+x)^{10} \, dx$

8 $\int \ln x \, dx$ **9** $\int 3 \, e^{-3x} \, dx$ **10** $\int \sin\left(\frac{x}{2}\right) dx$

11 $\int \frac{x}{x+3} \, dx$ **12** $\int \cos x - \tan x \, dx$ **13** $\int \sin^2 2x \, dx$

14 $\int \sqrt{5x+1} \, dx$ **15** $\int \frac{1}{\sqrt{x-4}} \, dx$ **16** $\int 3x(2x-1) \, dx$

Find the exact values of the following definite integrals:

17 $\int_0^{1} e^{x} + e^{-x} \, dx$ **18** $\int_1^{2} (e^{x}+1)^2 \, dx$ **19** $\int_1^{3} \frac{(e^{x}+1)}{e^{x}} \, dx$

20 $\int_1^{3} (3x-2)^2 \, dx$ **21** $\int_0^{1} (5x+1)^3 \, dx$ **22** $\int_{-2}^{3} (4x-3)^4 \, dx$

23 $\int_1^{2} \frac{4}{(2x-1)^3} \, dx$ **24** $\int_0^{\frac{1}{3}} \frac{4}{(3x+2)^2} \, dx$ **25** $\int_{0.8}^{2} \frac{4}{(5x-2)} \, dx$

26 $\int_0^{1} x \, e^{3x} \, dx$ **27** $\int_0^{\pi} x \cos x \, dx$ **28** $\int_1^{2} x \ln x \, dx$

29 $\int_5^{8} x\sqrt{9-x} \, dx$ **30** $\int_1^{4} \frac{2}{x} \, dx$ **31** $\int_{\frac{\pi}{6}}^{\frac{\pi}{2}} \cot x \, dx$

32 **a** If $u^2 = x+1$ then show that $2u\frac{du}{dx} = 1$.

b Use the substitution $u^2 = x+1$ to find $\int_0^{3} \frac{x}{\sqrt{x+1}} \, dx$.

c Show that you get the same value for $\int_0^{3} \frac{x}{\sqrt{x+1}} \, dx$ when you use the substitution $u = x+1$.

33 Use the substitution $u = \ln x$ to find $\int_1^{e} \frac{\ln x}{x} \, dx$.

34 **a** Show that $1 + \cot^2 x = \operatorname{cosec}^2 x$.

b Use this to find $\int_{\frac{\pi}{4}}^{\frac{\pi}{3}} \cot^2 x \, dx$.

35 **a** Calculate $\frac{d}{dx}\left((1 - x^2)^{\frac{3}{2}}\right)$ and so find $\int x\sqrt{1 - x^2} \, dx$.

b Calculate $\int x^3\sqrt{1 - x^2} \, dx$ by considering $x^3\sqrt{1 - x^2} = x^2(x\sqrt{1 - x^2})$.

36 Show that $\int_1^5 \frac{6x + 11}{(2x + 3)(x + 2)} \, dx = \ln\left(\frac{1183}{75}\right)$.

37 Use the substitution $x = \sin u$ to show that $\int \frac{1}{\sqrt{1 - x^2}} \, dx = \sin^{-1} x + c$

38 Evaluate $\int_0^{\ln 4} \left(\frac{e^{2x} + e^x}{e^x} - 1\right) dx$

39 Find $\int \sec 2x \, dx$

10.9 Differential equations

A first order differential equation contains $\frac{dy}{dx}$.

Consider the differential equation $\frac{dy}{dx} = 2x + 1$.

Integrating both sides with respect to x,

$y = x^2 + x + c$, where c is a constant.

This is the *general solution* of the differential equation.

If we are given that $y = 10$, when $x = 2$

then $10 = 2^2 + 2 + c$

$c = 4$

and $y = x^2 + x + 4$

This is the *particular solution* of the differential equation.

EXERCISE 9

1 Find the general solution of each equation.

a $\frac{dy}{dx} = \frac{x + 1}{x}$

b $\frac{dy}{dx} = 3x^2 - 1$

c $\frac{dy}{dx} = e^x - 1$

d $\frac{dy}{dx} = \cos x$

e $\frac{dy}{dx} = \frac{e^x + 3}{e^x}$

f $\frac{dy}{dx} = \frac{\cos x}{\sin x}$

2 Solve the following differential equations:

a $\dfrac{dy}{dx} = (2x + 1)^2$, given that $y = 1$ when $x = 0$.

b $\dfrac{dy}{dx} = \sqrt{x - 1}$, given that $y = 2$ when $x = 2$.

c $\dfrac{dy}{dx} = \sin 2x$, given that $y = 1$ when $x = \pi$.

d $\dfrac{dy}{dx} = \dfrac{1}{2x - 1}$, given that $y = 1$ when $x = 1$.

Separating the variables

Consider the equation $\dfrac{dy}{dx} = x^2 y$

Write this as $\dfrac{1}{y}\dfrac{dy}{dx} = x^2$

Integrating both sides with respect to x,

$$\int \frac{1}{y}\frac{dy}{dx}\,dx = \int x^2\,dx$$

From the work in section 10.3, we can write

$$\int \frac{1}{y}\,dy = \int x^2\,dx$$

$$\therefore \ \ln y = \frac{x^3}{3} + c$$

Until now the symbols dx and dy have no meaning when written on their own. They have generally appeared as $\dfrac{dy}{dx}$ or perhaps $\int f(x)dx$. When solving differential equations it is helpful to think of dx and dy as 'factors' which can be separated and then writing integral signs.

Example 1

Find the general solution of the equation $x\dfrac{dy}{dx} = 1$

$$x\frac{dy}{dx} = 1$$

Separating the variables, $\int dy = \int \frac{1}{x}\,dx$

Integrating, $\quad y = \ln x + c$

Or you can write $c = \ln A$, where A is a constant,

so that $\quad y = \ln x + \ln A$

$\quad y = \ln Ax.$

Example 2

Find the general solution of the equation $x\frac{dy}{dx} = y$.

Separating the variables, $\int \frac{1}{y}\,dy = \int \frac{1}{x}\,dx$

$$\ln y = \ln x + c$$

or $$\ln y = \ln x + \ln A$$

$$\ln y = \ln Ax$$

$$y = Ax.$$

Example 3

Solve the equation $x^2\frac{dy}{dx} = y$, given that $y = 1$ when $x = \frac{1}{2}$.

Separating the variables, $\int \frac{1}{y}\,dy = \int \frac{1}{x^2}\,dx$

$$\ln y = -\frac{1}{x} + c$$

When $y = 1, x = \frac{1}{2}$ $\quad \ln 1 = \frac{-1}{\frac{1}{2}} + c$

$$0 = -2 + c$$

$$c = 2$$

$$\therefore \quad \ln y = -\frac{1}{x} + 2$$

or $$y = e^{2-\frac{1}{x}}$$

(This is the *particular* solution of the equation.)

Example 4

Solve the equation $\frac{dy}{dx} = y\sin x$, given that $y = 1$ when $x = 0$.

Separating the variables, $\int \frac{1}{y}\,dy = \int \sin x\,dx$

$$\ln y = -\cos x + c$$

When $y = 1, x = 0 \Rightarrow$ $\ln 1 = -\cos 0 + c$

$$0 = -1 + c$$

$$c = 1$$

$$\therefore \quad \ln y = -\cos x + 1$$

$$\ln y = 1 + \cos x$$

$$y = e^{1-\cos x}$$

EXERCISE 10

1 Find the general solutions of the following equations.

a $\dfrac{dy}{dx} = x^2 y$ b $\dfrac{dy}{dx} = \dfrac{x}{y}$ c $\dfrac{dy}{dx} = \dfrac{y}{x}$

d $x\dfrac{dy}{dx} = x + 1$ e $2y\dfrac{dy}{dx} = 3x^2$ f $\dfrac{dy}{dx} = \dfrac{e^{4x}}{y}$

g $\dfrac{dy}{dx} = e^x y^2$ h $\dfrac{dy}{dx} = \dfrac{\cos x}{y}$ i $\dfrac{dy}{dx} = \dfrac{1}{1+x}$

j $\dfrac{dy}{dx} = 2y$ k $\dfrac{dy}{dx} = 3x^2(3 + y)$ l $x\dfrac{dy}{dx} = \sec y$

m $\dfrac{dy}{dx} = 4x\,e^{-y}$ n $\dfrac{dy}{dx} - 2xy = 0$ o $\dfrac{dy}{dx} = y^2$

2 Solve the equation $x\dfrac{dy}{dx} = 2y$, given that $y = 4$ when $x = 1$.

Express y in terms of x.

3 Solve the following differential equations, given the initial conditions:

a $\dfrac{dx}{dt} = 3x$, given $x = 10$, when $t = 0$

b $\dfrac{dy}{dt} = -10y$, given $y = 1000$, when $t = 0$

c $\dfrac{dx}{dt} = x - 5$, given $x = 20$ when $t = 0$

d $\dfrac{dy}{dt} = y + 8$, given $y = 10$ when $t = 0$.

4 Solve the following differential equations, giving x or y as a function of t.

a $\dfrac{dx}{dt} = 5x$, where $x = 100$, when $t = 0$

b $\dfrac{dy}{dt} = -2y$, where $y = 1000$, when $t = 0$

c $\dfrac{dx}{dt} = 3x$, where $x = x_0$, when $t = 0$

d $\dfrac{dy}{dt} = -4y$, where $y = y_0$, when $t = 0$

e $\dfrac{dx}{dt} = kx$, where $x = x_0$, when $t = 0$

f $\dfrac{dy}{dt} = ky$, where $y = y_0$, when $t = 0$.

5 Find the solution of following differential equations.

a $\dfrac{dy}{dx} = x\,e^y$, given that $y = 0$ when $x = 2$

b $\dfrac{dy}{dx} = y \sin x$, given that $y = 1$ when $x = \dfrac{\pi}{2}$

c $\dfrac{dy}{dx} = 2xy + 5x$, given that $y = -1$ when $x = 0$

d $y\dfrac{dy}{dx} = \tan x$, given that $y = 1$ when $x = \dfrac{\pi}{4}$

e $\dfrac{dy}{dx} = \sin x \cos^2 y$, given that $y = \dfrac{\pi}{4}$ when $x = 0$

f $\dfrac{dy}{dx} = xy + x$, given that $y = 1$ when $x = 1$

6 **a** Solve the differential equation

$$\frac{dx}{dt} = \frac{20 - x}{10}$$

given that $x = 5$ at $t = 0$.

b Find the value of t for which $x = 10$, giving your answer to three significant figures.

7 Solve the differential equation $\dfrac{dy}{dx} = y^2\,e^{-2x}$, given that $y = 1$ when $x = 0$.

Give your answer in a form expressing y in terms of x.

8 **a** Express $\dfrac{1}{(1 + x)(2 + x)}$ in partial fractions.

b Hence find the solution of the differential equation

$$\frac{dy}{dx} = \frac{y}{(1 + x)(2 + x)}, \; x > -1,$$

given that $y = 1$ when $x = 2$.

Express your answer in the form $y = f(x)$.

9 Given x and y are positive, $\dfrac{dy}{dx} + \dfrac{y}{x^2} = 0$ and $y = e$ when $x = 1$, show that the solution may be written in the form $y = e^{\frac{1}{x}}$.

10 Show that the general solution of the differential equation

$(1 + x)(1 + 2x)\dfrac{dy}{dx} = x \tan y$ may be written in the form $\sin y = k\left(\dfrac{1 + x}{\sqrt{1 + 2x}}\right)$.

Formation of differential equations

If the rate of increase of n is proportional to n, $\frac{dn}{dt} = kn$, where k is a positive constant.

If the rate of decrease of n is proportional to n, $\frac{dn}{dt} = -kn$, where k is a positive constant.

Example 1

Set up and then solve a differential equation for a population P whose rate of growth is proportional to P at that time. The initial population is P_0 and the populations doubles after 10 years.

$$\frac{dP}{dt} \propto P \Rightarrow \frac{dP}{dt} = kP$$

Separating the variables, $\int \frac{1}{P} dP = \int k \, dt$

$$\ln P = kt + c$$

When $t = 0$, $P = P_0$.

$$\ln P_0 = 0 + c \qquad [1]$$

$\therefore$ In [1], $\ln P = kt + \ln P_0$

$$\ln P - \ln P_0 = kt$$

$$\ln\left(\frac{P}{P_0}\right) = kt$$

$$\frac{P}{P_0} = e^{kt} \text{ or } P = P_0 e^{kt}$$

Now, when $t = 10$, $P = 2P_0$

$$2P_0 = P_0 e^{k \times 10}$$

$$e^{10k} = 2$$

$$10k = \ln 2$$

$$k = \frac{1}{10} \ln 2 \approx 0.0693 \quad \text{[3 significant figures]}$$

Finally $P = P_0 e^{0.0693t}$

Example 2

A time t minutes the rate of change of temperature of an object as it cools is proportional to the temperature T °C of the object at that time.

a Given that $T = 60$ °C when $t = 0$, show that $T = 60 e^{-kt}$ where k is a positive constant.

b Given also that $T = 40$ °C, when $t = 5$ minutes, find the temperature of the object after 20 minutes.

a We have $\dfrac{dT}{dt} = -kT$

Separating the variables, $\int \frac{1}{T} dT = \int -k\, dt$

$$\ln T = -kt + c \qquad [1]$$

When $t = 0$, $T = 60$

$\therefore \quad \ln 60 = 0 + c \Rightarrow c = \ln 60$

In [1]: $\qquad \ln T = -kt + \ln 60$

$$\ln T - \ln 60 = -kt$$

$$\ln\left(\frac{T}{60}\right) = -kt$$

$$\frac{T}{60} = e^{-kt}$$

$$T = 60\, e^{-kt} \qquad [2]$$

b When $t = 5$, $T = 40$

In [2]: $\qquad 40 = 60\, e^{-k \times 5}$

$$e^{5k} = \frac{60}{40} = \frac{3}{2}$$

$$5k = \ln\left(\frac{3}{2}\right)$$

$$k = \frac{1}{5}\ln\left(\frac{3}{2}\right) \approx 0.081\,093\ldots$$

To find the temperature after 20 minutes we use $k = 0.081093$ and $t = 20$ in equation [2]

$T = 60\, e^{-0.081093 \times 20}$

$T = 11.9\ °C$ (correct to 3 significant figures).

After 20 minutes the temperature of the object is 11.9 °C.

EXERCISE 11

1 In a sample of radioactive material, the rate of decay (or decrease) of the number N of radioactive nuclei is proportional to the number of radioactive nuclei as shown by the differential equation

$\dfrac{dN}{dt} = -kN$ where k is a positive constant.

Solve this equation, given that $N = N_0$ at $t = 0$, to show that $N = N_0\, e^{-kt}$.

2 In a biological experiment, the rate of increase in the number of yeast cells, N, is equal to kN, where k is a positive constant. At time $t = 0$, the number of yeast cells $= N_0$.

a Showing all your working, set up a differential equation and solve it to show $N = N_0\, e^{kt}$.

b Sketch a graph to show the relation between N and t.

3 A body moves in a viscous medium, which causes the velocity v to decrease at a rate proportional to the velocity.

a By forming and then solving an appropriate differential equation, show that $v = v_0 e^{-kt}$ where $v = v_0$ when $t = 0$.

b Show the time taken for the body to decrease its speed to $\frac{1}{2}v_0$ is $\frac{1}{k}\ln 2$.

4 The population P of a country is increasing yearly at a rate proportional to the population at that time. Form and solve an appropriate differential equation. Given that the population at the beginning of the year 1960 was 50 million, and at the beginning of 1980 was 60 million, find the values of the constants in your equation.

5 In a biological experiment the rate of increase of the number of bacteria, N, is proportional to the number of bacteria present at time t minutes.

Initially the number of bacteria was 100, and the rate of increase was 5 per minute.

a Set up and solve an appropriate differential equation.

b Find the time for the number of bacteria to increase to 1000.

c Find the number of bacteria after 1 hour.

6 At time t minutes the rate of change of temperature T °C of a body as it cools is proportional to the temperature of the body at that time. The initial temperature of the body was 80 °C.

a Show that $T = 80\,e^{-kt}$, where k is a positive constant.

b Given also that $T = 60$ when $t = 5$, find the time taken for the temperature to fall to 40 °C.

c Find the value of T when $t = 10$.

7 A model to estimate the value V of a car assumes the rate of decrease of V at time t is proportional to V.

a Showing all your working, set up and solve a differential equation to show

$$V = V_0 e^{-kt}$$

where V_0 and k are positive constants. Given that the car cost £10 000 and that its value decreased to £6000 in two years, find

b the value of the car after one year

c the age of the car in years when its value was £1000.

8 The rate at which liquid leaks from a container is proportional to the volume of liquid in the container.

The initial volume of liquid was 20 litres and 5 hours later the volume had fallen to 10 litres. Form a differential equation and solve it to find the time for the volume to fall to 5 litres.

9 Water flows out of a tank such that the depth h metres of water in the tank falls at a rate which is proportional to the square root of h.

a Show that the general solution of the differential equation may be written as $h = (c - kt)^2$ where c and k are constants.

b Given that at time $t = 0$ the depth of water in the tank is 4 m and that 10 minutes later the depth of water has reduced to 1 m, find the time which it takes for the tank to empty.

10 Liquid is poured into a container at a constant rate of 20 cm^3 s^{-1}. After t seconds liquid is leaking from the container at a rate of $\frac{v}{10}$ cm^3 s^{-1}, where v cm^3 is the volume of liquid in the container at that time.

a Show that

$$-10\frac{dv}{dt} = v - 200$$

Given that $v = 500$ when $t = 0$

b Find a solution of the differential equation in the form $v = f(t)$.

c Find the limiting value of v as $t \to \infty$.

10.10 Numerical integration

The trapezium rule

Many functions cannot be integrated exactly. We can obtain an approximate answer by splitting the area under the curve into several trapeziums as illustrated below. When many trapeziums are used (say 100, 1000 …) a computer can be used to give an answer to a high degree of accuracy.

The trapezium rule in a general case

Consider a curve $y = f(x)$. The shaded area can be approximated using five trapeziums of width h.

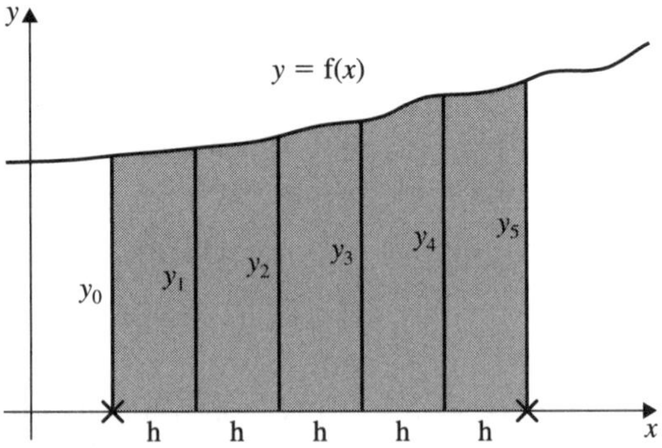

$$\text{Area} \simeq \frac{h}{2}(y_0 + y_1) + \frac{h}{2}(y_1 + y_2) + \frac{h}{2}(y_2 + y_3) + \frac{h}{2}(y_3 + y_4) + \frac{h}{2}(y_4 + y_5)$$

$$\simeq \frac{h}{2}[y_0 + 2(y_1 + y_2 + y_3 + y_4) + y_5]$$

↑ first ordinate; ↑ 'middle' ordinates; ↑ last ordinate

This is the trapezium rule which can shorten the working when several trapeziums are used.

Example 1

a Find an approximate value for the area under the curve $y = e^{\frac{x}{2}}$ between $x = 0$ and $x = 4$.

b Calculate the *exact* value for the area.

c Calculate the percentage error associated with using the trapezium rule in this case.

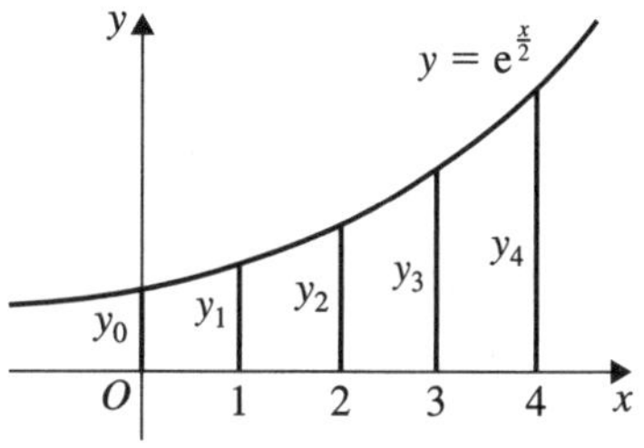

a Calculate the values of y at $x = 0, 1, 2, 3, 4$

x	0	1	2	3	4
y	1	1.649	2.718	4.482	7.389

Using the trapezium rule, with $h = 1$,

$$\text{Area} = \tfrac{1}{2}[1 + 2(1.649 + 2.718 + 4.482) + 7.389]$$

$$= 13.0 \quad \text{correct to three significant figures.}$$

b $\int_0^4 e^{\frac{x}{2}}\,dx = \left[2e^{\frac{x}{2}}\right]_0^4 = 2e^2 - 2e^0$

The exact value for the area is $2e^2 - 2$ (= 12.778 11...)

c $\text{Percentage error} = \dfrac{\text{error}}{\text{exact value}} \times 100\%$

$$= \frac{13.0 - (2e^2 - 2)}{(2e^2 - 2)} \times 100$$

$$= 1.74\%$$

EXERCISE 12

1 **a** Find an approximate value for the area under the curve $y = x^2$ between $x = 1$ and $x = 4$.
(Divide the area into three trapeziums as shown.)

b State, with a reason, whether your estimate is above or below the actual value for the area.

c Confirm your result by working out $\int_1^4 x^2\,dx$.

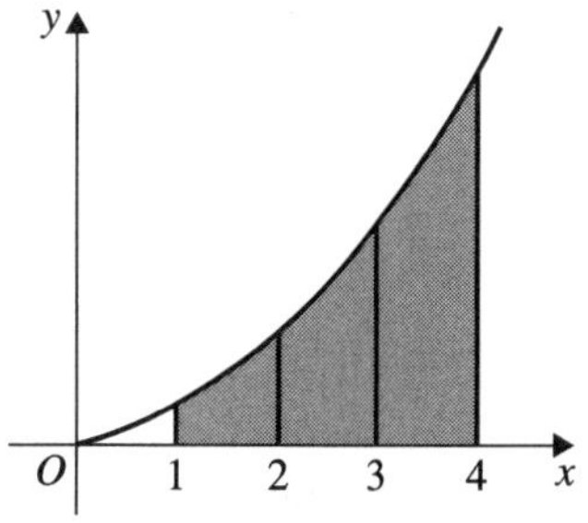

2 **a** Using the trapezium rule, find an approximate value for the following integral, using four equally spaced ordinates.

$$\int_1^4 \frac{4}{x}\,dx$$

b Work out the exact value of the integral and the error (difference between the actual value and the approximate value).

3 **a** Use the trapezium rule, with four trapeziums of width one unit, to evaluate approximately

$$\int_0^4 e^x\,dx.$$

b Evaluate the integral exactly and find the percentage error involved in using the trapezium rule.

$$\left(\text{Percentage error} = \frac{\text{error}}{\text{actual value}} \times 100\%\right)$$

4 Using the number of trapeziums indicated, find an approximate value for

$$\int_0^{\frac{\pi}{2}} \sin x\,dx.$$

a Using three trapeziums, width $\frac{\pi}{6}$.

b Using six trapeziums, width $\frac{\pi}{12}$.

c Work out the exact value for the integral and find the percentage error in each case.

5 Use the trapezium rule, with three trapeziums of equal width, to find an estimate for:

a $\int_0^{0.6} x^2\,dx$

b $\int_0^{\frac{\pi}{2}} \ln(1 + \cos x)\,dx$

c $\int_0^{\frac{\pi}{2}} \sin^2 x\,dx$

6 Using the trapezium rule with five equally spaced ordinates, find an approximate value for:

a $\int_{-1}^{1} e^{-2x}\,dx$

b $\int_{-\frac{\pi}{3}}^{\frac{\pi}{3}} \sec x\,dx$ (x in radians)

c $\int_0^2 (1 + \sin x)\,dx$

7 In the statistics of the Normal Distribution, it can be shown that approximately 95% of the distribution lies within ± 2 standard deviations of the mean.

This is given by the integral

$$\int_{-2}^{2} \frac{1}{\sqrt{2\pi}} e^{-\frac{x^2}{2}}\,dx \approx 0.95$$

Use the trapezium rule with four trapeziums to estimate the value of the integral.

Simpson's rule

Simpson's rule provides a more accurate approximation for the value of a definite integral. With Simpson's rule the top of each strip is replaced by a quadratic curve rather than a straight line as in the trapezium rule.

With Simpson's rule we use an even number of strips so that there is an odd number of ordinates (y values).

By Simpson's rule,

$$\int_a^b \mathrm{f}(x) \approx \frac{1}{3}h\left[y_0 + 4y_1 + 2y_2 + 4y_3 + 2y_4 + \ldots + 2y_{n-2} + 4y_{n-1} + y_n\right]$$

Example 1

The y values for a function $y = \mathrm{f}(x)$ are given below.

Use Simpson's rule with 7 ordinates to find an approximate value for $\int_1^7 \mathrm{f}(x)\,\mathrm{d}x$.

x	1	2	3	4	5	6	7
y	3	4	6	7	5	2	−1

We have $h = 1$, $y_0 = 3$, $y_1 = 4$ and so on.

$$\int_1^7 \mathrm{f}(x)\,\mathrm{d}x \approx \frac{1}{3}\left[3 + 4(4) + 2(6) + 4(7) + 2(5) + 4(2) + (-1)\right]$$

$$\approx \frac{76}{3} = 25.3 \text{ to three significant figures}$$

Exercise 13

Give your answers correct to four significant figures.

1 The tables give corresponding values for two functions $y = \mathrm{f}(x)$. In each case use Simpson's rule to find an approximate value for $\int_2^6 \mathrm{f}(x)\,\mathrm{d}x$.

a

x	2	3	4	5	6
y	1	4	5	6	3

b

x	2	3	4	5	6
y	10	12	16	27	20

2 The table gives corresponding values of x and y for a function $y = g(x)$. Use Simpson's rule to find an approximate value for $\int_0^{12} g(x)\mathrm{d}x$. [Note that '$h$' = 2]

x	0	2	4	6	8	10	12
y	−4	−1	0	2	5	7	12

3 In the last exercise we obtained an approximate value of 57.99 using the trapezium rule with 5 ordinates for $\int_0^4 e^x \, dx$.

a Use Simpson's rule with 5 ordinates to obtain a more accurate value for the same integral.

b Compare the two answers above with the exact answer which is $e^4 - 1$ ($= 53.598\ldots$)

4 Use Simpson's rule with 5 ordinates to find approximate values for each of the following.

a $\int_0^4 \frac{x}{1+x} \, dx$ **b** $\int_1^5 \frac{1}{1+x^2} \, dx$ **c** $\int_0^4 e^{x^2} \, dx$

5 Use Simpson's rule with 7 ordinates to find an approximate value for $\int_0^3 \sin\sqrt{x} \, dx$. [The limits of integration are in radians.]

Review Exercise 10

1 Calculate the following indefinite integrals:

a $\int \sin x \, dx$ **b** $\int \cos 2x \, dx$ **c** $\int \sin\frac{x}{4} \, dx$

d $\int \sec^2 x \, dx$ **e** $\int e^{7x} \, dx$ **f** $\int e^{-3x} \, dx$

g $\int \cos(5x + 4) \, dx$ **h** $\int (2x + 3)^3 \, dx$ **i** $\int (1 + 4x)^{-2} \, dx$

j $\int \frac{1}{x} \, dx$ **k** $\int \frac{5}{x} \, dx$ **l** $\int \frac{3}{2x+9} \, dx$

m $\int \frac{5}{x-7} \, dx$ **n** $\int \cos x + \sin x \, dx$ **o** $\int \sqrt{6x - 1} \, dx$

2 Show that $\int_{a-h}^{a} (a^2 - x^2) \, dx = \frac{h^2}{3}(3a - h)$

3 Evaluate the following.

a $\int_{\frac{\pi}{6}}^{\frac{\pi}{4}} \cos 2x \, dx$ **b** $\int_1^3 \frac{1}{x+1} \, dx$ **c** $\int_0^5 \sqrt{(3x + 1)} \, dx$

4 A curve has equation $y = 6x - e^{3x}$.

a Show that the x-coordinate of the stationary point on the curve is $\frac{1}{3}\ln 2$. Find the corresponding y-coordinate in the form $a \ln 2 + b$ where a and b are integers to be determined.

b Find an expression for $\frac{d^2y}{dx^2}$ and hence determine the nature of the stationary point.

c Show that the area of the region enclosed by the curve, the x-axis and the lines $x = 0$ and $x = 1$ is $\frac{1}{3}(10 - e^3)$.

5 a Find $\int \frac{2}{x+1} + \frac{3}{x+2}\,dx$

b Use part a to find $\int \frac{10x + 14}{x^2 + 3x + 2}\,dx$

6 Calculate the following, using natural logarithms:

a $\int \frac{3x^2}{x^3 + 5}\,dx$ **b** $\int \frac{2x}{4x^2 - 7}\,dx$ **c** $\int \frac{e^{2x}}{1 + e^{2x}}\,dx$

d $\int \frac{\sin x}{\cos x}\,dx$ **e** $\int \frac{f'(x)}{f(x)}\,dx$ **f** $\int \frac{\cos 2x}{2 \sin x \cos x}\,dx$

7 Calculate the following definite integrals by using the given substitution:

a $\int_0^1 x(2x + 1)^3\,dx$ $u = 2x + 1$

b $\int_2^3 \frac{2x + 1}{(x + 1)^3}\,dx$ $u = x + 1$

c $\int_{\frac{1}{3}}^{\frac{2}{3}} (x + 1)(3x - 1)^2\,dx$ $u = 3x - 1$

d $\int_{-1}^{1} (x - 1)^8 x\,dx$ $u = x - 1$

e $\int_1^5 \frac{3x + 1}{\sqrt{(2x - 1)}}\,dx$ $u^2 = 2x - 1$

f $\int_{\frac{\pi}{2}}^{\pi} \cos x\, e^{\sin x}\,dx$ $u = \sin x$

g $\int_0^2 3x^2 e^{x^3}\,dx$ $u = x^3$

h $\int_0^1 \frac{x}{(x^2 + 1)^3}\,dx$ $u = x^2 + 1$

8 Evaluate $\int_0^1 x\sqrt{4 - 3x^2}\,dx$. Give your answer as a fraction.

9 Find the exact value of $\int_{-1}^{2} x\sqrt{x + 2}\,dx$

10 Use the substitution $u = \tan x$ to show that

$$\int_0^{\frac{\pi}{6}} \sec^4 x \tan x\,dx = \frac{7}{36}$$

11 Calculate the following indefinite integrals:

a $\int xe^{2x}\,dx$ **b** $\int x \sin x\,dx$

c $\int 3x \ln x\,dx$ **d** $\int x^2 \ln x\,dx$

12 Evaluate $\int_0^2 x\,e^x\,dx$

13 Evaluate $\int_0^2 x\,e^{x^2}\,dx$

14 Show that $\int_0^{\frac{\pi}{8}} x \sin 2x\,dx = \dfrac{4 - \pi}{16\sqrt{2}}$

15 **a** Express $\cos 2x$ in terms of $\cos x$.

b Use this to show that $\cos^2 x = \frac{1}{2}(1 + \cos 2x)$.

c Hence find the exact value of $\int_0^{\frac{\pi}{2}} \cos^2 x\,dx$.

16 Find

a $\int \sin^2 x\,dx$ **b** $\int \tan^2 x\,dx$

17 **a** Find the area enclosed by the curve $y = \sqrt{x}$, the x-axis and the lines $x = 1$ and $x = 4$.

b Find the volume generated when this area is rotated through one revolution about the x-axis.

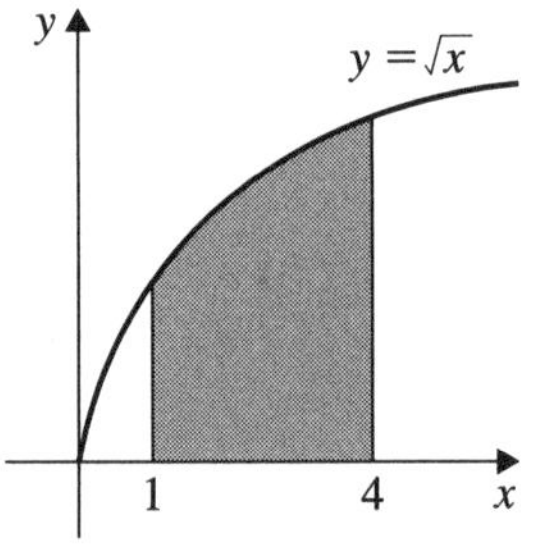

18 **a** Find, in terms of r and h, the volume of revolution when the area enclosed between the line $y = \dfrac{r}{h}x$, the x-axis and the line $x = h$ is rotated through 360° about the x-axis.

b What is this shape?

19 The area bounded by the curve $y = 2\cos x$, the x-axis and the lines $x = 0$ and $x = \dfrac{\pi}{4}$ is rotated through one revolution about the x-axis. Find the volume generated.

20 The diagram shows part of the curve $y = \dfrac{2}{x}$ and part of the line $y = 3 - x$ intersecting at points A and B.

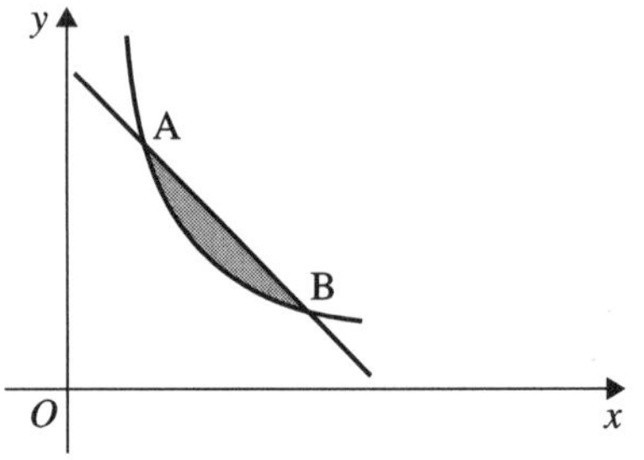

Find

a the coordinates of A and B,

b the area of the shaded region,

c the volume generated when the shaded region is rotated through 360° about the x-axis.

21 The graph shows the curve

$$y = \sqrt{x \cos x},\ 0 \leqslant x \leqslant \frac{\pi}{2}.$$

The maximum point on the curve is A.

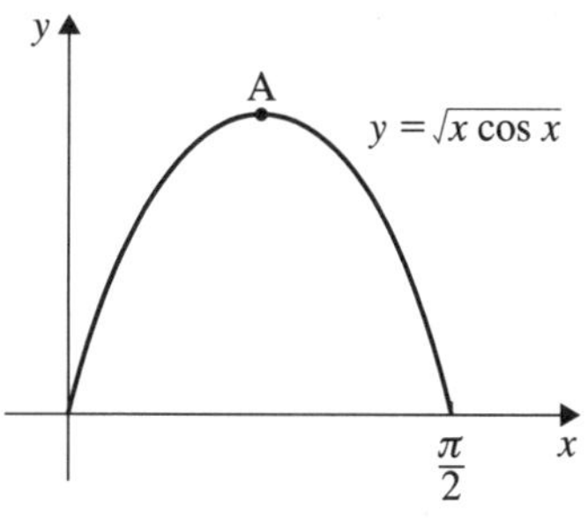

a Show that the x-coordinate of the point A satisfies $x = \cot x$

The finite region enclosed by the curve and the x-axis is rotated through 2π radians about the x-axis.

b Find the exact value of this volume.

22 Solve the differential equations.

a $\dfrac{dy}{dx} = y^{-\frac{1}{2}}$, given $y = 9$ when $x = 4$

b $\dfrac{dy}{dx} = \dfrac{20}{y}$, given $y = 2$ when $x = 0$

c $\dfrac{dy}{dx} = 2xy$, given $y = 1$ when $x = 2$

d $\dfrac{dy}{dx} = xy + x$, given $y = 1$ when $x = 1$

e $x\dfrac{dy}{dx} = x + 1$, given $y = 3$ when $x = 1$

f $y \cos^2 x \dfrac{dy}{dx} = y^2 + 1$, given $y = 2$ when $x = 0$

23 If $\dfrac{dx}{dt} = k(a - x)^2$ and $x = 0$ when $t = 0$, where k is a constant, show that

$$t = \frac{x}{ka(a - x)}.$$

24 The gradient of a curve at any point (x, y) on the curve is directly proportional to the product of x and y.
The curve passes through the point $(1, 1)$ and the gradient at this point is 4.
Form a differential equation and solve this equation to express y in terms of x.

25 During the initial stages some micro-organisms are growing at a rate $\left(\dfrac{dN}{dt}\right)$ proportional to the number (N) of micro-organisms present.

a Set up and solve a differential equation relating $\dfrac{dN}{dt}$ with N.

b Showing all your working show that

$$N = N_0 e^{kt},$$

where N_0 = the number of micro-organisms at $t = 0$ and k is a constant of proportionality. (P.T.O.)

c Given the number of micro-organisms increases by 50% in 10 hours, find k.

d Find the time for the number of micro-organisms to double.

26 Calculate the exact area enclosed between $y = 15 - \frac{50}{x}$ and $y = x$.

27 Show that $\int x(2x - 1)^7 \, dx = \frac{(2x - 1)^8(16x + 1)}{288} + c$

28 Find the following integrals

a $\int \sec x \tan x \, dx$ **b** $\int -\operatorname{cosec}^2 3x \, dx$ **c** $\int \ln 2x \, dx$

d $\int x^2 \cos x \, dx$ **e** $\int \frac{x - 1}{(x + 2)^2} \, dx$

29 Using the trapezium rule, find an approximate value for the following using four trapeziums in each case.

a $\int_0^4 e^{x^2} \, dx$ **b** $\int_2^6 \ln x \, dx$ **c** $\int_0^{\frac{\pi}{2}} \sin 2x \, dx$

EXAMINATION EXERCISE 10

1 Show that $\int_3^{12} \frac{2}{x} \, dx = \ln 16.$ [OCR]

2 Evaluate

$$\int_0^1 \left(e^{2x} + x^{\frac{1}{2}} \right) dx,$$

giving your answer in the form $pe^2 + q$, where p and q are rational numbers. [AQA]

3 Prove that

$$\int_A^{A^3} \frac{1}{x - 1} \, dx = \ln(A^2 + A + 1)$$

for values of the constant A such that $A > 1$. [OCR]

4 Show that the area of the region bounded by the curve $y = e^{-2x} + \frac{3}{x} + 3$, the x-axis, and the lines $x = 1$ and $x = 2$ is

$$\frac{e^2 - 1}{2e^4} + 3(\ln 2 + 1)$$ [AQA]

5 Fig. 4 shows the graph of $f(x) = \dfrac{x}{x-1}$.

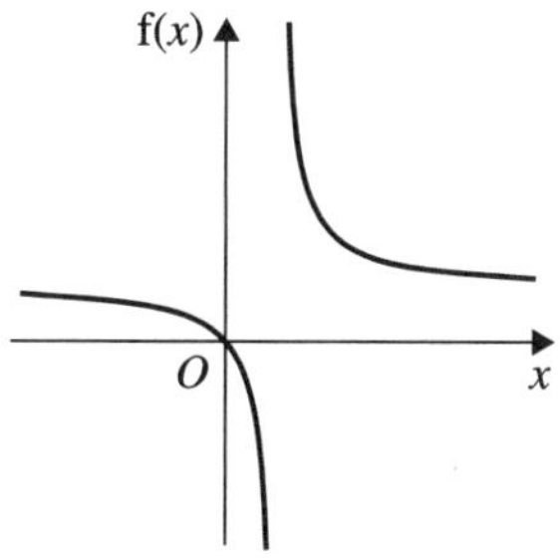

i State the value of x for which the function $f(x)$ is undefined.

ii Find the derivative $f'(x)$, simplifying your answer. How can you deduce that the gradient of the graph is always negative?

The area enclosed by the graph, the x-axis and the lines $x = 2$ and $x = 3$ is denoted by A.

iii Using the substitution $u = x - 1$, show that

$$A = \int_1^2 \left(1 + \frac{1}{u}\right) du.$$

Find the exact value of A.

iv Show algebraically that $f^{-1}(x) = f(x)$. What feature of the graph illustrates this result? [MEI]

6 Find $\int x \sec^2 x \, dx$. [OCR]

7 The function f is defined by

$$f(x) = \frac{2x}{(1+2x)^3}, \quad x \neq -\frac{1}{2}.$$

Using the substitution $u = 1 + 2x$, or otherwise, find $\int f(x)\, dx$. [AQA]

8 The function f is defined for $x \geqslant 0$ by

$f(x) = x^{\frac{1}{2}} + 2.$

a i Find $f'(x)$.

ii Hence find the gradient of the curve $y = f(x)$ at the point for which $x = 4$.

b i Find $\int f(x)\, dx$.

ii Hence show that $\int_0^4 f(x)\, dx = \dfrac{40}{3}$.

c Show that $f^{-1}(x) = (x - 2)^2$. (P.T.O.)

d The diagram shows a symmetrical shaded region A bounded by:

parts of the coordinate axes;

the curve $y = f(x)$ for $0 \leqslant x \leqslant 4$; and

the curve $y = f^{-1}(x)$ for $2 \leqslant x \leqslant 4$.

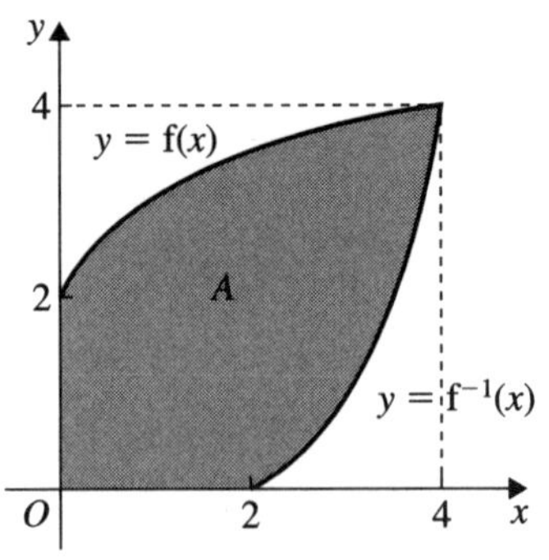

i Write down the equation of the line of symmetry of A.

ii Calculate the area of A. [AQA]

9 Use the substitution $u = \tan x$ to find the value of

$$\int_0^{\frac{1}{6}\pi} \frac{1 + \tan^2 x}{1 - \tan^2 x}\,dx,$$

giving your answer in logarithmic form. [OCR]

10 Use the substitution $u = 1 + \sin x$ and integration to show that

$$\int \sin x \cos x(1 + \sin x)^5\,dx = \frac{1}{42}(1 + \sin x)^6[6 \sin x - 1] + \text{constant}.$$

[EDEXCEL]

11 **a** Use the identity for $\cos(A + B)$ to prove that $\cos 2A = 2\cos^2 A - 1$.

b Use the substitution $x = 2\sqrt{2}\sin\theta$ to prove that

$$\int_2^{\sqrt{6}} \sqrt{(8 - x^2)}\,dx = \frac{1}{3}(\pi + 3\sqrt{3} - 6).$$

[EDEXCEL]

12 **a** Show that

$$\int_0^6 \frac{1}{2 + u}\,du = \ln n,$$

where n is an integer to be found.

b Use the substitution $x = u^2$ to show that

$$\int_0^{36} \frac{1}{\sqrt{x}(2 + \sqrt{x})}\,dx = \ln m,$$

where m is an integer to be found. [AQA]

13 **i** Use the derivative of $\cos x$ to prove that

$$\frac{d}{dx}(\sec x) = \sec x \tan x.$$

ii Use the substitution $u = \sec x$ to find the exact value of

$$\int_0^{\frac{1}{3}\pi} \sec^3 x \tan^3 x\,dx.$$

[OCR]

14 Use integration by parts to find

$$\int_0^{\frac{1}{2}} x\,e^{2x}\,dx.$$

[AQA]

15 a i Find $\int \sin 2x \, dx$.

ii Hence or otherwise show that

$$\int_0^{\frac{1}{3}\pi} x \sin x \cos x \, dx = \frac{2\pi + 3\sqrt{3}}{48}.$$

[MEI]

16 The diagram shows the curve with equation

$$y = 4e^{2x} - 18x.$$

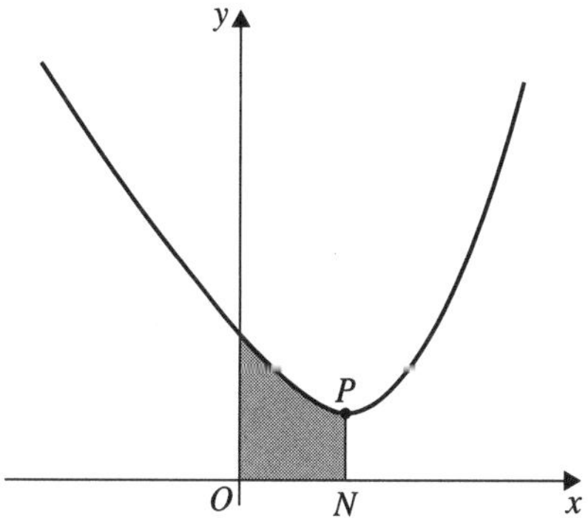

The minimum point of the curve is P. N is the point on the x-axis such that NP is parallel to the y-axis. The region shaded in the diagram is bounded by the curve and the lines $y = 0$, $x = 0$ and NP.

Show that the exact area of the shaded region can be written in the form

$$a - b(\ln \tfrac{3}{2})^2,$$

where the values of the constants a and b are to be stated. [OCR]

17 i Show that $\dfrac{2u^2}{u^2-1} = 2 + \dfrac{2}{u^2-1}$. Hence express $\dfrac{2u^2}{u^2-1}$ in partial fractions.

ii Using the substitution $u = \sqrt{x}$, show that $\displaystyle\int_4^9 \frac{\sqrt{x}}{x-1} \, dx = \int_2^3 \frac{2u^2}{u^2-1} \, du$.

Deduce that $\displaystyle\int_4^9 \frac{\sqrt{x}}{x-1} \, dx = \ln 3 - \ln 2 + 2$.

iii Use integration by parts, and the result of part (**ii**), to show that

$$\int_4^9 \frac{\ln(x-1)}{\sqrt{x}} \, dx = 20 \ln 2 - 6 \ln 3 - 4.$$

[MEI]

18 A function $f(x)$ is defined for $0 \leqslant x < 1$ by

$$f(x) = \frac{1+x}{1-x^3}.$$

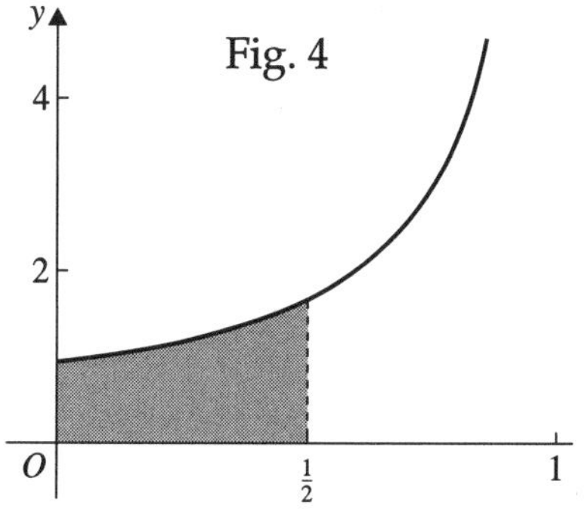

Fig. 4 shows the graph of $y = f(x)$. The area of the shaded region enclosed by the graph of $y = f(x)$, the y-axis, the x-axis and the line $x = \frac{1}{2}$ is denoted by A.

i Find the first three non-zero terms in the expansion of $\dfrac{1}{1-x^3}$ for $|x| < 1$. Deduce that, for small positive values of x, $f(x) \approx 1 + x + x^3 + x^4$.

(P.T.O.)

ii Find an estimate of the area A by calculating $\int_0^{\frac{1}{2}} (1 + x + x^3 + x^4)\,dx$, giving your answer to 3 significant figures.

iii Show that, for $0 \leqslant x < 1$,

$$\frac{2}{1-x} + \frac{1+2x}{1+x+x^2} = 3f(x).$$

iv Differentiate $\ln(1 + x + x^2)$ with respect to x. Using your answer, and the result in part **iii**, show that the exact value of A is $\frac{1}{3}\ln 7$.

v Explain why the approximation to A found in part **ii** must be an underestimate. [MEI]

19 The graph below shows the region R enclosed by the curve $y = x - \dfrac{1}{x}$, the x-axis and the line $x = 2$.

Find the exact volume of the solid formed when the region R is rotated through 2π radians about the x-axis. [AQA]

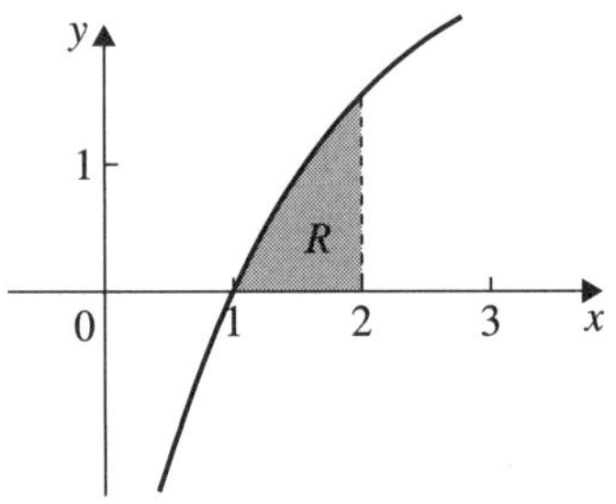

20 The diagram shows the curve $y = \dfrac{1}{\sqrt{(4x-1)}}$

The region R (shaded in the diagram) is enclosed by part of the curve and by the lines $x = 1$, $x = 2$ and $y = 0$.

i Find the exact area of R.

ii The region R is rotated through four right angles about the x-axis.
Find the exact volume of the solid formed. [OCR]

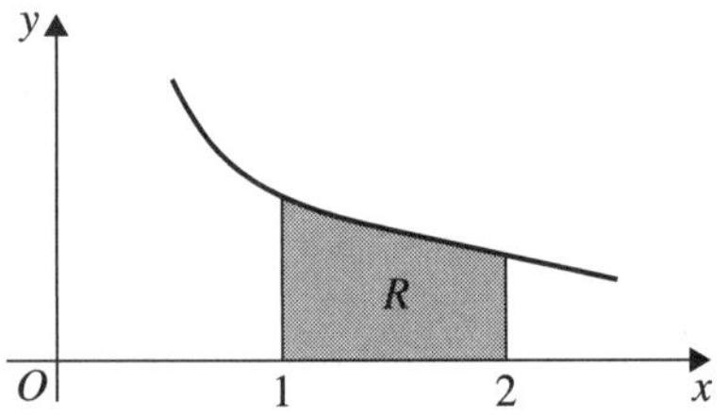

21 Fig. 3 shows part of the curve $y = x - 2\sin x$. P is a minimum point on the curve.

i Show that the x-coordinate of P is $\dfrac{\pi}{3}$.

ii Evaluate $\int_0^{\frac{1}{2}\pi} x \sin x\,dx$.

iii Show that $\int_0^{\frac{1}{2}\pi} \sin^2 x\,dx = \dfrac{\pi}{4}$.

Fig. 3

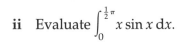

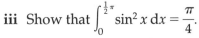

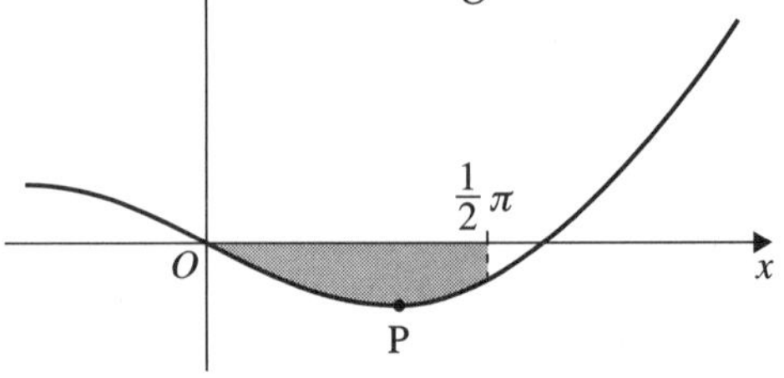

iv The shaded region bounded by the curve, the x-axis and the line $x = \frac{1}{2}\pi$ is rotated through 360° about the x-axis. Using your results from parts **(ii)** and **(iii)**, find the volume of the solid of revolution formed, giving your answer in terms of π. [MEI]

22 a Use integration by parts to show that

$$\int_0^{\frac{\pi}{4}} x\sec^2 x\,\mathrm{d}x = \tfrac{1}{4}\pi - \tfrac{1}{2}\ln 2.$$

The finite region R, bounded by the equation $y = x^{\frac{1}{2}}\sec x$, the line $x = \dfrac{\pi}{4}$ and the x-axis is shown in Fig. 1. The region R is rotated through 2π radians about the x-axis.

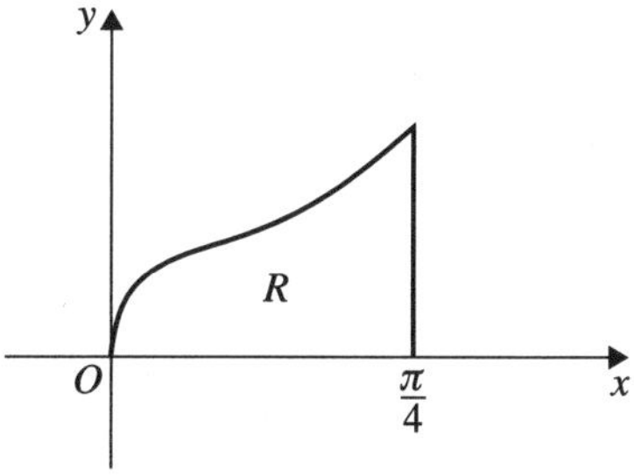

b Find the volume of the solid of revolution generated. [EDEXCEL]

23 Figure 2 shows the curve with equation $y = x^{\frac{1}{2}}\mathrm{e}^{-2x}$.

Fig. 2

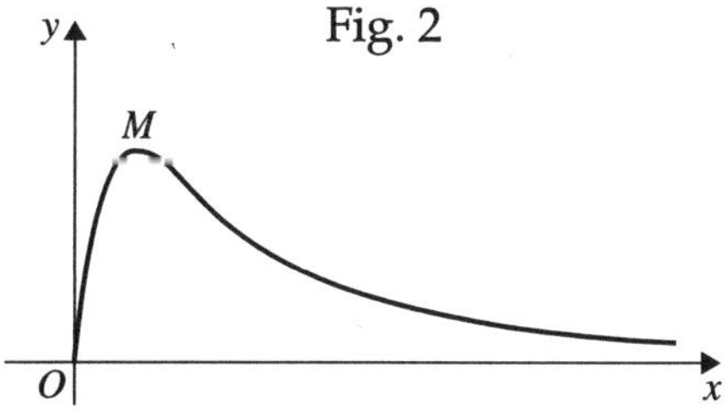

a Find the x-coordinate of M, the maximum point of the curve.

The finite region enclosed by the curve, the x-axis and the line $x = 1$ is rotated through 2π about the x-axis.

b Find, in terms of π and e, the volume of the solid generated. [EDEXCEL]

24 Find the solution of the differential equation

$$\frac{\mathrm{d}y}{\mathrm{d}x} = \frac{x^4 - 1}{x^2y^2},$$

given that $y = 2$ when $x = 1$. [OCR]

25 a Solve the differential equation $\dfrac{\mathrm{d}y}{\mathrm{d}x} = \dfrac{1}{y^2}$ giving the general solution for y in terms of x.

b Find the particular solution of this differential equation for which $y = -1$ when $x = 1$. [AQA]

26 i Find $\int x^2 \sin x\,\mathrm{d}x$.

ii Find $\dfrac{\mathrm{d}}{\mathrm{d}x}(\tan x - x)$, giving your answer in its simplest form.

iii Find the general solution of the differential equation

$$\frac{\mathrm{d}y}{\mathrm{d}x} = x^2 \sin x \cot^2 y.$$ [OCR]

27 a Find $\int x\mathrm{e}^x\,\mathrm{d}x$. Hence show that the solution of the differential equation

$$\mathrm{e}^y\frac{\mathrm{d}y}{\mathrm{d}x} = -x\mathrm{e}^x,$$

for which $y = 0$ when $x = 0$, is $y = x + \ln(1 - x)$. [MEI]

28 Liquid is poured into a container at a constant rate of 30 $\text{cm}^3\,\text{s}^{-1}$. At time t seconds liquid is leaking from the container at a rate of $\frac{2}{15}V$ $\text{cm}^3\,\text{s}^{-1}$, where V cm^3 is the volume of liquid in the container at that time.

a Show that

$$-15\frac{\mathrm{d}V}{\mathrm{d}t} = 2V - 450.$$

Given that $V = 1000$ when $t = 0$,

b find the solution of the differential equation, in the form $\mathrm{V} = \mathrm{f}(t)$.

c find the limiting value of V as $t \to \infty$. [EDEXCEL]

29 A rectangular water tank rests on a horizontal surface. The tank is emptied in such a way that the depth of water decreases at a rate which is proportional to the square root of the depth of the water.

The depth of the water is h metres at time t hours. Initially there is a depth of 1 metre of water in the tank. The depth of water is 0.5 metres after 2 hours.

a i Write down a differential equation for h.

ii Hence show that $2\sqrt{h} = 2 - kt$, where k is a constant.

iii Find the value of k giving your answer to three decimal places.

b Find how long it will take to empty the tank completely, giving your answer to the nearest minute. [AQA]

30 A radioactive isotope decays in such a way that the rate of change of the number N of radioactive atoms present after t days, is proportional to N.

a Write down a differential equation relating N and t.

b Show that the general solution may be written as $N = A\mathrm{e}^{-kt}$, where A and k are positive constants.

Initially the number of radioactive atoms present is 7×10^{18} and 8 days later the number present is 3×10^{17}.

c Find the value of k.

d Find the number of radioactive atoms present after a further 8 days. [EDEXCEL]

31 The variables x and t are such that the rate of increase of $\ln x$ with respect to t is proportional to x.

i Express this statement as an equation, and hence show that the rate of increase of x with respect to t is proportional to x^2.

ii Find the general solution of the differential equation

$$\frac{\mathrm{d}x}{\mathrm{d}t} = kx^2,$$

where k is a constant, expressing x in terms of t in your answer.

iii Given that $x = 1$ when $t = 0$ and that $x = 2$ when $t = 1$, find the value of t near which x becomes very large. [OCR]

32 The following is a table of values for $y = \sqrt{(1 + \sin x)}$, where x is in radians.

x	0	0.5	1	1.5	2
y	1	1.216	p	1.413	q

a Find the value of p and the value of q.

b Use the trapezium rule and all the values of y in the completed table to obtain an estimate of I, where

$$I = \int_0^2 \sqrt{(1 + \sin x)}\, dx.$$

[EDEXCEL]

33 Use the trapezium rule with 3 intervals, each of width 1, to find an approximate value of

$$\int_2^5 x \ln x\, dx.$$

Give your answer correct to 3 significant figures. [OCR]

34 a Find $\int (4x + 3)^{10}\, dx$.

b Use the trapezium rule with two intervals to estimate the value of

$$\int_0^6 \ln (x^2 + 1)\, dx,$$

giving your answer correct to 3 significant figures. [OCR]

35 Use the trapezium rule with five ordinates (four strips) to find an approximation to

$$\int_0^2 e^{(1-x)^2}\, dx,$$

giving your answer to two decimal places. [AQA]

36 i Use the trapezium rule, with two intervals each of width 5, to show that

$$\int_5^{15} \frac{3}{x+5}\, dx \approx \frac{17}{8}.$$

ii Show that the exact value of $\int_5^{15} \frac{3}{x+5}\, dx$ is $\ln 8$.

iii Use a sketch graph to show how you can deduce from parts **i** and **ii** that

$$\ln 8 < \frac{17}{8}.$$

[OCR]

PART 11

Vectors

11.1 Vector rules

A vector quantity has both magnitude and direction. A translation is described by a vector and a vector can also represent physical quantities such as velocity, force, acceleration, etc. The symbol for a vector is shown in a text book by a bold letter, e.g. **a**, **x**. In your own work on paper you should show vectors by drawing a line underneath, e.g. $\underline{a}$ or $\underline{x}$.

On a coordinate grid, the magnitude and direction of the vector can be shown by a column vector, e.g. $\begin{pmatrix}2\\1\end{pmatrix}, \begin{pmatrix}1\\-3\end{pmatrix}$ where the upper number shows the distance in the x direction and the lower number shows the distance in the y direction.

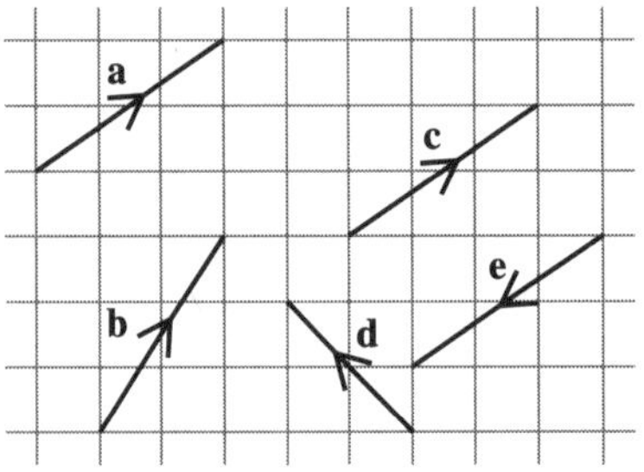

$\mathbf{a} = \begin{pmatrix}3\\2\end{pmatrix}$ $\mathbf{d} = \begin{pmatrix}-2\\2\end{pmatrix}$

$\mathbf{b} = \begin{pmatrix}2\\3\end{pmatrix}$ $\mathbf{e} = \begin{pmatrix}-3\\-2\end{pmatrix}$

$\mathbf{c} = \begin{pmatrix}3\\2\end{pmatrix}$

Equal vectors

Two vectors are equal if they have the same length *and* the same direction. The actual position of the vector on the diagram or in space is of no consequence.

Thus in the example, vectors **a** and **c** are equal because they have the same magnitude and direction. Even though vector **b** also has the same length (magnitude) as **a** and **c**, it is not equal to **a** or **c** because it acts in a different direction. Likewise, vector **e** has the same length as vector **c** but acts in the reverse direction and so cannot equal **c**.
Equal vectors have identical column vectors.

Addition of vectors

Vectors **a** and **b** are represented by the line segments shown below, they can be added by using the 'nose-to-tail' method to give a single equivalent vector.

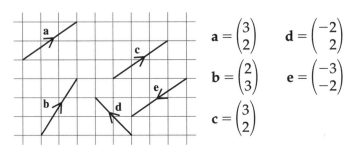

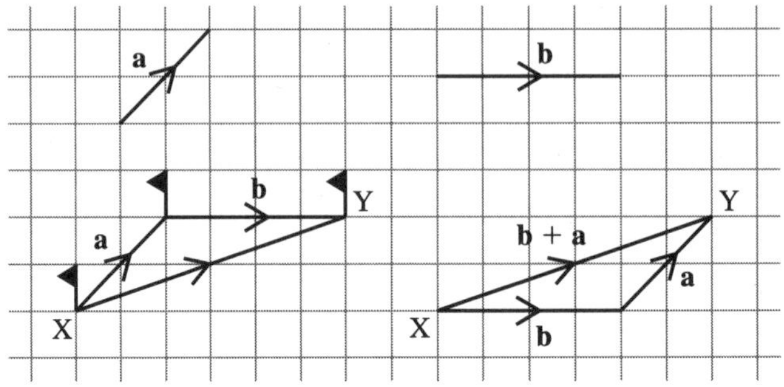

The 'tail' of vector **b** is joined to the 'nose' of vector **a**.
Alternatively the tail of **a** can be joined to the 'nose' of vector **b**.

In both cases the vector $\overrightarrow{XY}$ has the same length and direction and therefore $\mathbf{a} + \mathbf{b} = \mathbf{b} + \mathbf{a}$.

In the first diagram the flag is moved by translation **a** and then translation **b**. The translation $\mathbf{a} + \mathbf{b}$ is the equivalent or *resultant* translation.

Multiplication by a scalar

A scalar quantity has magnitude but no direction (e.g. mass, volume, temperature). Ordinary numbers are scalars.

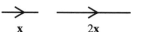

When vector **x** is multiplied by 2, the result is 2**x**.

When **x** is multiplied by −3 the result is −3**x**.

Note
1 The negative sign reverses the direction of the vector.
2 The result $\mathbf{a} - \mathbf{b}$ is $\mathbf{a} + -\mathbf{b}$.
So, subtracting **b** is equivalent to adding the negative of **b**.

Exercise 1

In Questions **1** and **2**, use the diagram below. $\overrightarrow{LM} = \mathbf{a}$, $\overrightarrow{LQ} = \mathbf{b}$.

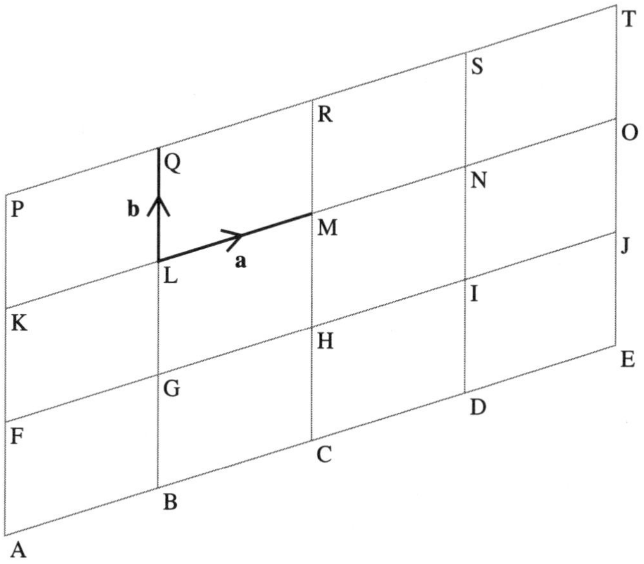

1 Write these vectors in terms of **a** and **b**.

a $\overrightarrow{GN}$ **b** $\overrightarrow{CO}$

c $\overrightarrow{TN}$ **d** $\overrightarrow{FT}$

e $\overrightarrow{KC}$ **f** $\overrightarrow{CJ}$

2 From your answers to Question **1**, find the vector which is:

a parallel to $\overrightarrow{LR}$

b 'opposite' to $\overrightarrow{LR}$

c parallel to $\overrightarrow{CJ}$ with twice the magnitude

d parallel to the vector $(\mathbf{a} - \mathbf{b})$.

In Questions **3** to **6**, write each vector in terms of **a**, **b**, or **a** and **b**.

3 **a** $\overrightarrow{BA}$
b $\overrightarrow{AC}$
c $\overrightarrow{DB}$
d $\overrightarrow{AD}$

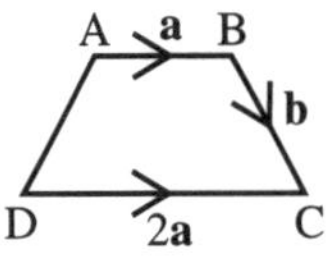

4 **a** $\overrightarrow{ZX}$
b $\overrightarrow{YW}$
c $\overrightarrow{XY}$
d $\overrightarrow{XZ}$

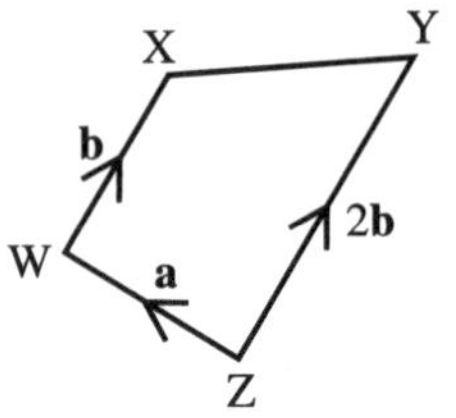

5 **a** $\overrightarrow{MK}$
b $\overrightarrow{NL}$
c $\overrightarrow{NK}$
d $\overrightarrow{KN}$

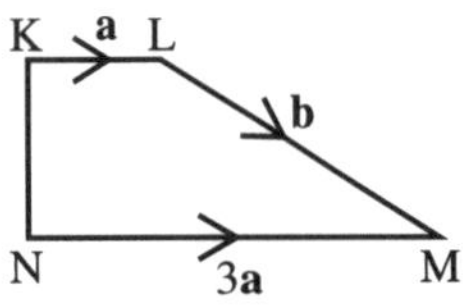

6 **a** $\overrightarrow{FE}$
b $\overrightarrow{BC}$
c $\overrightarrow{FC}$
d $\overrightarrow{DA}$

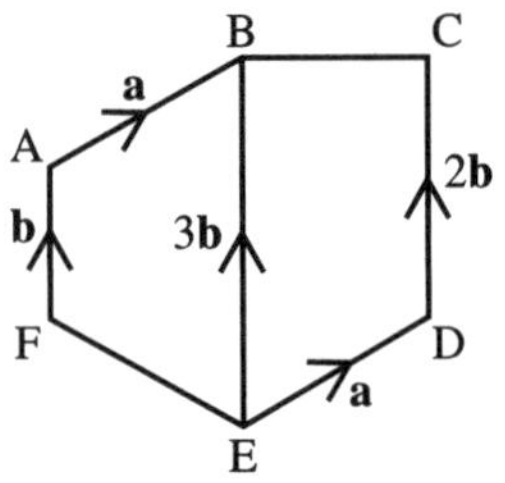

7 The points A, B and C lie on a straight line and the vector $\overrightarrow{AB}$ is $\mathbf{a} + 2\mathbf{b}$. Which of the following vectors is possible for $\overrightarrow{AC}$:

a $3\mathbf{a} + 6\mathbf{b}$ **b** $4\mathbf{a} + 4\mathbf{b}$ **c** $\mathbf{a} - 2\mathbf{b}$ **d** $5\mathbf{a} + 10\mathbf{b}$?

8 Find three pairs of parallel vectors from those below.

$\mathbf{a} + 3\mathbf{b}$	$\mathbf{a} - \mathbf{b}$	$6\mathbf{a} - 3\mathbf{b}$	$2\mathbf{a} + 6\mathbf{b}$	$3\mathbf{a} - 3\mathbf{b}$	$2\mathbf{a} - \mathbf{b}$	$\mathbf{a} + \mathbf{b}$
A	B	C	D	E	F	G

11.2 Vector geometry

Example 1

In the diagram, OA = AP and BQ = 3OB. N is the mid-point of PQ; $\overrightarrow{OA} = \mathbf{a}$ and $\overrightarrow{OB} = \mathbf{b}$.

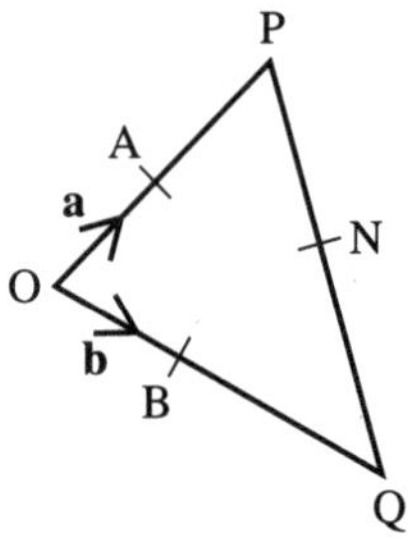

Express each of the following vectors in terms of **a**, **b**, or **a** and **b**.

a $\overrightarrow{AP}$ **b** $\overrightarrow{AB}$ **c** $\overrightarrow{OQ}$ **d** $\overrightarrow{PO}$

e $\overrightarrow{PQ}$ **f** $\overrightarrow{PN}$ **g** $\overrightarrow{ON}$ **h** $\overrightarrow{AN}$

a $\overrightarrow{AP} = \mathbf{a}$

b $\overrightarrow{AB} = -\mathbf{a} + \mathbf{b}$

c $\overrightarrow{OQ} = 4\,\mathbf{b}$

d $\overrightarrow{PO} = -2\,\mathbf{a}$

e $\overrightarrow{PQ} = \overrightarrow{PO} + \overrightarrow{OQ}$
$= 2\mathbf{a} + 4\mathbf{b}$

f $PN = \frac{1}{2}\overrightarrow{PQ}$
$= -\mathbf{a} + 2\mathbf{b}$

g $\overrightarrow{ON} = \overrightarrow{OP} + \overrightarrow{PN}$
$= 2\mathbf{a} + (-\mathbf{a} + 2\mathbf{b})$
$= \mathbf{a} + 2\mathbf{b}$

h $\overrightarrow{AN} = \overrightarrow{AP} + \overrightarrow{PN}$
$= \mathbf{a} + (-\mathbf{a} + 2\mathbf{b})$
$= 2\mathbf{b}$

EXERCISE 2

In Questions **1** and **2**, $\overrightarrow{OA} = \mathbf{a}$ and $\overrightarrow{OB} = \mathbf{b}$. Copy each diagram and use the information given to express the following vectors in terms of **a**, **b** or **a** and **b**.

a $\overrightarrow{AP}$ **b** $\overrightarrow{AB}$ **c** $\overrightarrow{OQ}$ **d** $\overrightarrow{PO}$ **e** $\overrightarrow{PQ}$

f $\overrightarrow{PN}$ **g** $\overrightarrow{ON}$ **h** $\overrightarrow{AN}$ **i** $\overrightarrow{BP}$ **j** $\overrightarrow{QA}$

1 A, B and N are mid-points of OP, OB and PQ respectively.

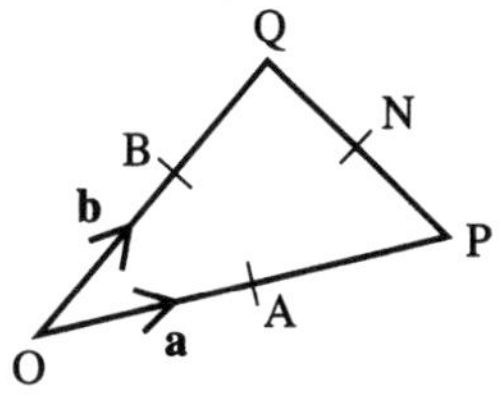

2 A and N are mid-points of OP and PQ; BQ = 2OB.

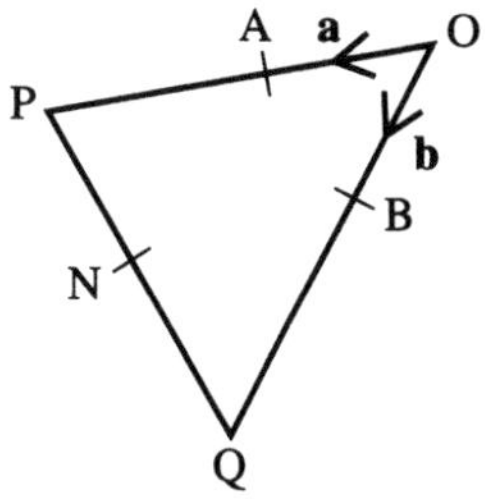

3 In $\triangle XYZ$, the mid-point of YZ is M. If $\overrightarrow{XY} = \mathbf{s}$ and $\overrightarrow{ZX} = \mathbf{t}$, find $\overrightarrow{XM}$ in terms of **s** and **t**.

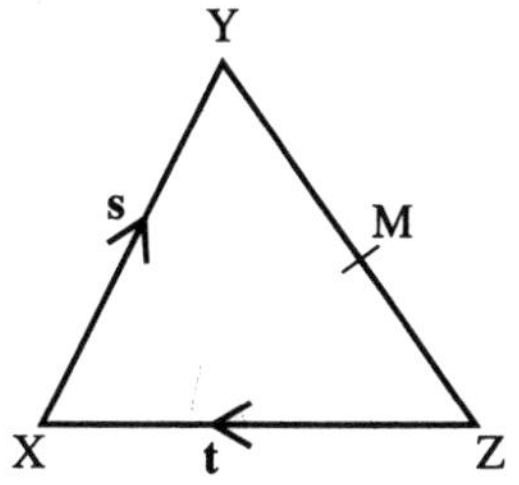

4 In $\triangle AOB$, AM : MB = 2 : 1. If $\overrightarrow{OA} = \mathbf{a}$, and $\overrightarrow{OB} = \mathbf{b}$ find $\overrightarrow{OM}$ in terms of **a** and **b**.

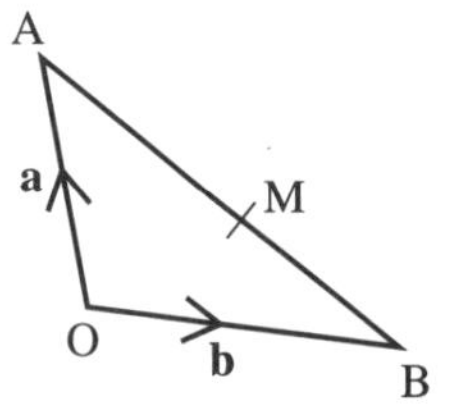

5 O is any point in the plane of the square ABCD. The vectors $\overrightarrow{OA}$, $\overrightarrow{OB}$, and $\overrightarrow{OC}$, are **a**, **b** and **c** respectively. Find the vector $\overrightarrow{OD}$, in terms of **a**, **b** and **c**.

6 ABCDEF is a regular hexagon with $\overrightarrow{AB}$, representing the vector **m** and $\overrightarrow{AF}$, representing the vector **n**. Find the vector representing $\overrightarrow{AD}$.

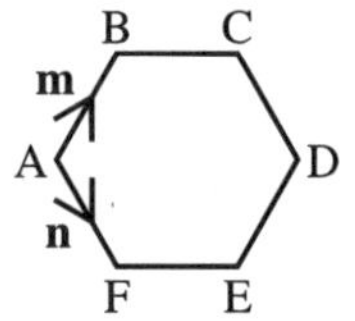

7 ABCDEF is a regular hexagon with centre O. $\overrightarrow{FA} = \mathbf{a}$ and $\overrightarrow{FB} = \mathbf{b}$.

Express the following vectors in terms of **a** and/or **b**.

a $\overrightarrow{AB}$ **b** $\overrightarrow{FO}$ **c** $\overrightarrow{FC}$

d $\overrightarrow{BC}$ **e** $\overrightarrow{AO}$ **f** $\overrightarrow{FD}$

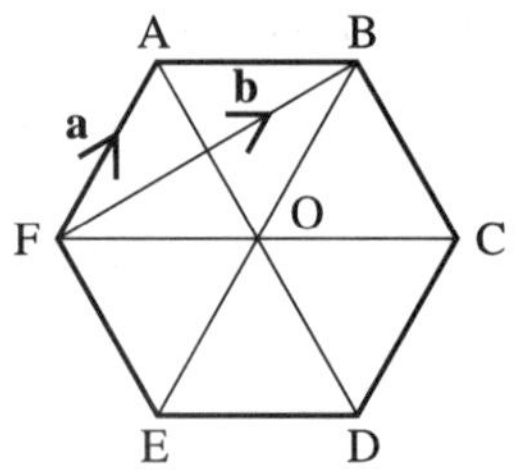

8 In the diagram, M is the mid-point of CD, BP : PM = 2 : 1, $\overrightarrow{AB} = \mathbf{x}$, and $\overrightarrow{AC} = \mathbf{y}$ and $\overrightarrow{AD} = \mathbf{z}$.

Express the following vectors in terms of **x**, **y** and **z**.

a $\overrightarrow{DC}$ **b** $\overrightarrow{DM}$ **c** $\overrightarrow{AM}$

d $\overrightarrow{BM}$ **e** $\overrightarrow{BP}$ **f** $\overrightarrow{AP}$

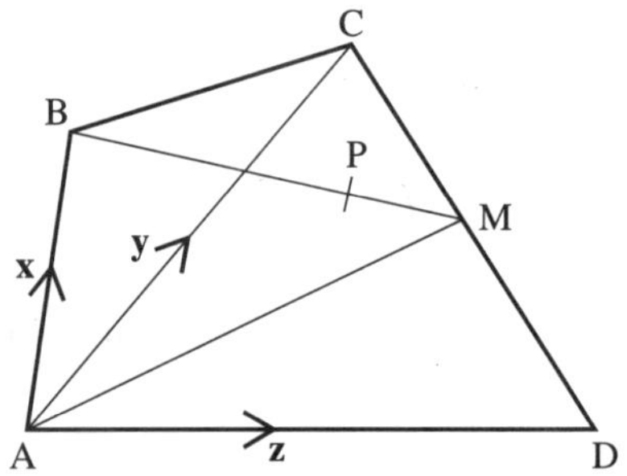

11.3 Position vectors and unit vectors

The *position vector* of a point P is the vector from the origin O to the point P. The usual notation is $\overrightarrow{OP}$ or **p**. The position vector can be described in terms of its *components*.

A *unit vector* is a vector of length one unit in a given direction. We denote a unit vector in the x direction by **i** and a unit vector in the y direction by **j**.

In two dimensions, $\overrightarrow{OP} = \begin{pmatrix} 4 \\ 3 \end{pmatrix}$ or $\overrightarrow{OP} = 4\mathbf{i} + 3\mathbf{j}$.

When we work in three dimensions we use **k** for a unit vector in the plane at right angles to the plane containing **i** and **j**.

Fig. 1

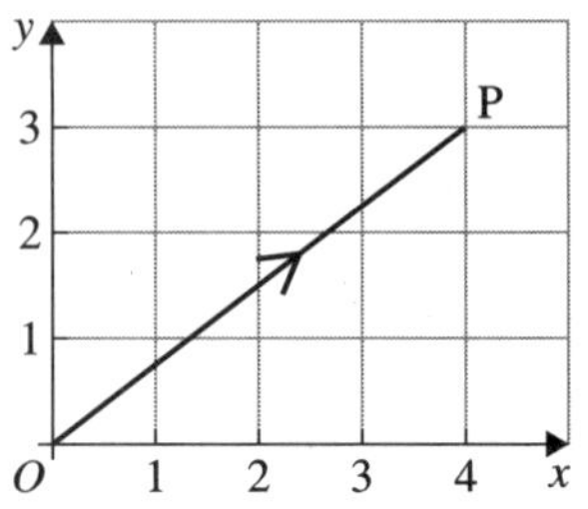

The diagram shows the vector $\overrightarrow{OA}$, where $\overrightarrow{OA} = 12\mathbf{i} + 3\mathbf{j} + 4\mathbf{k}$ or, as a column vector, $\overrightarrow{OA} = \begin{pmatrix} 12 \\ 3 \\ 4 \end{pmatrix}$

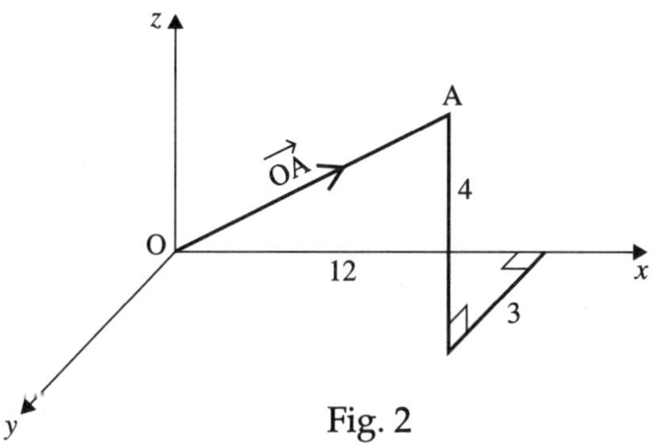

Fig. 2

Many people choose to write vectors as column vectors to save writing **'i'**, **'j'** and **'k'** many times. Either notation is acceptable.

Magnitude of a vector

If a vector $\mathbf{r} = a\,\mathbf{i} + b\,\mathbf{j}$, the magnitude of the vector is, by Pythagoras Theorem, $\sqrt{a^2 + b^2}$.
In Figure 1 above the magnitude of $\overrightarrow{OP}$ is $\sqrt{4^2 + 3^2} = 5$ units. We write $|\overrightarrow{OP}| = 5$.

In three dimensions, the magnitude of a vector $s = a\,\mathbf{i} + b\,\mathbf{j} + c\,\mathbf{k}$ is $\sqrt{a^2 + b^2 + c^2}$.

In Figure 2, $|\overrightarrow{OA}| = \sqrt{12^2 + 3^2 + 4^2} = 13$ units.

A unit vector in the direction $\overrightarrow{OA}$ is $\dfrac{\overrightarrow{OA}}{|\overrightarrow{OA}|} = \dfrac{12}{13}\mathbf{i} + \dfrac{3}{13}\mathbf{j} + \dfrac{4}{13}\mathbf{k} = \dfrac{1}{13}(12\mathbf{i} + 3\mathbf{j} + 4\mathbf{k})$

The distance between two points

Points A and B have position vectors **a** and **b** relative to the origin O.

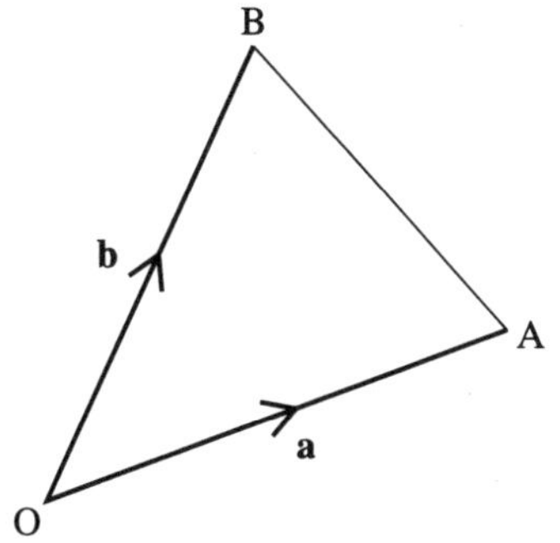

$$\begin{aligned}\overrightarrow{AB} &= \overrightarrow{AO} + \overrightarrow{OB} \\ &= -\overrightarrow{OA} + \overrightarrow{OB} \\ &= -\mathbf{a} + \mathbf{b} \\ &= \mathbf{b} - \mathbf{a}\end{aligned}$$

$$|\overrightarrow{AB}| = |\mathbf{b} - \mathbf{a}|$$

In general the distance d between the points (x_1, y_1, z_1) and (x_2, y_2, z_2) is given by

$$d^2 = (x_1 - x_2)^2 + (y_1 - y_2)^2 + (z_1 - z_2)^2.$$

Example 1

Find the magnitudes of the vectors $\mathbf{m} = 3\mathbf{i} - 2\mathbf{j}$ and $\mathbf{n} = \begin{pmatrix} 1 \\ 3 \\ -5 \end{pmatrix}$

$$|\mathbf{m}| = \sqrt{3^2 + (-2)^2} = \sqrt{13}$$
$$|\mathbf{n}| = \sqrt{1^2 + 3^2 + (-5)^2} = \sqrt{35}$$

Example 2

Find a unit vector in the direction of the vector $3\mathbf{i} - 2\mathbf{j} + \mathbf{k}$.

$$|3\mathbf{i} - 2\mathbf{j} + \mathbf{k}| = \sqrt{3^2 + (-2)^2 + 1^2} = \sqrt{14}$$

The unit vector required $= \dfrac{1}{\sqrt{14}}(3\mathbf{i} - 2\mathbf{j} + \mathbf{k})$.

Example 3

Find the distance between the points A(7, 4, 9) and B(2, 0, −1).

$$\text{Distance} = \sqrt{[(7-2)^2 + (4-0)^2 + (9-(-1))^2]}$$
$$= \sqrt{141}.$$

Example 4

If $\mathbf{m} = 2\mathbf{i} + 7\mathbf{j} + \mathbf{k}$ and $\mathbf{n} = \mathbf{i} + 2\mathbf{j} + 3\mathbf{k}$, find

a $\mathbf{m} + \mathbf{n}$ **b** $|\mathbf{m} + \mathbf{n}|$ **c** $\mathbf{m} - \mathbf{n}$ **d** $2\mathbf{m}$

a $\mathbf{m} + \mathbf{n} = (2\mathbf{i} + 7\mathbf{j} + \mathbf{k}) + (\mathbf{i} + 2\mathbf{j} + 3\mathbf{k})$
$= 3\mathbf{i} + 9\mathbf{j} + 4\mathbf{k}$

b $|\mathbf{m} + \mathbf{n}| = \sqrt{3^2 + 9^2 + 4^2} = \sqrt{106}$

c $\mathbf{m} - \mathbf{n} = (2\mathbf{i} + 7\mathbf{j} + \mathbf{k}) - (\mathbf{i} + 2\mathbf{j} + 3\mathbf{k}) = \mathbf{i} + 5\mathbf{j} - 2\mathbf{k}$

d $2\mathbf{m} = 2(2\mathbf{i} + 2\mathbf{j} + \mathbf{k}) = 4\mathbf{i} + 14\mathbf{j} + 2\mathbf{k}$

EXERCISE 3

1 Find the distance between the following pairs of points:

a A(1, 2) B(4, 6)

b A(−3, −1) B(12, 4)

c A(−2, 5) B(3, −1)

d A(3, −2) B(−2, 4)

2 Find the magnitude of the following vectors and hence find a unit vector in their respective directions.

a $4\mathbf{i} + 3\mathbf{j}$ **b** $-5\mathbf{i} + 12\mathbf{j}$ **c** $7\mathbf{i} - 24\mathbf{j}$

3 Find the distance between the following pairs of points:

a A(1, 2, 4) B(4, 8, 6) **b** A(3, −1, −4) B(11, 3, −3)

c A(2, −1, 3) B(0, 3, −2) **d** A(−4, 0, 2) B(1, −1, 3)

4 Find a unit vector in the direction of the following vectors:

a $2\mathbf{i} + 2\mathbf{j} + \mathbf{k}$

b $4\mathbf{i} - 2\mathbf{j} + 4\mathbf{k}$

c $3\mathbf{i} + 6\mathbf{j} - 2\mathbf{k}$

d $6\mathbf{i} - 3\mathbf{j} + 6\mathbf{k}$

5 If $3\mathbf{i} + a\mathbf{j}$ is parallel to $6\mathbf{i} + 10\mathbf{j}$, find the value of a.

6 Find a vector that is parallel to $3\mathbf{i} + 4\mathbf{j}$ and has magnitude 20 units.

7 Find a vector that is parallel to $5\mathbf{i} - 12\mathbf{j}$ and has magnitude 52 units.

8 Find the number a such that $a\begin{pmatrix}1\\3\end{pmatrix} + \begin{pmatrix}-1\\2\end{pmatrix} = \begin{pmatrix}2\\11\end{pmatrix}$.

9 Find the number b such that $\begin{pmatrix}0\\3\end{pmatrix} + b\begin{pmatrix}2\\-1\end{pmatrix} = \begin{pmatrix}8\\-1\end{pmatrix}$.

10 Express each of the vectors as a column vector.

a $2\mathbf{i} + 3\mathbf{j} + \mathbf{k}$ **b** $\mathbf{i} - \mathbf{j} + 4\mathbf{k}$ **c** $2\mathbf{j} + \mathbf{k}$

d $-\mathbf{i} + \mathbf{j} + 4\mathbf{k}$ **e** $2\mathbf{j}$ **f** $2\mathbf{i} + 3\mathbf{k}$

11 If $\mathbf{p} = \begin{pmatrix}2\\3\end{pmatrix}$, $\mathbf{q} = \begin{pmatrix}4\\1\end{pmatrix}$ and $\mathbf{r} = \begin{pmatrix}-2\\2\end{pmatrix}$, find the number n such that $\mathbf{p} + n\mathbf{q} = \mathbf{r}$.

12 Given $\overrightarrow{OP} = 3\mathbf{i} + 7\mathbf{j}$ and $\overrightarrow{OQ} = \mathbf{i} - \mathbf{j}$, find

a $\overrightarrow{PQ}$

b $\overrightarrow{QP}$

13 Given $\overrightarrow{OS} = \mathbf{i} + 2\mathbf{j} + \mathbf{k}$ and $\overrightarrow{OT} = 3\mathbf{i} + 8\mathbf{j} + \mathbf{k}$, find:

a $\overrightarrow{ST}$

b $\overrightarrow{SM}$, where M is the mid-point of $\overrightarrow{ST}$.

14 If $\overrightarrow{OA} = \begin{pmatrix}3\\2\end{pmatrix}$ and $\overrightarrow{AB} = \begin{pmatrix}4\\1\end{pmatrix}$, find $\overrightarrow{OB}$.

15 If $\overrightarrow{OM} = \mathbf{i} + 2\mathbf{j} + \mathbf{k}$ and $\overrightarrow{MN} = 2\mathbf{i} - \mathbf{j} + 3\mathbf{k}$, find $\overrightarrow{ON}$.

16 If $\mathbf{s} = \begin{pmatrix} 4 \\ 1 \\ 2 \end{pmatrix}$ and $\mathbf{t} = \begin{pmatrix} -1 \\ 2 \\ 3 \end{pmatrix}$, find

a $\mathbf{s} + \mathbf{t}$

b $|\mathbf{s} + \mathbf{t}|$

c vector $\mathbf{u}$ such that $\mathbf{s} + \mathbf{u} = \mathbf{t}$.

17 If $\mathbf{a} = \begin{pmatrix} -2 \\ -3 \\ 3 \end{pmatrix}$, $\mathbf{b} = \begin{pmatrix} -1 \\ 1 \\ -12 \end{pmatrix}$ and $\mathbf{c} = \begin{pmatrix} -2 \\ 3 \\ 0 \end{pmatrix}$ then find the following:

a $3\mathbf{a}$

b $5\mathbf{a} + 2\mathbf{b}$

c $2\mathbf{a} - \mathbf{b} + \mathbf{c}$

d the magnitude of $\mathbf{b} + \mathbf{c}$

e a unit vector parallel to $\mathbf{a} + \mathbf{c}$

18 If A, B and C are three points with co-ordinates (1, −2, 3), (4, 2, 7) and (16, 5, 11) respectively then find the following:

a the vector $\overrightarrow{AB}$

b the distance BC

c the co-ordinates of the point D such that $\overrightarrow{AD} = 2\overrightarrow{AB}$

d the co-ordinates of the point E such that $\overrightarrow{AE} = 2\overrightarrow{AC}$

e the ratio of DE : BC.

11.4 The vector equation of a line

Vectors describe journeys. The vector equation of a line describes a journey that starts at the origin O and finishes at **some point** on the line.

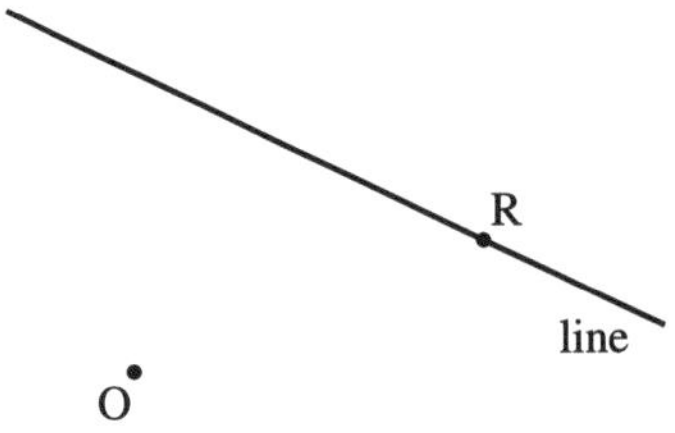

We want the vector $\overrightarrow{OR}$ (the position vector of a variable point R). This journey can be described in two parts:

Part 1: Get from O to a point on the line, say point A.
Part 2: Get from the point A to the point R.

So $\overrightarrow{OR} = \overrightarrow{OA} + \overrightarrow{AR}$.

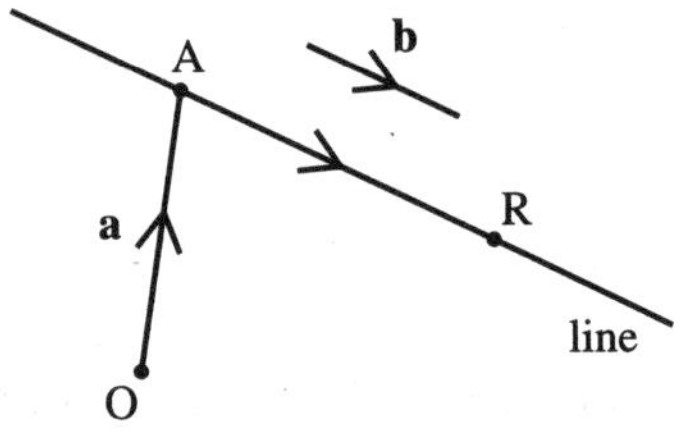

The vector that describes the journey from O to meet the line at A is **a**. Now $\overrightarrow{AR}$ describes the journey from A to R. It is sufficient to know a vector that is *parallel* to the line in order to find $\overrightarrow{AR}$. This is because parallel vectors have the same direction. This parallel vector is usually called **b**. Once we know **b**, we can work out $\overrightarrow{AR}$ by multiplying **b** by an appropriate number λ (remember $\lambda\mathbf{b}$ is in the same direction as **b**, but its length is multiplied by λ). So $\overrightarrow{AR} = \lambda\mathbf{b}$.

Hence $\overrightarrow{OR} = \overrightarrow{OA} + \overrightarrow{AR}$ becomes

$$\overrightarrow{OR} = \mathbf{a} + \lambda\mathbf{b}$$

If the position vector of point R is **r** then

$$\mathbf{r} = \mathbf{a} + \lambda\mathbf{b}$$

This is the vector equation for a line which passes through a point with position vector **a** and which is parallel to a vector **b**. R is *any* point on the line.

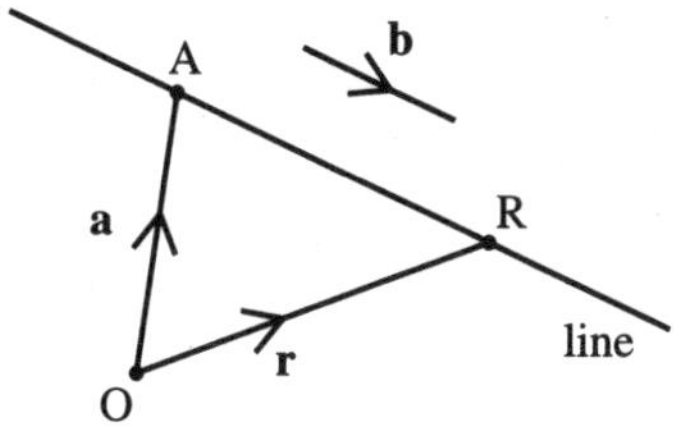

Any point on the line is found with the appropriate value of λ. This sometimes causes confusion. Remember that λ can be *any* value [e.g. -7, $2\frac{1}{2}$, 1001, 2.003 etc.].

Suppose the equation of a line is $\mathbf{r} = \begin{pmatrix}1\\2\\3\end{pmatrix} + \lambda\begin{pmatrix}4\\1\\2\end{pmatrix}$

With $\lambda = 1$, $r_1 = \begin{pmatrix}1\\2\\3\end{pmatrix} + 1\begin{pmatrix}4\\1\\2\end{pmatrix} = \begin{pmatrix}5\\3\\5\end{pmatrix}$

With $\lambda = 2$, $r_2 = \begin{pmatrix}1\\2\\3\end{pmatrix} + 2\begin{pmatrix}4\\1\\2\end{pmatrix} = \begin{pmatrix}9\\4\\7\end{pmatrix}$

With $\lambda = -5$, $r_{-5} = \begin{pmatrix}1\\2\\3\end{pmatrix} - 5\begin{pmatrix}4\\1\\2\end{pmatrix} = \begin{pmatrix}-19\\-3\\-7\end{pmatrix}$

The points with position vectors $\begin{pmatrix}5\\3\\5\end{pmatrix}$, $\begin{pmatrix}9\\4\\7\end{pmatrix}$ and $\begin{pmatrix}-19\\-3\\-7\end{pmatrix}$ all lie on the given line.

When you have two different lines in three dimensions, either

a the lines intersect in one point,
b the lines are parallel or
c the lines do not intersect and are *skew*.

Note. Consider the line $\mathrm{r} = \begin{pmatrix} a \\ b \\ c \end{pmatrix} + \lambda\begin{pmatrix} l \\ m \\ n \end{pmatrix}$. If the variable point on the line has coordinates (x, y, z), where $\mathbf{r} = \begin{pmatrix} x \\ y \\ z \end{pmatrix}$, the equation of the line can be written as $\begin{pmatrix} x \\ y \\ z \end{pmatrix} = \begin{pmatrix} a \\ b \\ c \end{pmatrix} + \lambda\begin{pmatrix} l \\ m \\ n \end{pmatrix}$.

Example 1

Find the vector equation of the straight line which passes through the points with position vectors **c** and **d**.

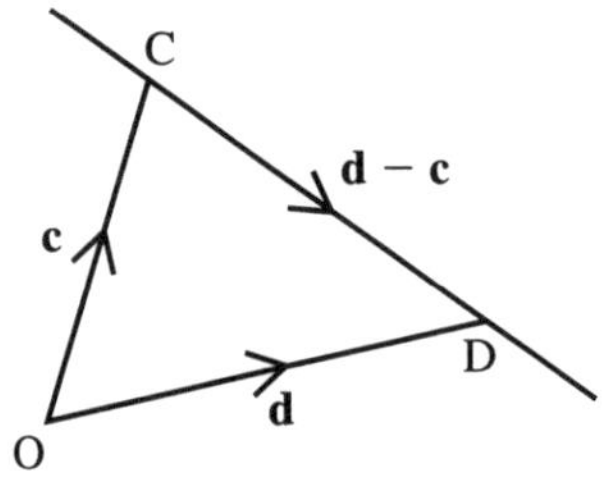

The direction of the line is given by the vector $\mathbf{d} - \mathbf{c}$.

We can take either **c** or **d** as the position vector of a point on the line.
Taking **c**, the equation of the line is

$$\mathbf{r} = \mathbf{c} + \lambda(\mathbf{d} - \mathbf{c})$$

Example 2

Write down the vector equation of the straight line that is parallel to the vector $3\mathbf{i} + \mathbf{j}$ and which passes through the point with position vector $\mathbf{i} - 4\mathbf{j}$.

The equation of the line is $\mathbf{r} = \mathbf{i} - 4\mathbf{j} + \lambda(3\mathbf{i} + \mathbf{j})$.

Example 3

a Write the vector equation of the line L that is parallel to the vector $\begin{pmatrix} 2 \\ 1 \\ 4 \end{pmatrix}$ and which passes through the point with position vector $\begin{pmatrix} 1 \\ -2 \\ 3 \end{pmatrix}$.

b Show that the point with position vector $\begin{pmatrix} 7 \\ 1 \\ 15 \end{pmatrix}$ lies on line L.

a The equation of line L is $\mathbf{r} = \begin{pmatrix} 1 \\ -2 \\ 3 \end{pmatrix} + \lambda \begin{pmatrix} 2 \\ 1 \\ 4 \end{pmatrix}$

[or $\mathbf{r} = \mathbf{i} - 2\mathbf{j} + 3\mathbf{k} + \lambda(2\mathbf{i} + \mathbf{j} + 4\mathbf{k})$]

b If $\begin{pmatrix} 7 \\ 1 \\ 15 \end{pmatrix}$ lies on the line there must be some value of λ such that

$$\begin{pmatrix} 7 \\ 1 \\ 15 \end{pmatrix} = \begin{pmatrix} 1 \\ -2 \\ 3 \end{pmatrix} + \lambda \begin{pmatrix} 2 \\ 1 \\ 4 \end{pmatrix}.$$

Equating the coefficients of **i**, **j** and **k** gives

$$7 = 1 + 2\lambda,$$
$$1 = -2 + \lambda$$
$$15 = 3 + 4\lambda \text{ respectively.}$$

From the first equation $\lambda = 3$. This value of λ satisfies the other two equations.

Thus the point with position vector $\begin{pmatrix} 7 \\ 1 \\ 15 \end{pmatrix}$ does lie on line L.

Example 4

a Show that the lines $\mathbf{r} = 3\mathbf{i} + 5\mathbf{j} + \mathbf{k} + \lambda(2\mathbf{i} + 3\mathbf{j} + 4\mathbf{k})$ and $\mathbf{r} = \mathbf{i} + 16\mathbf{j} + 7\mathbf{k} + \mu(4\mathbf{i} - \mathbf{j} + 3\mathbf{k})$ intersect.

b Find the position vector of the point of intersection.

a Points on the two lines can be written as

$$(3 + 2\lambda)\mathbf{i} + (5 + 3\lambda)\mathbf{j} + (1 + 4\lambda)\mathbf{k} \text{ and}$$
$$(1 + 4\mu)\mathbf{i} + (16 - \mu)\mathbf{j} + (7 + 3\mu)\mathbf{k} \text{ respectively.}$$

For these to be the *same* point we must have

$3 + 2\lambda = 1 + 4\mu$ [1] (equating **i** components)
$5 + 3\lambda = 16 - \mu$ [2] (equating **j** components)
$1 + 4\lambda = 7 + 3\mu$ [3] (equating **k** components)

Solving equations [1] and [2], we obtain $\lambda = 3$ and $\mu = 2$.
Substitute these values into equation [3]

L.H.S. $= 1 + 4 \times 3 = 13$
R.H.S. $= 7 + 3 \times 2 = 13$

So the equations [1], [2] and [3] are consistent which shows that the lines *do* intersect.

b With $\lambda = 3$, $\mu = 2$ we obtain the vector $9\mathbf{i} + 14\mathbf{j} + 13\mathbf{k}$ which is the position vector of the point of intersection.

Example 5

a Find a vector equation for a line L which passes through the point (1, 2, 6) and which is parallel to the line with vector equation $\mathbf{r} = \begin{pmatrix} 1 \\ -4 \\ 2 \end{pmatrix} + \lambda \begin{pmatrix} 1 \\ 3 \\ 2 \end{pmatrix}$.

b Find the coordinates of the point where line L meets the xy-plane.

a The direction of the given line is the vector $\begin{pmatrix} 1 \\ 3 \\ 2 \end{pmatrix}$.

The equation of the line L is $\mathbf{r} = \begin{pmatrix} 1 \\ 2 \\ 6 \end{pmatrix} + \mu \begin{pmatrix} 1 \\ 3 \\ 2 \end{pmatrix}$.

b The **k** component of any point on the xy-plane is zero. So for line L we have $6 + 2\mu = 0$

$\Rightarrow \quad \mu = -3$

The line L meets the xy-plane at $\begin{pmatrix} 1 \\ 2 \\ 6 \end{pmatrix} - 3\begin{pmatrix} 1 \\ 3 \\ 2 \end{pmatrix} = \begin{pmatrix} -2 \\ -7 \\ 0 \end{pmatrix}$.

That is at the point $(-2, -7, 0)$.

EXERCISE 4

1 The diagram shows two points A(−2, 1) and B(4, 4) together with the origin O.

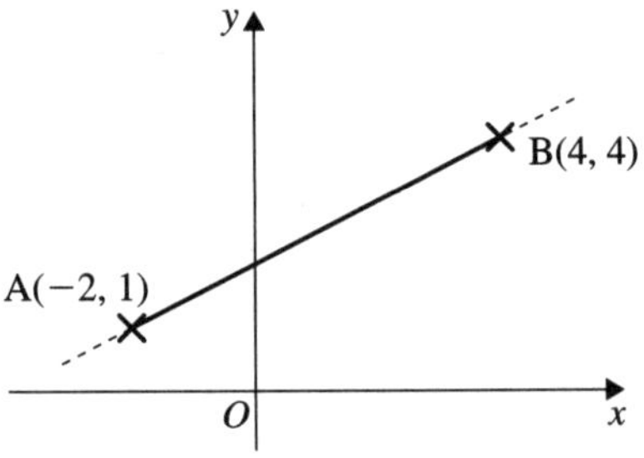

a Write down a vector representing each of $\overrightarrow{OA}$ and $\overrightarrow{OB}$.

b Hence find vector $\overrightarrow{AB}$.

c Using the equation $\mathbf{r} = \mathbf{a} + \lambda\mathbf{m}$, find a vector equation of the line through A and B.
[Note: **a** can be either of the fixed points A and B and **m** is a vector in the direction AB].

2 With respect to an origin O, the position vectors of points A and B are given. In each case write down the vector $\overrightarrow{AB}$ and find a vector equation of the line which passes through points A and B.

a A(3**i** + 2**j**), B(5**i** + 6**j**)

b A(−3**i** + 4**j**), B(2**i** + 5**j**)

c A(−**i** − 2**j**), B(−4**i** + 3**j**)

d A(2**i** + 3**j**), B(−4**i** − 2**j**)

3 Find the vector equation of a line which passes through the point P(1, −3) and is parallel to the vector −2**i** + 5**j**.

4 Write down the vector equation of a line which passes through the point (0, 2) with a direction vector 3**i** − 4**j**.

5 State which of the following lines are parallel to the line

$r = 3\mathbf{i} - 2\mathbf{j} + \lambda(2\mathbf{i} - \mathbf{j})$.

l_1: $r = 2\mathbf{i} + 3\mathbf{j} + \lambda(6\mathbf{i} - 3\mathbf{j})$
l_2: $r = 6\mathbf{i} - 4\mathbf{j} + \lambda(2\mathbf{i} + \mathbf{j})$
l_3: $r = 3\mathbf{i} - 2\mathbf{j} + \lambda(\mathbf{i} - \mathbf{j})$
l_4: $r = 3\mathbf{i} + 4\mathbf{j} + \lambda(4\mathbf{i} + 2\mathbf{j})$

6 Show that the following points lie on the given lines:

a P(7, 4) $r = 3\mathbf{i} + 2\mathbf{j} + \lambda(2\mathbf{i} + \mathbf{j})$

b P(9, −5) $r = 3\mathbf{i} - 2\mathbf{j} + \lambda(2\mathbf{i} - \mathbf{j})$

c P(5, −8) $r = \begin{pmatrix} 2 \\ 1 \end{pmatrix} + \lambda\begin{pmatrix} 1 \\ -3 \end{pmatrix}$

d P(−1, −2) $r = \begin{pmatrix} -5 \\ 4 \end{pmatrix} + \mu\begin{pmatrix} 2 \\ -3 \end{pmatrix}$

7 The diagram shows two lines intersecting at point P where $r_1 = r_2$.

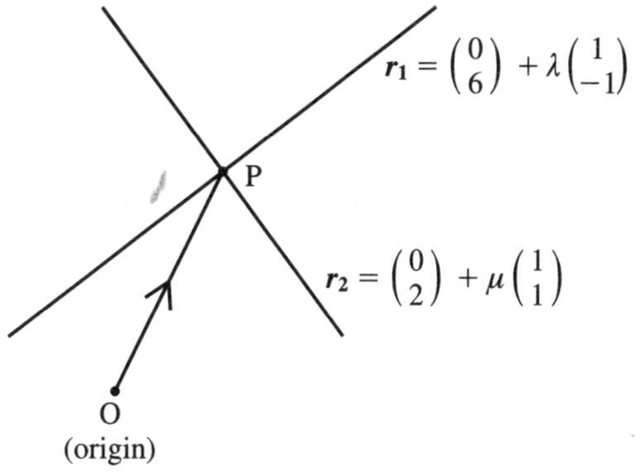

a By putting $r_1 = r_2$ and equating the coefficients of **i** and **j**, form two simultaneous equations and solve for λ and μ.

b Hence find the position vector $\overrightarrow{OP}$ of the point of intersection of the two lines.

8 Find the position vector of the point of intersection of the following pairs of lines:

a $r = \begin{pmatrix} 0 \\ 2 \end{pmatrix} + \lambda\begin{pmatrix} 3 \\ 1 \end{pmatrix}$ $r = \begin{pmatrix} 2 \\ 0 \end{pmatrix} + \mu\begin{pmatrix} 1 \\ 3 \end{pmatrix}$

b $r = (1 + 2\lambda)\mathbf{i} + (3 + \lambda)\mathbf{j}$ $r = (3 + \mu)\mathbf{i} + (2 + \mu)\mathbf{j}$

c $r = -3\mathbf{i} - 4\mathbf{j} + \lambda(\mathbf{i} + 2\mathbf{j})$ $r = 4\mathbf{i} + 2\mathbf{j} + \mu(3\mathbf{i} - 2\mathbf{j})$

d $r = \begin{pmatrix} -3 \\ -4 \end{pmatrix} + \lambda\begin{pmatrix} 1 \\ 2 \end{pmatrix}$ $r = \begin{pmatrix} 4 \\ 2 \end{pmatrix} + \mu\begin{pmatrix} 3 \\ -2 \end{pmatrix}$

9 The diagram shows a line joining points A and B whose position vectors are $\overrightarrow{OA} = 2\mathbf{i} - 2\mathbf{j} + \mathbf{k}$ and $\overrightarrow{OB} = \mathbf{i} + 3\mathbf{j} + 2\mathbf{k}$.

a Find the direction vector $\overrightarrow{AB}$.

b Using the equation $\mathbf{r} = \mathbf{a} + \lambda\mathbf{m}$, where $\mathbf{a}$ is the position vector of a fixed point and $\mathbf{m}$ is a direction vector, find a vector equation of the line AB.

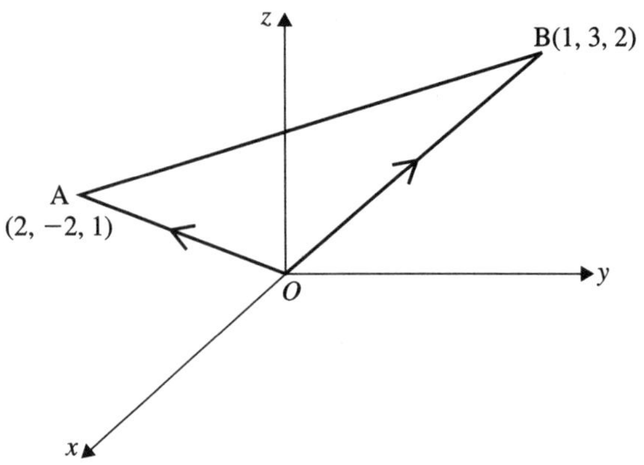

10 Find a vector equation of the line through the following pairs of points:

a A(3, 2, 1) B(1, 4, 2)

b A(1, 4, 0) B(3, 2, 7)

c A(3, −2, 1) B(−2, 0, 4)

d A(1, −2, 1) B(−3, −1, 2)

11 Find the vector equation of the line which passes through point P(2, −1, 0) and is parallel to the vector $-3\mathbf{i} + \mathbf{j} + 2\mathbf{k}$.

12 State which of the following lines are parallel to the line $r = 3\mathbf{i} - 2\mathbf{j} + \mathbf{k} + \lambda(\mathbf{i} - 2\mathbf{j} + 3\mathbf{k})$

l_1: $\mathbf{r} = -\mathbf{i} + \mathbf{j} + 2\mathbf{k} + \lambda(2\mathbf{i} - 4\mathbf{j} + 6\mathbf{k})$

l_2: $\mathbf{r} = 6\mathbf{i} - 4\mathbf{j} + 2\mathbf{k} + \lambda(-\mathbf{i} + 2\mathbf{j} + 3\mathbf{k})$

l_3: $\mathbf{r} = (2 - \lambda)\mathbf{i} + (-4 + 2\lambda)\mathbf{j} + (1 - 3\lambda)\mathbf{k}$

l_4: $\begin{pmatrix} x \\ y \\ z \end{pmatrix} = \begin{pmatrix} 3 \\ -2 \\ 1 \end{pmatrix} + \lambda\begin{pmatrix} 1 \\ 2 \\ 3 \end{pmatrix}$

13 Show that the following points lie on the given lines:

a P(9, 5, 8), $\mathbf{r} = \mathbf{i} + 3\mathbf{j} + 2\mathbf{k} + \lambda(4\mathbf{i} + \mathbf{j} + 3\mathbf{k})$

b P(1, −1, 7), $\mathbf{r} = (2 - \lambda)\mathbf{i} + (-3 + 2\lambda)\mathbf{j} + (4 + 3\lambda)\mathbf{k}$

c P(2, −3, 2), $\begin{pmatrix} x \\ y \\ z \end{pmatrix} = \begin{pmatrix} 4 \\ 1 \\ 0 \end{pmatrix} + \lambda\begin{pmatrix} -1 \\ -2 \\ 1 \end{pmatrix}$

14 Show that the lines with equations $\mathbf{r} = 3\mathbf{i} - 2\mathbf{j} + \mathbf{k} + \lambda(2\mathbf{i} + 4\mathbf{j} + 3\mathbf{k})$ and $\mathbf{r} = \mathbf{i} + 2\mathbf{k} + \mu(2\mathbf{i} + \mathbf{j} + \mathbf{k})$ intersect and find the point of intersection.

a Check the direction vectors are not equal so the lines are not parallel.

b Equate the coefficients of **i** and **j** to find λ and μ.

c Show that with these values of λ and μ, the coefficients of **j** are equal.

d Find the point of intersection of the lines.

15 Show that the following lines intersect and find the position vector of the point of intersection.

a $\mathbf{r} = 4\mathbf{i} - 3\mathbf{j} + 2\mathbf{k} + \lambda(\mathbf{i} + 4\mathbf{j} + 3\mathbf{k})$
$\mathbf{r} = 3\mathbf{i} - \mathbf{j} + 5\mathbf{k} + \mu(\mathbf{i} + 2\mathbf{j} + \mathbf{k})$

b $\mathbf{r} = \begin{pmatrix} -3 \\ 4 \\ -2 \end{pmatrix} + \lambda\begin{pmatrix} 2 \\ 1 \\ 1 \end{pmatrix}$

$\mathbf{r} = \begin{pmatrix} 2 \\ 3 \\ -1 \end{pmatrix} + \mu\begin{pmatrix} 1 \\ 4 \\ 2 \end{pmatrix}$

c $\mathbf{r} = (2 + \lambda)\mathbf{i} + (1 + 3\lambda)\mathbf{j} + (4 - 2\lambda)\mathbf{k}$
$\mathbf{r} = (7 - \mu)\mathbf{i} + (-8 + 3\mu)\mathbf{j} + (-6 + 2\mu)\mathbf{k}$

16 If A, B and C have position vectors $\mathbf{a} = \begin{pmatrix} 1 \\ 5 \\ 2 \end{pmatrix}$, $\mathbf{b} = \begin{pmatrix} -3 \\ 7 \\ 4 \end{pmatrix}$ and $\mathbf{c} = \begin{pmatrix} 7 \\ 9 \\ -5 \end{pmatrix}$ respectively then find a vector equation of the straight line which passes through

a A and B.

b C and the origin.

c B and the midpoint of AC.

17 Find a vector equation of the straight line which:

a Passes through (1, 4, 7) and is parallel to the z-axis.

b Passes through (3, 5, −4) and is parallel to the vector $2\mathbf{i} + \mathbf{j} + 3\mathbf{k}$.

c Passes through (2, −7, 1) and (4, −3, 2).

d Passes through (3, 2, −5) and is parallel to the straight line with vector equation $\mathbf{r} = (1 + 2\lambda)\mathbf{i} + (3 - 4\lambda)\mathbf{j} + 5\lambda\mathbf{k}$.

18 Find the coordinates of the point where the line
$\mathbf{r} = \begin{pmatrix} 4 \\ 1 \\ 6 \end{pmatrix} + \lambda\begin{pmatrix} 1 \\ 3 \\ -2 \end{pmatrix}$ meets the xy-plane.

19 Find the coordinates of the point where the line
$\mathbf{r} = \begin{pmatrix} 4 \\ 3 \\ -2 \end{pmatrix} + \lambda\begin{pmatrix} 8 \\ 10 \\ 4 \end{pmatrix}$ meets the yz-plane.

20 Line l cuts the xy-plane at (4, 5, 0) and the yz-plane at (0, 15, 4).

a Find a vector equation of line l.

b Find the coordinates of the points where line l meets the xz-plane.

21 In parts **a**, **b** and **c** you are given a pair of lines. Determine whether the lines are parallel, skew or intersect at a point.

a $\mathbf{r} = \begin{pmatrix} 2 \\ 3 \\ 8 \end{pmatrix} + \lambda \begin{pmatrix} 1 \\ 4 \\ 0 \end{pmatrix}, \quad \mathbf{r} = \begin{pmatrix} 5 \\ 1 \\ 3 \end{pmatrix} + \mu \begin{pmatrix} -1 \\ 10 \\ 2 \end{pmatrix}$

b $\mathbf{r} = \begin{pmatrix} 1 \\ 2 \\ 2 \end{pmatrix} + \lambda \begin{pmatrix} -3 \\ 0 \\ 4 \end{pmatrix}, \quad \mathbf{r} = \begin{pmatrix} 1 \\ -1 \\ 0 \end{pmatrix} + \mu \begin{pmatrix} 2 \\ 1 \\ -2 \end{pmatrix}$

c $\mathbf{r} = \begin{pmatrix} 1 \\ 4 \\ 3 \end{pmatrix} + \lambda \begin{pmatrix} 2 \\ 3 \\ -1 \end{pmatrix}, \quad \mathbf{r} = \begin{pmatrix} 1 \\ -5 \\ 2 \end{pmatrix} + \mu \begin{pmatrix} 6 \\ 9 \\ -3 \end{pmatrix}$

11.5 The scalar product

The scalar product of two vectors **a** and **b** is defined as

$$\mathbf{a}.\mathbf{b} = |\mathbf{a}|\,|\mathbf{b}|\cos\theta$$

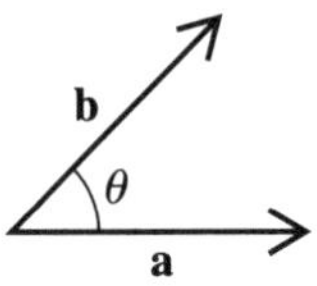

where θ is the angle between the two vectors.
The angle θ can be acute or obtuse and the vectors are drawn with their 'tails' drawn from the same point.

This product is called the *scalar* product because the result is a scalar quantity (a number).
It is important to write the scalar product with a dot.
For **a.b**, we say 'a dot b' and the product is commonly known as the 'dot product'.

The unit vectors **i**, **j** and **k** are mutually perpendicular.

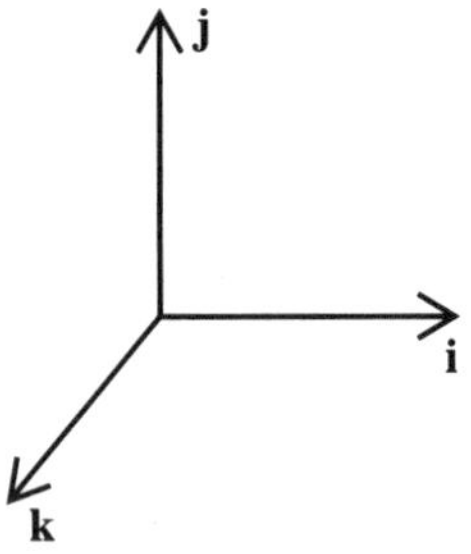

$$\therefore \quad \mathbf{i}.\mathbf{i} = \mathbf{j}.\mathbf{j} = \mathbf{k}.\mathbf{k} = 1 \times 1 \times \cos 0° = 1$$

and $\mathbf{i}.\mathbf{j} = \mathbf{i}.\mathbf{k} = \mathbf{j}.\mathbf{k} = 1 \times 1 \times \cos 90° = 0$

The scalar product is commutative (**a.b** = **b.a**) and also distributive over addition [**a.(b** + **c)** = **a.b** + **a.c**]

Now consider $\mathbf{a} = l\mathbf{i} + m\mathbf{j} + n\mathbf{k}$ and $\mathbf{b} = p\mathbf{i} + q\mathbf{j} + r\mathbf{k}$.

$$\begin{aligned} \mathbf{a}.\mathbf{b} &= (l\mathbf{i} + m\mathbf{j} + n\mathbf{k}).(p\mathbf{i} + q\mathbf{j} + r\mathbf{k}) \\ &= l\mathbf{i}.p\mathbf{i} + l\mathbf{i}.q\mathbf{j} + l\mathbf{i}.r\mathbf{k} + m\mathbf{j}.p\mathbf{i} + m\mathbf{j}.q\mathbf{j} + m\mathbf{j}.r\mathbf{k} + n\mathbf{k}.p\mathbf{i} + n\mathbf{k}.q\mathbf{j} + n\mathbf{k}.r\mathbf{k} \\ &= lp + mq + nr \text{ [since } i.i. = j.j = k.k = 1 \text{ and } i.j = i.k = j.k = 0] \end{aligned}$$

> Remember
> $(l\mathbf{i} + m\mathbf{j} + n\mathbf{k}).(p\mathbf{i} + q\mathbf{j} + r\mathbf{k}) = lp + mq + nr$
>
> or $\begin{pmatrix} l \\ m \\ n \end{pmatrix} . \begin{pmatrix} p \\ q \\ r \end{pmatrix} = lp + mq + nr$

2 Find the modulus ($|\mathbf{a}|$, $|\mathbf{b}|$ etc) for each of the following vectors:

$$\mathbf{a} = 3\mathbf{i} + 4\mathbf{j},\ \mathbf{b} = -5\mathbf{i} + 12\mathbf{j},\ \mathbf{c} = 7\mathbf{i} - 24\mathbf{j},\ \mathbf{d} = -3\mathbf{i} - 4\mathbf{j}$$

3 Using the rule $\mathbf{a}.\mathbf{b} = |\mathbf{a}||\mathbf{b}|\cos\theta$, find the cosine of the angle between the following pairs of vectors.

a $\mathbf{a} = 3\mathbf{i} + 4\mathbf{j}$, $\mathbf{b} = 5\mathbf{i} + 12\mathbf{j}$

b $\mathbf{a} = 3\mathbf{i} - 4\mathbf{j}$, $\mathbf{b} = -7\mathbf{i} + 24\mathbf{j}$

4 Find the angle in degrees between each of the following pairs of vectors:

a $\mathbf{a} = 3\mathbf{i} + 4\mathbf{j}$ $\quad \mathbf{b} = -5\mathbf{i} + 12\mathbf{j}$ $\qquad$ **b** $\mathbf{a} = \begin{pmatrix} 1 \\ 1 \end{pmatrix}$ $\quad \mathbf{b} = \begin{pmatrix} -1 \\ 1 \end{pmatrix}$

c $\mathbf{a} = 2\mathbf{i} + \mathbf{j}$ $\quad \mathbf{b} = \mathbf{i} + 2\mathbf{j}$ $\qquad$ **d** $\mathbf{a} = \begin{pmatrix} -3 \\ 2 \end{pmatrix}$ $\quad \mathbf{b} = \begin{pmatrix} 4 \\ 1 \end{pmatrix}$

5 a Sketch a diagram and use basic trigonometry to find the angle which each of the following vectors makes with the **i** direction:

$$\mathbf{a} = \mathbf{i} + \mathbf{j},\ \mathbf{b} = \sqrt{3}\mathbf{i} + \mathbf{j}$$

b Using the scalar product, find the angle between vectors **a** and **b**.

c Confirm that this angle is the difference between the angles found in part **a**.

6 Find the angle in degrees between each of the following pairs of lines:

a $\mathbf{r} = 3\mathbf{i} - 2\mathbf{j} + \lambda(\mathbf{i} - \mathbf{j})$, $\ \mathbf{r} = -4\mathbf{i} + \mathbf{j} + \mu(\sqrt{3}\mathbf{i} + \mathbf{j})$

b $\mathbf{r} = \mathbf{i} + 2\mathbf{j} + \lambda(2\mathbf{i} - \mathbf{j})$, $\ \mathbf{r} = -3\mathbf{i} - \mathbf{j} + \mu(3\mathbf{i} + 2\mathbf{j})$

c $\mathbf{r} = \begin{pmatrix} 0 \\ 3 \end{pmatrix} + \lambda\begin{pmatrix} -3 \\ 1 \end{pmatrix}$, $\ \mathbf{r} = \begin{pmatrix} 2 \\ 2 \end{pmatrix} + \mu\begin{pmatrix} 1 \\ 2 \end{pmatrix}$

7 Show the following pairs of lines are perpendicular.

a $\mathbf{r} = 3\mathbf{i} - 2\mathbf{j} + \lambda(4\mathbf{i} + 3\mathbf{j})$
$\mathbf{r} = -\mathbf{i} + 4\mathbf{j} + \mu(-3\mathbf{i} + 4\mathbf{j})$

b $\mathbf{r} = \begin{pmatrix} 5 \\ 0 \end{pmatrix} + \lambda\begin{pmatrix} 1 \\ 2 \end{pmatrix}$
$\mathbf{r} = \begin{pmatrix} 3 \\ -1 \end{pmatrix} + \mu\begin{pmatrix} -6 \\ 3 \end{pmatrix}$

c $\mathbf{r} = (3 + 2\lambda)\mathbf{i} + (2 + \lambda)\mathbf{j}$
$\mathbf{r} = (-5 + 2\mu)\mathbf{i} + (1 - 4\mu)\mathbf{j}$

8 The points A, B, C and D have position vectors $3\mathbf{i} + 4\mathbf{j}$, $4\mathbf{i} + 6\mathbf{j}$, $\mathbf{i} + 5\mathbf{j}$ and $7\mathbf{i} + 2\mathbf{j}$ respectively. Show that AB is perpendicular to CD.

9 If $\mathbf{a} = \begin{pmatrix} 5 \\ -12 \end{pmatrix}$ find a unit vector perpendicular to **a**.

10 Find the value of t such that $\mathbf{a} = \begin{pmatrix} 5 \\ 4 \end{pmatrix}$ is perpendicular to $\mathbf{b} = \begin{pmatrix} 4 \\ t \end{pmatrix}$.

11 Two vectors are given by $\mathbf{a} = 3\mathbf{i} - 6\mathbf{j} + 2\mathbf{k}$ and $\mathbf{b} = 8\mathbf{i} + \mathbf{j} + 4\mathbf{k}$.

a Find the scalar product $\mathbf{a}.\mathbf{b}$.

b Find the modulus of each vector.

c Find the angle in degrees between the two vectors.

12 Find the angle in degrees between each of the following pairs of vectors:

a $\mathbf{a} = 2\mathbf{i} + 2\mathbf{j} + \mathbf{k}, \mathbf{b} = 9\mathbf{i} + 6\mathbf{j} + 2\mathbf{k}$ **b** $\mathbf{a} = 2\mathbf{i} + 3\mathbf{j} + 6\mathbf{k}, \mathbf{b} = 4\mathbf{i} - 4\mathbf{j} + 2\mathbf{k}$

c $\mathbf{a} = -8\mathbf{i} + \mathbf{j} + 4\mathbf{k}, \mathbf{b} = 6\mathbf{i} + 3\mathbf{j} + 6\mathbf{k}$ **d** $\mathbf{a} = 2\mathbf{i} + 3\mathbf{j} - \mathbf{k}, \mathbf{b} = \mathbf{i} + 2\mathbf{k}$

e $\mathbf{a} = 3\mathbf{i} - 2\mathbf{j} + 4\mathbf{k}, \mathbf{b} = \mathbf{i} + 5\mathbf{j} + 2\mathbf{k}$

13 Show that the vectors **a** and **b** are perpendicular:

a $\mathbf{a} = \mathbf{i} + 2\mathbf{j} + 3\mathbf{k}, \mathbf{b} = \mathbf{i} + \mathbf{j} - \mathbf{k}$

b $\mathbf{a} = \mathbf{i} - \mathbf{j} - 3\mathbf{k}, \mathbf{b} = \mathbf{i} - 2\mathbf{j} + \mathbf{k}$

14 Given that the vectors **a** and **b** are perpendicular, find the value of t.

a $\mathbf{a} = \mathbf{i} - 3\mathbf{j} + 2\mathbf{k}$ $\mathbf{b} = 4\mathbf{i} + 2\mathbf{j} + t\mathbf{k}$

b $\mathbf{a} = t\mathbf{i} + 2\mathbf{j} - \mathbf{k}$ $\mathbf{b} = 3\mathbf{i} - 4\mathbf{j} + t\mathbf{k}$

15 **a** Write down a unit vector in the positive x-direction.

b Find the angle which the following vectors make with the positive x-direction.

i $2\mathbf{i} + \mathbf{j} + 2\mathbf{k}$

ii $3\mathbf{i} - 6\mathbf{j} + 2\mathbf{k}$

iii $-8\mathbf{i} + 4\mathbf{j} + \mathbf{k}$

16 Find the angle in degrees between each of the following pairs of lines:

a $\mathbf{r} = \mathbf{i} - 2\mathbf{j} - 3\mathbf{k} + \lambda(3\mathbf{i} - 2\mathbf{j} + 6\mathbf{k}), \mathbf{r} = -2\mathbf{i} + \mathbf{j} + 4\mathbf{k} + \mu(8\mathbf{i} - 4\mathbf{j} - \mathbf{k})$

b $\mathbf{r} = \begin{pmatrix} 1 \\ -2 \\ 3 \end{pmatrix} + \lambda\begin{pmatrix} 2 \\ -3 \\ 4 \end{pmatrix}, \mathbf{r} = \begin{pmatrix} 0 \\ 1 \\ -4 \end{pmatrix} + \mu\begin{pmatrix} -1 \\ 2 \\ 5 \end{pmatrix}$

17 Show that the following lines are perpendicular:

$$\mathbf{r} = 3\mathbf{i} - 2\mathbf{j} - \mathbf{k} + \lambda(\mathbf{i} + 3\mathbf{j} - 2\mathbf{k}), \mathbf{r} = -3\mathbf{i} + 2\mathbf{j} + \mathbf{k} + \mu(4\mathbf{i} - 2\mathbf{j} - \mathbf{k})$$

EXERCISE 6

1 Three points A, B and C have position vectors:

$$\overrightarrow{OA} = 3\mathbf{i} - 2\mathbf{j} + \mathbf{k}$$
$$\overrightarrow{OB} = -\mathbf{i} + \mathbf{j} - 3\mathbf{k}$$
$$\overrightarrow{OC} = 2\mathbf{i} - 5\mathbf{j} + 3\mathbf{k}$$

Find the angles of the triangle ABC in degrees correct to 3 significant figures. Confirm that the sum of the angles is approximately equal to 180°.

2 If $\overrightarrow{AB} = \begin{pmatrix} 3 \\ 4 \\ 12 \end{pmatrix}$ and $\overrightarrow{AC} = \begin{pmatrix} 6 \\ 2 \\ z \end{pmatrix}$ then find two possible values of z (to 2 sf) such that the angle BAC is 60°.

3 **a** If $\mathbf{a} = \begin{pmatrix} t \\ t+2 \\ 2 \end{pmatrix}$ and $\mathbf{b} = \begin{pmatrix} 4 \\ t+3 \\ 1 \end{pmatrix}$ then find the two values of t for which $\mathbf{a}$ is perpendicular to $\mathbf{b}$.

b If $t = 3$ then find the angle (to the nearest degree) between $\mathbf{a}$ and the z-axis.

4 If A, B and C are three points with co-ordinates $(-2, 1, 2)$, $(2, 3, 5)$ and $(1, -4, z)$ respectively then find the following:

a the value of z given that CBA is a right angle.

b the size of the angle CBA given that $z = 2$.

c the equation of the line through A and B in the form $\mathbf{r} = \mathbf{a} + t\mathbf{b}$ where $\mathbf{a}$ and $\mathbf{b}$ are vectors to be determined.

d If $z = 3$ then find the co-ordinates of D where $ABCD$ is a parallelogram.

5 Relative to origin O, the position vectors of points A, B and C are $\begin{pmatrix} 4 \\ 7 \\ 7 \end{pmatrix}$, $\begin{pmatrix} 1 \\ 3 \\ 2 \end{pmatrix}$ and $\begin{pmatrix} 2 \\ 4 \\ 6 \end{pmatrix}$ respectively.

a Find the vectors $\overrightarrow{BC}$ and $\overrightarrow{BA}$.

b Show that $\cos A\hat{B}C = \frac{9}{10}$.

c Find the area of $\triangle$ABC.

6 The straight lines l and m are given by l: $r = \begin{pmatrix} 1 \\ -1 \\ 0 \end{pmatrix} + \lambda\begin{pmatrix} 2 \\ 1 \\ -2 \end{pmatrix}$ and m: $r = \begin{pmatrix} 1 \\ 2 \\ 2 \end{pmatrix} + \mu\begin{pmatrix} -3 \\ 0 \\ 4 \end{pmatrix}$.

a Show that l and m intersect and find the position vector of their point of intersection.

b Find the acute angle between the lines.

7 Points A and B have position vectors $\begin{pmatrix} 4 \\ -11 \\ 4 \end{pmatrix}$ and $\begin{pmatrix} 7 \\ 1 \\ 7 \end{pmatrix}$ respectively.

a Find a vector equation of the line through A and B in terms of a parameter λ.

b The point M on AB is such that OM is perpendicular to AB. Find the position vector of M.

8 The diagram shows the line $\mathbf{r} = 2\mathbf{i} + \lambda(3\mathbf{i} + \mathbf{j})$ which passes through the fixed point A(2, 0). Point P has coordinates (5, 4).

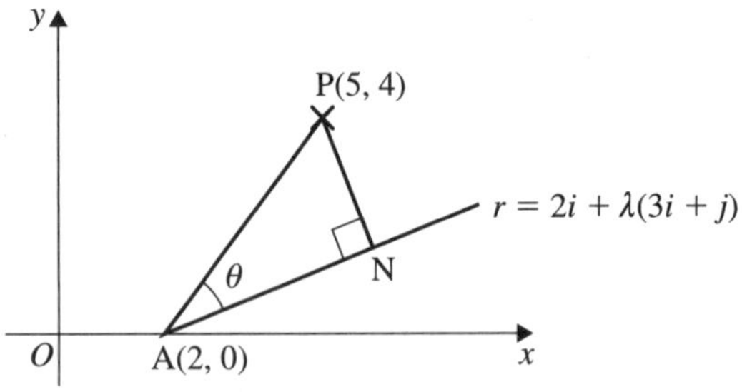

a Write down vector $\overrightarrow{AP}$.

b Using the scalar product rule, find the acute angle between $\overrightarrow{AP}$ and the line given.

c Find the distance AP.

d Using trigonometry, find the shortest distance of P from the line (this is PN in the diagram).

9 Find the shortest distance of the following points from the given lines.

a P(6, 1) $\mathbf{r} = \mathbf{i} + 2\mathbf{j} + \lambda(2\mathbf{i} + \mathbf{j})$

b P(2, 6) $\mathbf{r} = 4\mathbf{i} + 2\mathbf{j} + \lambda(\mathbf{i} + \mathbf{j})$

c P(5, 2) $\mathbf{r} = -4\mathbf{i} + 2\mathbf{j} + \lambda(3\mathbf{i} + \mathbf{j})$

10 A line with equation $\mathbf{r} = \mathbf{i} - 2\mathbf{j} + 3\mathbf{k} + \lambda(2\mathbf{i} + 3\mathbf{j} - \mathbf{k})$ passes through the point A with position vector $5\mathbf{i} + 4\mathbf{j} + \mathbf{k}$. A point P has position vector $7\mathbf{i} + 5\mathbf{j} + 3\mathbf{k}$.

a Write down vector $\overrightarrow{AP}$.

b Find the distance AP.

c Using the scalar product rule, find the acute angle θ between $\overrightarrow{AP}$ and the line. Visualising a right angled triangle,

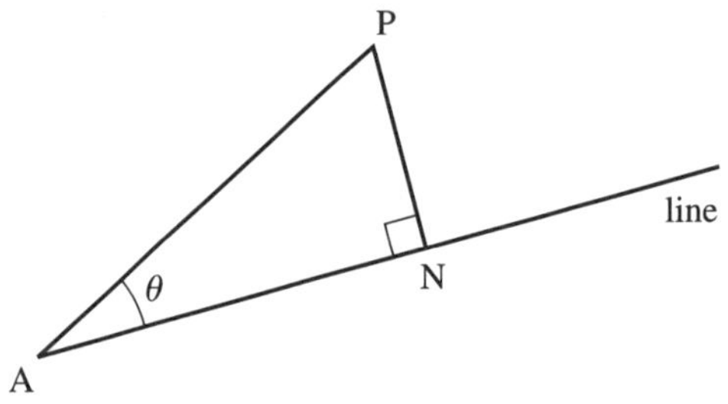

find the shortest distance (PN) from point P to the line.

11 A line has an equation $\mathbf{r} = \mathbf{i} - 2\mathbf{k} + \lambda(-2\mathbf{i} + \mathbf{j} + 3\mathbf{k})$. Point P has a position vector $3\mathbf{i} + 5\mathbf{j} + 6\mathbf{k}$.

a Write down the position vector of a point on the line.

b Following the method in question 10 find the shortest distance of point P from the line.

12 Find the shortest distance of the following points from the given lines:

a P(5, 1, 6) $\mathbf{r} = \mathbf{i} + 2\mathbf{k} + \lambda(2\mathbf{i} - \mathbf{j} + 3\mathbf{k})$

b P(8, 1, 8) $\mathbf{r} = 3\mathbf{i} + \mathbf{j} + 2\mathbf{k} + \lambda(\mathbf{i} - 3\mathbf{j} + 2\mathbf{k})$

c P(−3, 10, −12) $\mathbf{r} = 3\mathbf{i} - 4\mathbf{j} + 2\mathbf{k} + \lambda(2\mathbf{i} - \mathbf{j} + 4\mathbf{k})$.

13 If $\mathbf{a} = \begin{pmatrix} 2 \\ 1 \\ 3 \end{pmatrix}$, $\mathbf{b} = \begin{pmatrix} -4 \\ 2 \\ 1 \end{pmatrix}$ and $\mathbf{c} = \begin{pmatrix} 5 \\ 7 \\ 6 \end{pmatrix}$ then find the following:

a **a.b**

b **a.c**

c **b.c**

d which two vectors are perpendicular to one another.

e **a.(b + c)**

f an expression for **a.(b + c)** in terms of **a**, **b** and **c**.

14 If A, B and C have position vectors $\mathbf{a} = \begin{pmatrix} -1 \\ 4 \\ 5 \end{pmatrix}$, $\mathbf{b} = \begin{pmatrix} 3 \\ -1 \\ 2 \end{pmatrix}$ and $\mathbf{c} = \begin{pmatrix} 13 \\ 14 \\ 7 \end{pmatrix}$ respectively then find the following:

a **b − a**

b **c − b**

c the position vector of the point D which lies on BC and cuts BC in the ratio 2 : 3.

15* Find a vector which is perpendicular to each of the vectors $\begin{pmatrix} 1 \\ 3 \\ 4 \end{pmatrix}$ and $\begin{pmatrix} -1 \\ 2 \\ 1 \end{pmatrix}$.

16* Find a vector which is perpendicular to each of the vectors $\begin{pmatrix} 2 \\ -3 \\ 1 \end{pmatrix}$ and $\begin{pmatrix} 1 \\ 3 \\ 5 \end{pmatrix}$.

17* Find the vector equation of the straight line which:

Passes through A(3, 4, 6) and is perpendicular to AB where B is (11, −2, 6) and is also perpendicular to $\mathbf{i} - \mathbf{j} + \mathbf{k}$.

18* Line l has equation $\mathbf{r} = \begin{pmatrix} 1 \\ 3 \\ -1 \end{pmatrix} + \lambda \begin{pmatrix} 3 \\ 3 \\ 2 \end{pmatrix}$. Point P is on the line l where $\lambda = p$.

Point M has position vector $\begin{pmatrix} 3 \\ 5 \\ 15 \end{pmatrix}$.

a Write down the position vector of P, in terms of p.

b Show that $\overrightarrow{MP}.\begin{pmatrix} 3 \\ 3 \\ 2 \end{pmatrix} = 22p - 44$.

c Hence find the position vector of the foot of the perpendicular from M to line l.

19* Line L has equation $\begin{pmatrix} x \\ y \\ z \end{pmatrix} = \begin{pmatrix} 2 \\ -4 \\ 1 \end{pmatrix} + \lambda\begin{pmatrix} 4 \\ 1 \\ 3 \end{pmatrix}$. Point A is on line L where $\lambda = a$.

Point T has coordinates (5, 2, 21).

a Show that $\overrightarrow{TA}.\begin{pmatrix} 4 \\ 1 \\ 3 \end{pmatrix} = 26a - 78$.

b Hence find the coordinates of the foot of the perpendicular from T to line L.

REVIEW EXERCISE 11

1 Write each vector in terms of **c**, **d** or **c** and **d**.

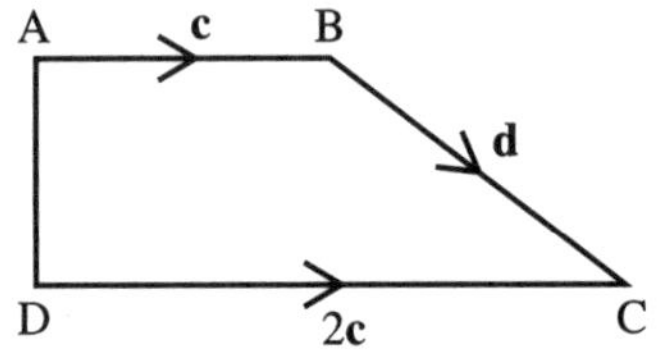

a $\overrightarrow{BA}$

b $\overrightarrow{AC}$

c $\overrightarrow{DB}$

d $\overrightarrow{AD}$

2 Write each vector in terms of **c**, **d** or **c** and **d**.

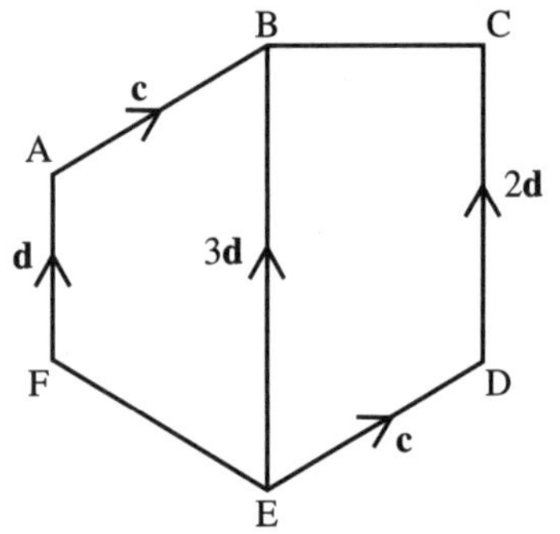

a $\overrightarrow{FE}$

b $\overrightarrow{BC}$

c $\overrightarrow{FC}$

d $\overrightarrow{DA}$

3 In the quadrilateral shown $\overrightarrow{OA} = 2\mathbf{a}$, $\overrightarrow{OB} = 2\mathbf{b}$, $\overrightarrow{OC} = 2\mathbf{c}$. Points P, Q, R and S are the mid-points of the sides shown.

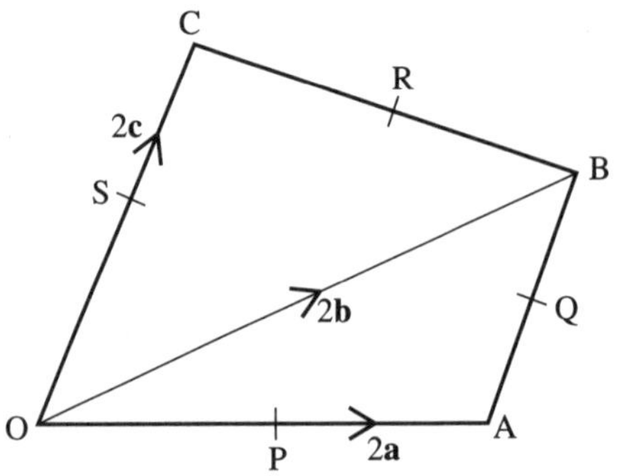

a Express in terms of **a**, **b** and **c**:

i $\overrightarrow{AB}$ **ii** $\overrightarrow{BC}$ **iii** $\overrightarrow{PQ}$ **iv** $\overrightarrow{QR}$ **v** $\overrightarrow{PS}$.

b Describe the relationship between QR and PS.

c What sort of quadrilateral is PQRS?

4 In the diagram, $\overrightarrow{OA} = \mathbf{a}$, $\overrightarrow{OB} = \mathbf{b}$, OC = CA, OB = BE and BD : DA = 1 : 2.

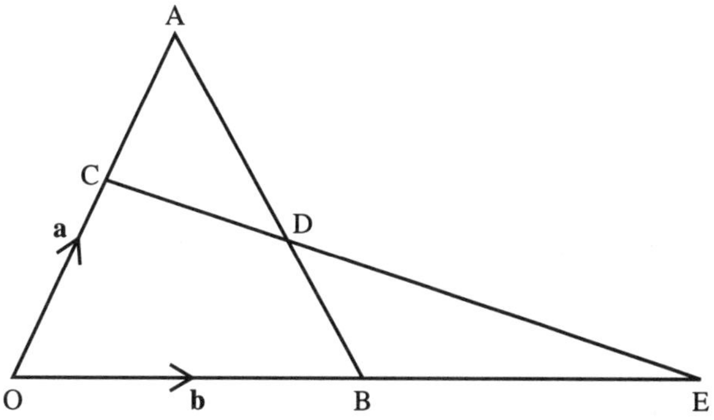

a Express in terms of **a** and **b**:

i $\overrightarrow{BA}$ **ii** $\overrightarrow{BD}$ **iii** $\overrightarrow{CD}$ **iv** $\overrightarrow{CE}$.

b Explain why points C, D and E lie on a straight line.

5 Find the magnitude of the following vectors.

a $5\mathbf{i} + 12\mathbf{j} + 0\mathbf{k}$ **b** $\mathbf{i} + \mathbf{j} + \mathbf{k}$ **c** $3\mathbf{i}$

d $\begin{pmatrix} 1 \\ -1 \\ \sqrt{7} \end{pmatrix}$ **e** $\begin{pmatrix} 4 \\ 3 \\ 2 \end{pmatrix}$

6 Find the distance between the following pairs of points:

a (1, 3, 6) and (3, 4, 9) **b** (−1, 2, 0) and (1, 2, 6)

7 Find a unit vector in the direction of:

a $\mathbf{i} + 2\mathbf{j} + 2\mathbf{k}$ **b** $6\mathbf{i} - 2\mathbf{j} + 3\mathbf{k}$

8 If $\overrightarrow{OP} = 2\mathbf{i} + 5\mathbf{j}$ and $\overrightarrow{OQ} = 3\mathbf{i} - \mathbf{j}$, find

a $\overrightarrow{PQ}$ **b** $\overrightarrow{QP}$

9 If $\mathbf{a} = \begin{pmatrix} 5 \\ 1 \\ 3 \end{pmatrix}$ and $\mathbf{b} = \begin{pmatrix} -2 \\ 2 \\ 3 \end{pmatrix}$, find

a $\mathbf{a} + \mathbf{b}$

b $|\mathbf{a} + \mathbf{b}|$

c vector **c** such that $\mathbf{a} + \mathbf{c} = \mathbf{b}$.

10 Find the vector equation of the straight line which:

a Passes through (2, 3, 5) and is parallel to $\begin{pmatrix} 3 \\ -1 \\ 7 \end{pmatrix}$.

b Passes through (2, 1, −3) and is parallel to the vector $3\mathbf{i} - \mathbf{j} + 5\mathbf{k}$.

c Passes through (7, 5, 2) and (1, −1, 4).

d Passes through (2, 0, −1) and is parallel to the straight line with vector equation $\mathbf{r} = (3 - 4\mu)\mathbf{i} + (2 + \mu)\mathbf{j} + (-3 + 5\mu)\mathbf{k}$.

11 Find whether the following pairs of straight lines intersect, are skew or are parallel. If they do intersect then find the point of intersection:

a $\mathbf{r} = \begin{pmatrix} 3 \\ 5 \\ 1 \end{pmatrix} + \lambda \begin{pmatrix} 2 \\ 3 \\ 4 \end{pmatrix}$ and $\mathbf{r} = \begin{pmatrix} 1 \\ 16 \\ 7 \end{pmatrix} + \mu \begin{pmatrix} 4 \\ -1 \\ 3 \end{pmatrix}$

b $\mathbf{r} = (2 + 3\lambda)\mathbf{i} + (7 + \lambda)\mathbf{j} + (4 + 5\lambda)\mathbf{k}$ and $\mathbf{r} = (4 + 2\mu)\mathbf{i} + (26 - 3\mu)\mathbf{j} + (5 - \mu)\mathbf{k}$

c $\mathbf{r} = \begin{pmatrix} 2 \\ 1 \\ 7 \end{pmatrix} + \lambda \begin{pmatrix} 2 \\ -3 \\ 1 \end{pmatrix}$ and $\mathbf{r} = \begin{pmatrix} 7 \\ 5 \\ 11 \end{pmatrix} + \mu \begin{pmatrix} 1 \\ 10 \\ 2 \end{pmatrix}$

12 Find **a.b** in the following:

a $\mathbf{a} = \begin{pmatrix} 2 \\ 1 \end{pmatrix}$ and $\mathbf{b} = \begin{pmatrix} 3 \\ 5 \end{pmatrix}$.

b $\mathbf{a} = \begin{pmatrix} 3 \\ -2 \end{pmatrix}$ and $\mathbf{b} = \begin{pmatrix} 2 \\ 3 \end{pmatrix}$.

c $\mathbf{a} = \begin{pmatrix} 2 \\ 1 \\ 5 \end{pmatrix}$ and $\mathbf{b} = \begin{pmatrix} 5 \\ 1 \\ 3 \end{pmatrix}$.

13 Find the value of t such that $\mathbf{a} = \begin{pmatrix} 5 \\ -2 \end{pmatrix}$ is perpendicular to $\mathbf{b} = \begin{pmatrix} 4 \\ t \end{pmatrix}$.

14 If $\mathbf{a} = \begin{pmatrix} 1 \\ 3 \\ 2 \end{pmatrix}$ then:

a find:

i the exact magnitude of **a** (denoted by $|\mathbf{a}|$)

ii **a.a**

b Show that $\mathbf{a}.\mathbf{a} = |\mathbf{a}|^2$

15 Find the angle (to 1 dp) between:

a $\mathbf{a} = \begin{pmatrix} 1 \\ 0 \end{pmatrix}$, $\mathbf{b} = \begin{pmatrix} 3 \\ 4 \end{pmatrix}$

b $\mathbf{a} = \begin{pmatrix} 0 \\ 2 \end{pmatrix}$, $\mathbf{b} = \begin{pmatrix} 5 \\ 12 \end{pmatrix}$

c $\mathbf{a} = \begin{pmatrix} 6 \\ 8 \end{pmatrix}$, $\mathbf{b} = \begin{pmatrix} 8 \\ 15 \end{pmatrix}$

d $\mathbf{a} = \begin{pmatrix} 7 \\ 24 \end{pmatrix}$, $\mathbf{b} = \begin{pmatrix} -3 \\ 4 \end{pmatrix}$

e $\mathbf{a} = \begin{pmatrix} 12 \\ 5 \\ 0 \end{pmatrix}$, $\mathbf{b} = \begin{pmatrix} 1 \\ 1 \\ 1 \end{pmatrix}$

f $\mathbf{a} = \begin{pmatrix} 1 \\ 2 \\ 2 \end{pmatrix}$, $\mathbf{b} = \begin{pmatrix} -6 \\ 2 \\ 3 \end{pmatrix}$

16 If $\mathbf{a} = 3\mathbf{i} + 5\mathbf{j} + 7\mathbf{k}$, $\mathbf{b} = 2\mathbf{i} - 3\mathbf{j} + \mathbf{k}$ and $\mathbf{c} = \mathbf{i} + 7\mathbf{j} - 2\mathbf{k}$ then find the following:

a $3\mathbf{a} + 2\mathbf{b}$

b $\mathbf{a} - \mathbf{b} + \mathbf{c}$

c $|\mathbf{a}|$

d **a.b**

e the angle **a** makes with the x-axis (to the nearest degree)

f the angle between **a** and **b** (to the nearest degree)

17 If A, B and C have co-ordinates (3, 4), (7, −4) and (14, 7) respectively then find:

a The vector equation of the line l_1 through A and B (use λ as the parameter).

b The vector equation of the line l_2 through C and perpendicular to AB (use μ as the parameter).

c The point of intersection of l_1 and l_2.

d The exact perpendicular distance between C and the line AB.

18 Three points A, B and C have position vectors $\mathbf{a} = \begin{pmatrix} 3 \\ 5 \\ 1 \end{pmatrix}$, $\mathbf{b} = \begin{pmatrix} 12 \\ 17 \\ 1 \end{pmatrix}$ and $\mathbf{c} = \begin{pmatrix} 7 \\ 9 \\ 3 \end{pmatrix}$ respectively.

a Find the vector equation of the line which passes through A and B.

b Find the vector equation of the line which passes through A and C.

c Calculate the angle CAB (to the nearest degree).

d Calculate the lengths AB and AC.

e Hence find the area of the triangle ABC (to 2 sf).

f Calculate the perpendicular distance (to 3 sf) from the point C to the line AB.

19 Three straight lines l_1, l_2 and l_3 have equations as follows:

$$l_1: \ \mathbf{r} = \begin{pmatrix} -2 \\ -1 \\ -5 \end{pmatrix} + s\begin{pmatrix} 3 \\ 4 \\ 12 \end{pmatrix}$$

$$l_2: \ \mathbf{r} = \begin{pmatrix} 11 \\ 13 \\ 12 \end{pmatrix} + t\begin{pmatrix} 2 \\ 2 \\ 1 \end{pmatrix}$$

$$l_3: \ \mathbf{r} = \begin{pmatrix} 10 \\ 11 \\ 1 \end{pmatrix} + u\begin{pmatrix} 3 \\ 2 \\ -9 \end{pmatrix}$$

a Find the co-ordinates of the following points:

i A, the point of intersection of l_1 and l_2.

ii B, the point of intersection of l_1 and l_3.

iii C, the point of intersection of l_2 and l_3.

b Find the angle enclosed between l_1 and l_2 (to 1 dp).

c Find the area of the triangle ABC (to 3 sf).

d Show that the exact area of ABC is $\dfrac{39\sqrt{5}}{2}$.

EXAMINATION EXERCISE 11

1 The diagram shows points O, A and B, with $\overrightarrow{OA} = \mathbf{a}$ and $\overrightarrow{OB} = \mathbf{b}$.

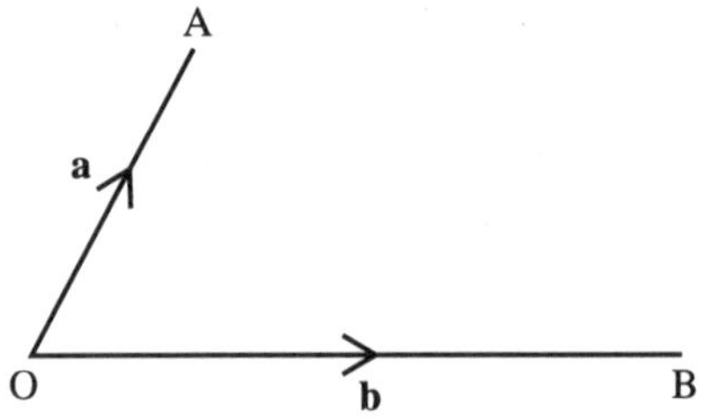

i Make a sketch of the diagram and mark the points C and D such that $\overrightarrow{OC} = \mathbf{a} + 2\mathbf{b}$ and $\overrightarrow{OD} = 2\mathbf{a} + \mathbf{b}$.

ii Express $\overrightarrow{DC}$ in terms of **a** and **b**, simplifying your answer.

iii Prove that ABCD is a parallelogram. [OCR]

2 The line L_1 passes through the point (3, 6, 1) and is parallel to the vector $2\mathbf{i} + 3\mathbf{j} - \mathbf{k}$. The line L_2 passes through the point (3, −1, 4) and is parallel to the vector $\mathbf{i} - 2\mathbf{j} + \mathbf{k}$.

i Write down vector equations for the lines L_1 and L_2.

ii Prove that L_1 and L_2 intersect, and find the coordinates of the point of intersection.

iii Calculate the acute angle between the lines. [OCR]

3 Two lines have equations

$$\mathbf{r} = \begin{pmatrix} 3 \\ 1 \\ -2 \end{pmatrix} + s\begin{pmatrix} 1 \\ -1 \\ 4 \end{pmatrix} \quad \text{and} \quad \mathbf{r} = \begin{pmatrix} 1 \\ 0 \\ 5 \end{pmatrix} + t\begin{pmatrix} -2 \\ -3 \\ 1 \end{pmatrix}.$$

i Find the acute angle between the directions of the two lines.

ii Prove that the lines are skew. [OCR]

4 Referred to an origin O, the points A, B and C have position vectors $(9\mathbf{i} - 2\mathbf{j} + \mathbf{k})$, $(6\mathbf{i} + 2\mathbf{j} + 6\mathbf{k})$ and $(3\mathbf{i} + p\mathbf{j} + q\mathbf{k})$ respectively, where p and q are constants.

a Find, in vector form, an equation of the line l which passes through A and B.

Given that C lies on l,

b find the value of p and the value of q,

c calculate, in degrees, the acute angle between OC and AB.

The point D lies on AB and is such that OD is perpendicular to AB.

d Find the position vector of D. [EDEXCEL]

5 The position vectors of the points A and B, relative to a fixed origin O, are $6\mathbf{i} + 4\mathbf{j} - \mathbf{k}$ and $8\mathbf{i} + 5\mathbf{j} - 3\mathbf{k}$ respectively.

i Find $\overrightarrow{AB}$.

ii Find the length AB.

iii Show that, for all values of the parameter λ, the point P with position vector

$$(8 + 2\lambda)\mathbf{i} + (5 + \lambda)\mathbf{j} - (3 + 2\lambda)\mathbf{k}$$

lies on the straight line through A and B.

iv Determine the value of λ for which $\overrightarrow{OP}$ is perpendicular to $\overrightarrow{AB}$. [OCR]

6 The equations of the lines l_1 and l_2 are

$$\mathbf{r} = \begin{pmatrix} 9 \\ 2 \\ -1 \end{pmatrix} + s\begin{pmatrix} 3 \\ -4 \\ 0 \end{pmatrix} \quad \text{and} \quad \mathbf{r} = \begin{pmatrix} 2 \\ 3 \\ 2 \end{pmatrix} + t\begin{pmatrix} a \\ 3 \\ b \end{pmatrix}$$

respectively. It is given that l_1 and l_2 are perpendicular and also that l_1 and l_2 intersect.

i Find the values of a and b.

ii Hence show that the point of intersection A has position vector $\begin{pmatrix} 6 \\ 6 \\ -1 \end{pmatrix}$.

iii There are two points on l_1 whose distance from A is 10. Find the position vector of **one** of these points. [OCR]

7 Relative to a fixed origin O, the point A has position vector $3\mathbf{i} + 2\mathbf{j} - \mathbf{k}$, the point B has position vector $5\mathbf{i} + \mathbf{j} + \mathbf{k}$, and the point C has position vector $7\mathbf{i} - \mathbf{j}$.

a Find the cosine of angle ABC.

b Find the exact value of the area of triangle ABC.

The point D has position vector $7\mathbf{i} + 3\mathbf{k}$.

c Show that AC is perpendicular to CD.

d Find the ratio AD : DB. [EDEXCEL]

8 The equations of the lines l_1 and l_2 are given by

$$l_1: \quad \mathbf{r} = \mathbf{i} + 3\mathbf{j} + 5\mathbf{k} + \lambda(\mathbf{i} + 2\mathbf{j} - \mathbf{k}),$$
$$l_2: \quad \mathbf{r} = -2\mathbf{i} + 3\mathbf{j} - 4\mathbf{k} + \mu(2\mathbf{i} + \mathbf{j} + 4\mathbf{k}),$$

where λ and μ are parameters.

a Show that l_1 and l_2 intersect and find the coordinates of Q, their point of intersection.

b Show that l_1 is perpendicular to l_2.

The point P with x-coordinate 3 lies on the line l_1 and the point R with x-coordinate 4 lies on the line l_2.

c Find, in its simplest form, the exact area of the triangle PQR. [EDEXCEL]

9 The line l_1 passes through the point A(2, 0, 2) and has equation

$$\mathbf{r} = \begin{pmatrix} 2 \\ 0 \\ 2 \end{pmatrix} + t\begin{pmatrix} 2 \\ 6 \\ -3 \end{pmatrix}.$$

a Show that the line l_2 which passes through the points B(4, 4, −5) and C(0, −8, 1) is parallel to the line l_1.

b P is the point on line l_1 such that angle CPA is a right angle.

i Find the value of the parameter t at P.

ii Hence show that the shortest distance between the lines l_1 and l_2 is $2\sqrt{5}$.

[AQA]

10 Referred to a fixed origin O, the points A and B have position vectors $(\mathbf{i} + 2\mathbf{j} - 3\mathbf{k})$ and $(5\mathbf{i} - 3\mathbf{j})$ respectively.

a Find, in vector form, an equation of the line l_1 which passes through A and B.

The line l_2 has equation $\mathbf{r} = (4\mathbf{i} - 4\mathbf{j} + 3\mathbf{k}) + \mu(\mathbf{i} - 2\mathbf{j} + 2\mathbf{k})$, where μ is a scalar parameter.

b Show that A lies on l_2.

c Find, in degrees, the acute angle between the lines l_1 and l_2.

The point C with position vector $(2\mathbf{i} - \mathbf{k})$ lies on l_2.

d Find the shortest distance from C to the line l_1. [EDEXCEL]

ANSWERS

EXERCISE 1 *page 1*

1 $\frac{3}{5}$ **2** $4a$ **3** $\frac{1}{3}$ **4** 4

5 $\frac{a}{2b}$ **6** 3 **7** $\frac{a}{2}$ **8** $2b$

9 $\frac{3}{4y}$ **10** $\frac{11y}{12x}$ **11** $\frac{2ya}{3}$ **12** $\frac{4m}{n}$

13 A/G, B/E, C/H, D/F

14 $\frac{a}{5b}$ **15** a **16** $\frac{7}{8}$ **17** $\frac{3}{4-x}$

18 $\frac{5+2x}{3}$ **19** $\frac{3x+1}{x}$ **20** $\frac{4+5a}{5}$ **21** $\frac{b}{3+2a}$

22 $\frac{3-x}{2}$ **23** $\frac{2x+1}{y}$ **24** $\frac{2x+5}{2}$ **25** $y+xy$

26 $\frac{5x+4}{3x-2}$ **27** $\frac{1+4a}{b+b^2}$ **28** $\frac{1+2x+2x^2}{x}$

29 $\frac{6n-3m}{2mn}$

30 a False **b** True **c** True **d** False

31 a $(x-3)(x+2)$ **b** $x-3$

32 a $x-4$ **b** $\frac{x-2}{x-1}$ **c** $\frac{x+5}{x+2}$

d $\frac{x}{x+1}$ **e** $\frac{x+4}{2(x-5)}$ **f** $\frac{x+5}{x-2}$

g $\frac{x}{2x-1}$ **h** $\frac{2x+1}{2x+3}$ **i** $\frac{x-1}{3x-4}$

33 $\frac{6x+3}{6x+2}$

34 a $\frac{x^2+1}{x^2}$ **b** $2x^2-1$ **c** $2x-1$

d $12x+1$ **e** $30x-2$ **f** $\frac{1-4x}{2}$

g $\frac{3x^2+1}{x^2+2}$ **h** $\frac{x+2}{x}$ **i** $\frac{x}{2x+3}$

j x^2-1 **k** $\frac{x+2}{x}$ **l** $\frac{x+3}{x+4}$

EXERCISE 2 *page 3*

1 a $\frac{3x}{5}$ **b** $\frac{3}{x}$ **c** $\frac{4x}{7}$ **d** $\frac{4}{7x}$

e $\frac{7x}{8}$ **f** $\frac{7}{8x}$ **g** $\frac{5x}{6}$ **h** $\frac{5}{6x}$

2 AF, BH, CD, EG

3 a $\frac{23x}{20}$ **b** $\frac{23}{20x}$ **c** $\frac{x}{12}$

d $\frac{1}{12x}$ **e** $\frac{5x+2}{6}$ **f** $\frac{7x+2}{12}$

4 a $\frac{1-2x}{20}$ **b** $\frac{2x-9}{15}$ **c** $\frac{3x+12}{14}$

5 a $\frac{9x+13}{10}$ **b** $\frac{3x+1}{x(x+1)}$

c $\frac{7x-8}{x(x-2)}$ **d** $\frac{8x+9}{(x-2)(x+3)}$

e $\frac{4x+11}{(x+1)(x+2)}$ **f** $\frac{x^2+4x+6}{2(x+2)}$

EXERCISE 3 *page 4*

1 a **2** $\frac{10m}{3}$ **3** $\frac{2y}{3x}$ **4** $\frac{15b}{4a}$

5 $\frac{4}{a^2}$ **6** $\frac{8}{3}$ **7** $\frac{x-1}{x+2}$ **8** $\frac{y^2}{5x}$

9 $\frac{x}{8}$ **10** $3(x+2)$

11 a 6 **b** 4 **c** $2a$

12 AF, BG, CE, DG

13 a a **b** $\frac{10m}{3}$ **c** $\frac{2xy}{9}$ **d** $\frac{a}{q}$

e $\frac{8}{3}$ **f** $\frac{y^2}{5x}$ **g** $2a$ **h** 1

14 a 1 **b** $\frac{ab^3}{x}$ **c** $\frac{x}{x-1}$ **d** $\frac{z^2}{y}$

15 a $\frac{x}{10}$ **b** $3x$ **c** 11

16 $\frac{1}{4}$

17 $\frac{1}{5}$

EXERCISE 4 *page 7*

1 a one-one **b** many-one
c one-one **d** one-one
e many-one **f** many-one
g many-one **h** many-one
i one-one **j** one-one

2 a $f(x) \geqslant 3$ **b** $f(x) \geqslant 2$
c $f(x) \geqslant 1$ **d** $f(x) \geqslant 0$
e $0 \leqslant f(x) \leqslant 1$ **f** $-\frac{9}{4} \leqslant f(x) \leqslant 0$
g $0 \leqslant f(x) \leqslant \frac{25}{4}$ **h** $0 \leqslant f(x) \leqslant 5$
i $0 < f(x) \leqslant 1$ **j** $0 < f(x) \leqslant 1$

3 a $(x+3)^2-5$ **b** $f(x) \geqslant -5$

4 a $f(x) \geqslant 3$ **b** $f(x) \geqslant -50$
c $0 < f(x) \leqslant \frac{1}{2}$ **d** $0 < f(x) \leqslant 1$

EXERCISE 5 *page 10*

1 a $x \to 4(x+5)$ **b** $x \to 4x+5$
c $x \to (4x)^2$ **d** $x \to 4x^2$
e $x \to x^2+5$ **f** $x \to 4(x^2+5)$
g $x \to [4(x+5)]^2$

2 a -2.5 **b** $\pm\sqrt{\frac{5}{3}}$

3 a $x \to 2(x-3)$ b $x \to 2x-3$
c $x \to x^2-3$ d $x \to (2x)^2$
e $x \to (2x)^2-3$ f $x \to (2x-3)^2$

4 a 2 b 11 c 6
d 2 e 1 f 64

5 a -3 b 2 c $1\frac{1}{2}$ d 5

6 a $x \to 2(3x-1)+1$ b $x \to 3(2x+1)-1$
c $x \to 2x^2+1$ d $x \to (3x-1)^2$
e $x \to 2(3x-1)^2+1$ f $x \to 3(2x^2+1)-1$

7 a 11 b 9 c 11
d 14 e 81 f -1

8 a 2 b 0, 2 c $\pm\sqrt{2}$

9 $x \to \dfrac{x+2}{5}$ 10 $x \to \dfrac{x}{5}+2$

11 $x \to \dfrac{x}{6}-2$ 12 $x \to \dfrac{3x-1}{2}$

13 $x \to \dfrac{4x}{3}+1$ 14 $x \to \dfrac{(x+6)/2-4}{3}$

15 $x \to \dfrac{2(x-10)-4}{5}$ 16 $x \to \dfrac{x-3}{-7}$

17 $x \to \dfrac{3x-12}{-5}$ 18 $x \to \dfrac{3(x-2)-4}{-1}$

19 $x \to \dfrac{4(5x+3)+1}{2}$ 20 $x \to \dfrac{12}{x}$

21 a $x \to \dfrac{x}{3}$ b $x \to x+5$ c $x \to \dfrac{x-1}{2}$
d $x \to 3(x-5)$ e $x \to \dfrac{x}{3}+5$ f $x \to \dfrac{x}{3}+5$

22 a $x \to 6x+1$ b $x \to \dfrac{2x}{3}+1$ c $x \to \dfrac{x-1}{6}$
d $x \to 2x-9$ e $x \to \dfrac{x+9}{2}$ f $x \to \dfrac{x+9}{2}$

23 a 7 b 21 c 5
d 5 e 2

24 a 1 b $x > -2\frac{1}{2}$; $x > 5$

25 a $x \to \dfrac{3x-5}{2}$ b 2

EXERCISE 6 *page 13*

1 a $2-x$ b $\dfrac{13}{2x}$ c $\dfrac{3-x}{5}$
d $\dfrac{7x+5}{x-2}$ e $\dfrac{7x+5}{3-2x}$

2

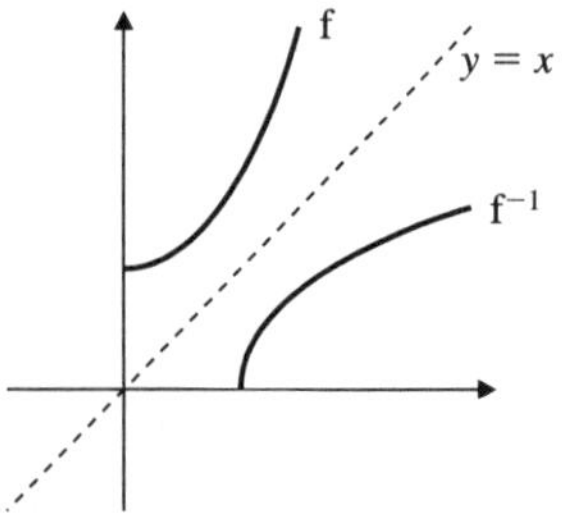

3 a $\dfrac{x+8}{3}, \dfrac{x+22}{3}$ b 4, 8

4 a $9x-4, \dfrac{x+1}{3}$ b Both are $\dfrac{x+4}{9}$

5 a $\dfrac{8x+10}{x-1}$ b $12\frac{1}{2}$
c 10 d 10

6 i $f(x) \geqslant -4$
ii domain $x \geqslant -4$, range $f^{-1}(x) \geqslant 1$
iii

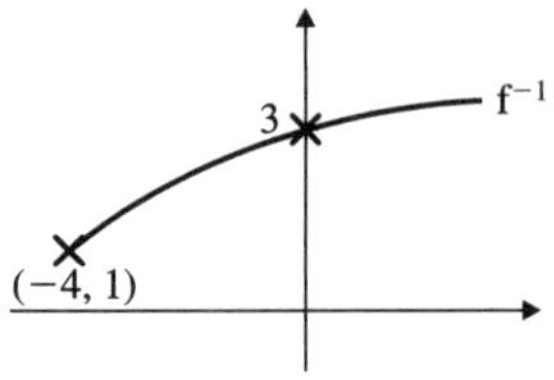

7 a $f^{-1}(x) = \dfrac{1}{2-x}, x \neq 2$
b $x = 1$

8 i domain $x \geqslant 0$, range $f^{-1}(x) \geqslant 2$
ii $f^{-1}(x) = 2+\sqrt{x}$
iii $x = 4$

9 a $a = 2$ $b = 0$
b $f^{-1}(x) = x^2+2, x \geqslant 0$; $g^{-1}(x) = \dfrac{1}{\sqrt{x}}, x \geqslant 0$

10 $f^{-1}(x) = \dfrac{x-1}{2}$; $g^{-1}(x) = \dfrac{5+3x}{x}, x \neq 0$

11 a $-\infty < x < \infty$ b $x \geqslant 0$
c $-90 < x \leqslant 90°$ d $x \geqslant 1$
e $x \leqslant -1$ f $x \geqslant -4$

EXERCISE 7 *page 17*

1 a

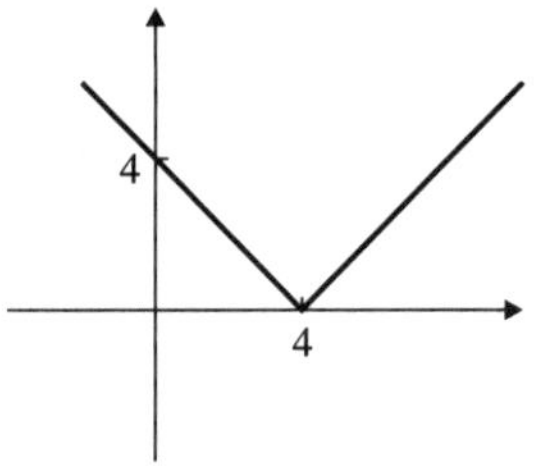

b

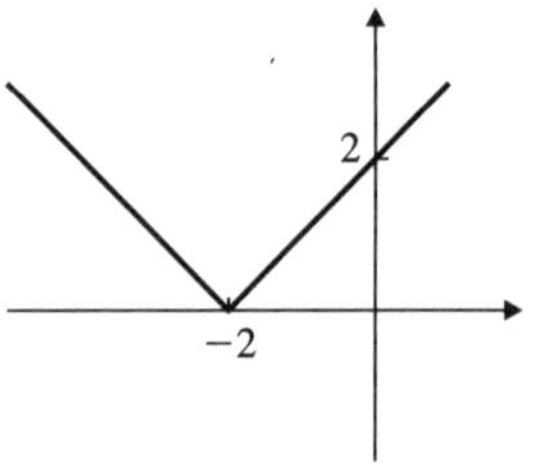

c

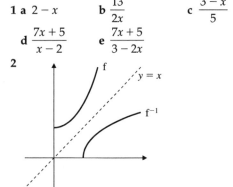

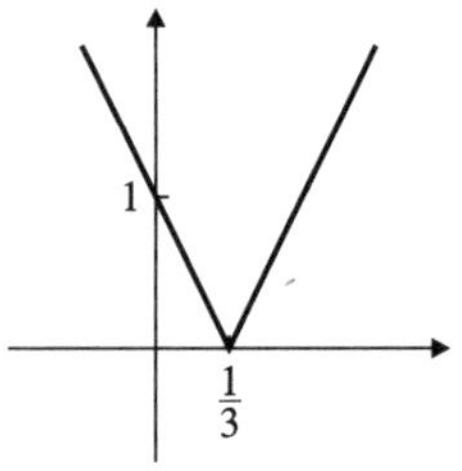

d

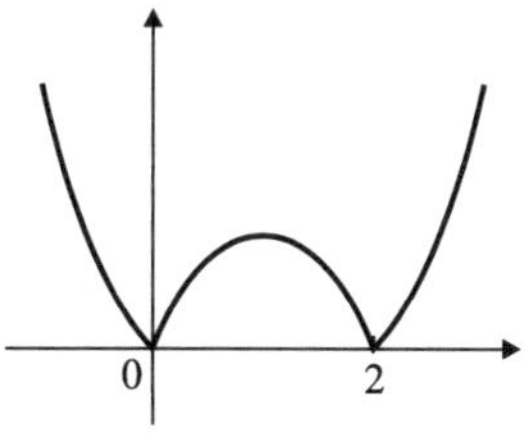

e

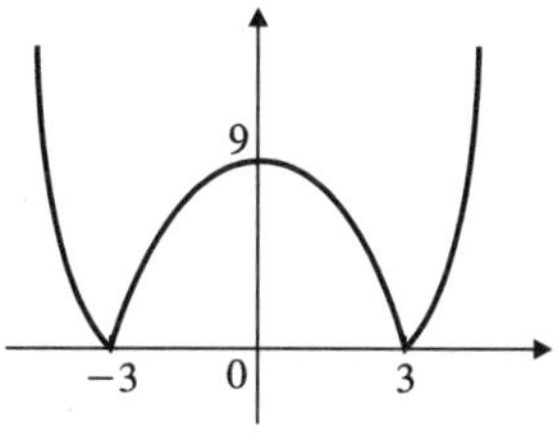

f

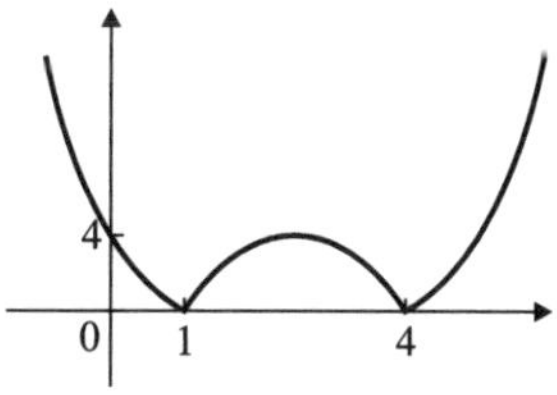

g

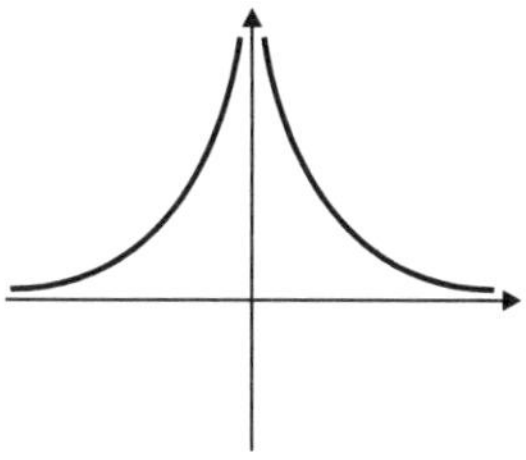

h

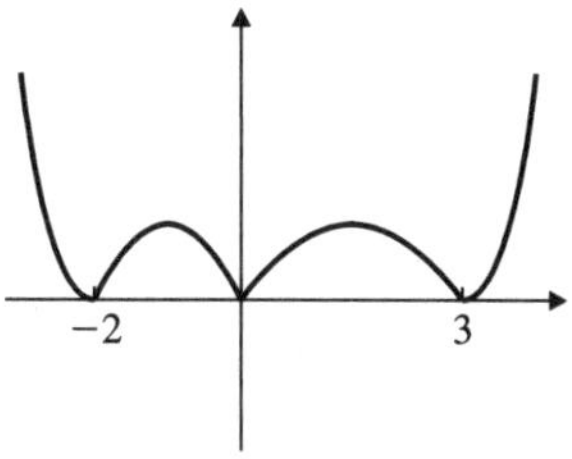

2 a $-3, 7$ **b** $0, 3$
c $-2, 3$ **d** $\pm 1, \pm 3$
e $\pm 5, \pm 7$ **f** $-7, 1$
g $-1, 2$ **h** $\pm \frac{1}{2}$

3 a $x < -1, x > 9$ **b** $-4 < x < 2$
c $x > 2, x < -5$ **d** all x
e $-\frac{14}{3} < x < 6$ **f** $-2 < x < 2$
g $\frac{1}{3} < x < 1$ **h** $x > -1$
i $0 < x < \frac{4}{3}$

4 $4 \pm \sqrt{11}$

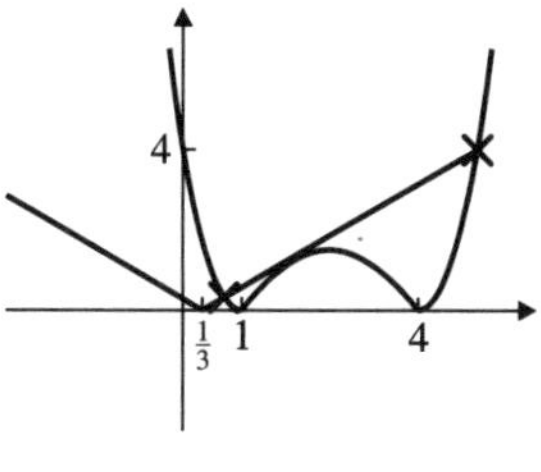

5 a

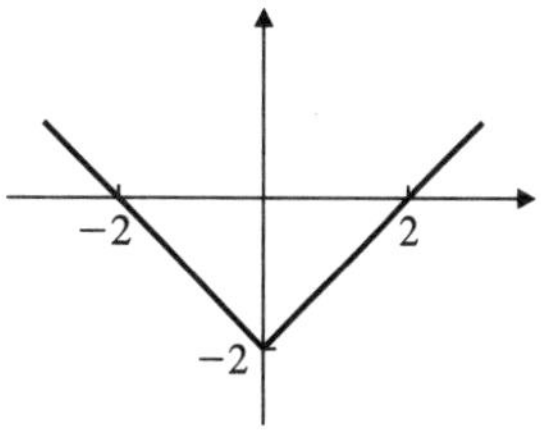

b

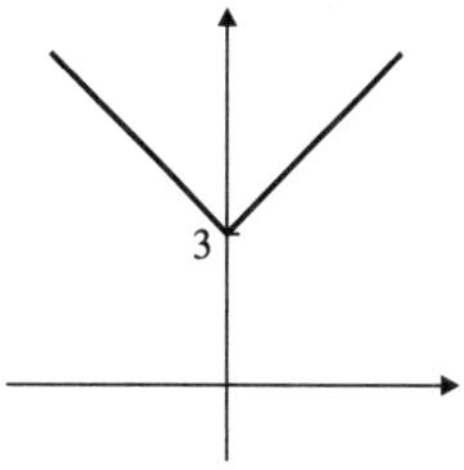

c

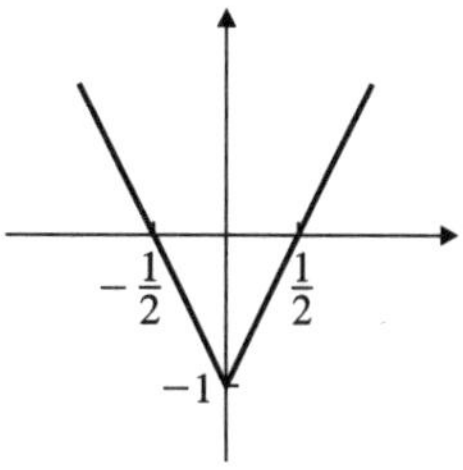

d

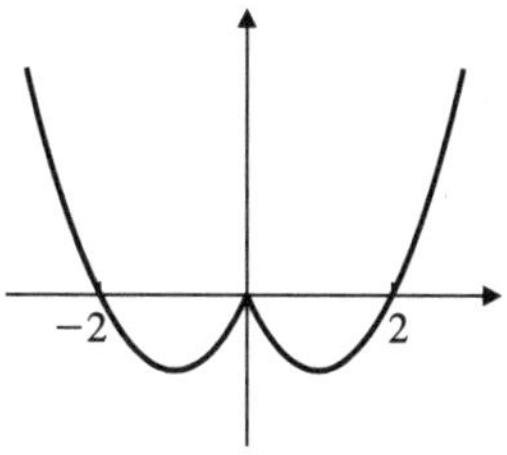

e

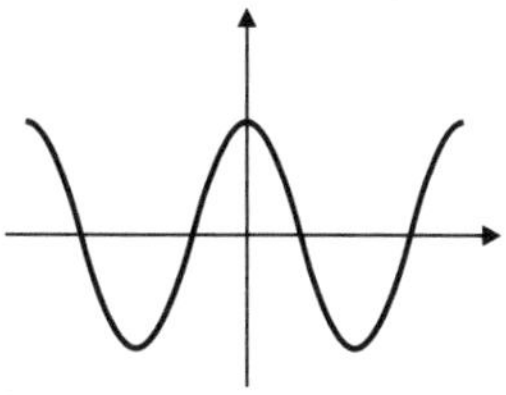

f

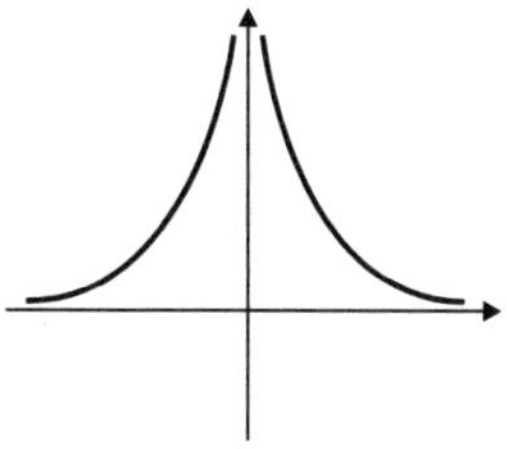

6 a

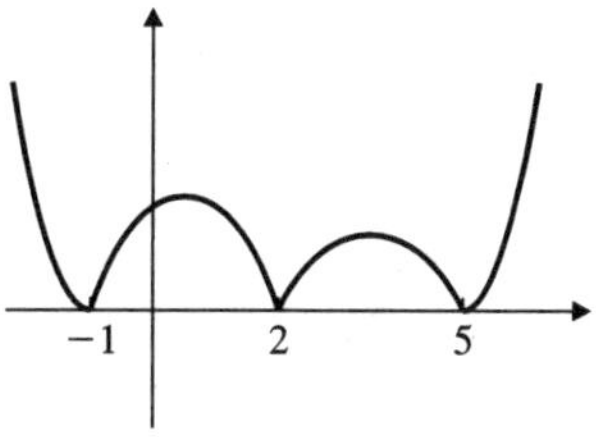

b

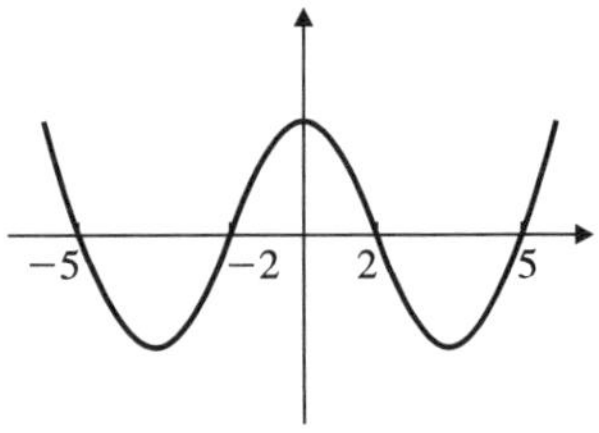

c

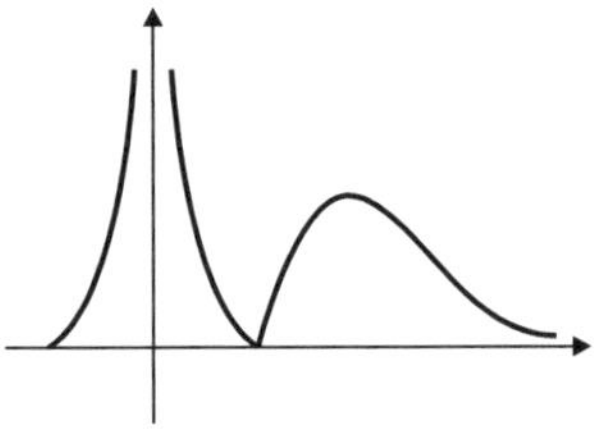

d

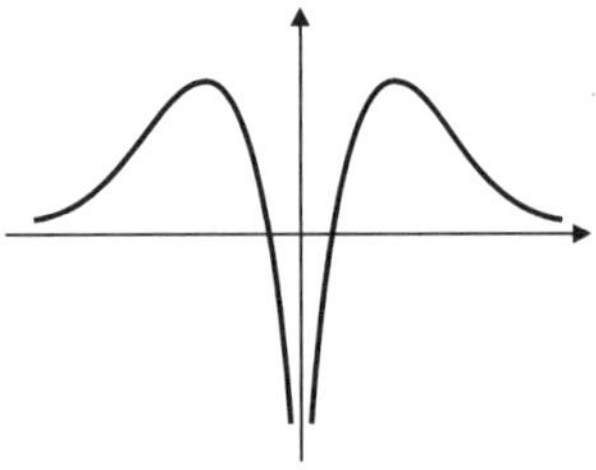

7

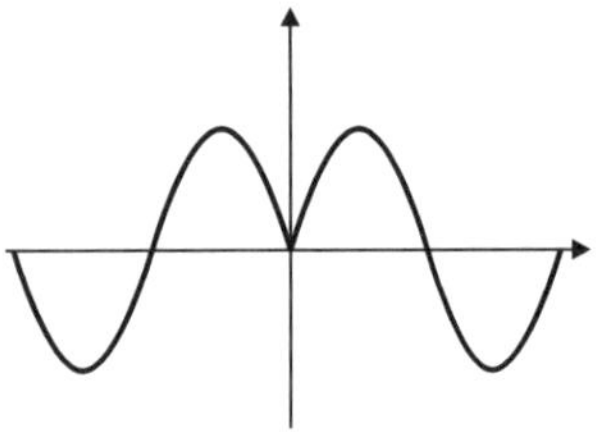

8 a

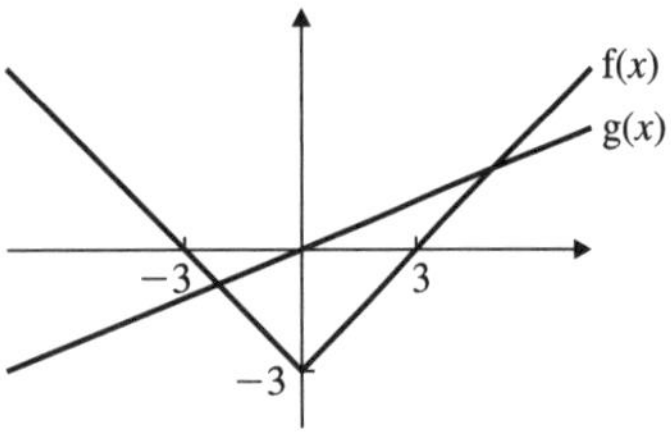

b $x > \frac{9}{2}, x < -\frac{9}{4}$

9 a

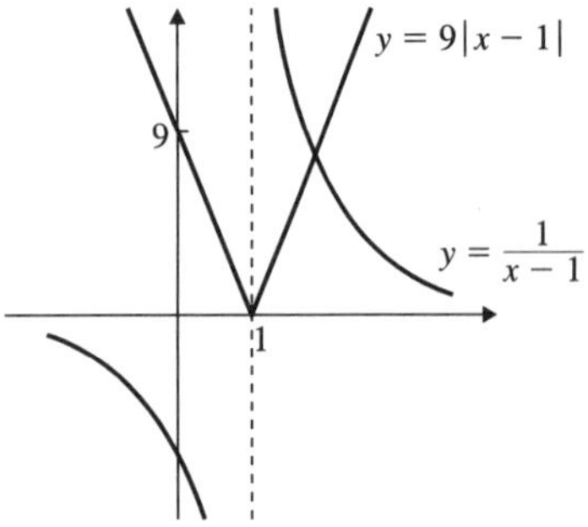

b $x > \frac{4}{3}, x < 1$

10 a i

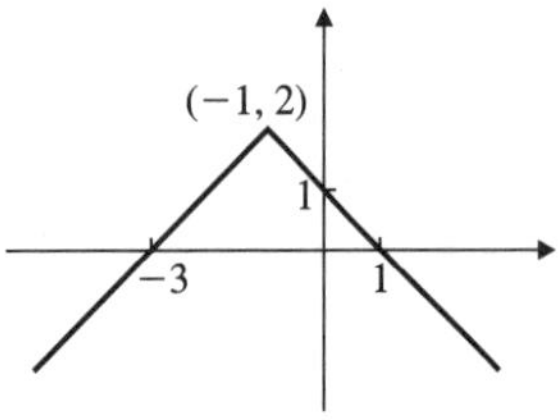

ii

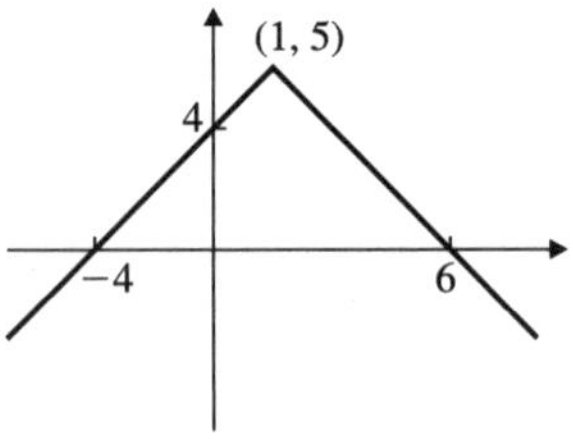

b i range $y < 2$ **ii** range $y < 5$

11 a

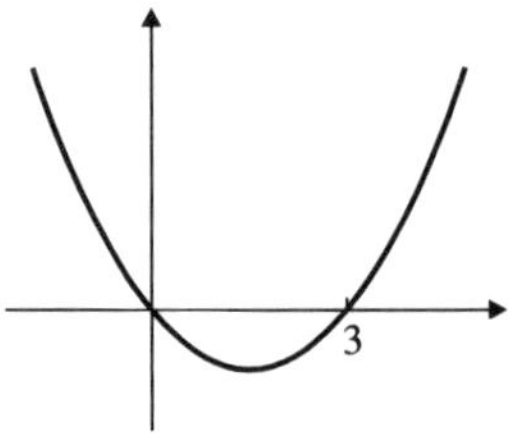

b

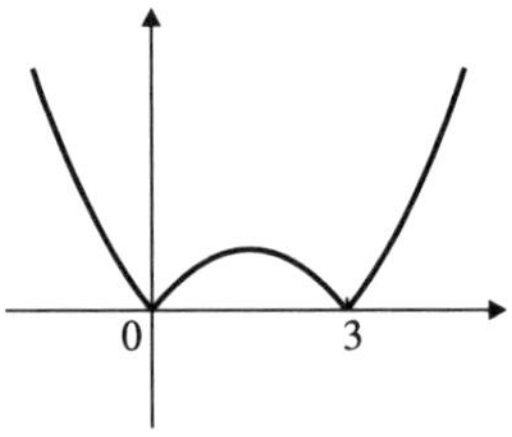

c

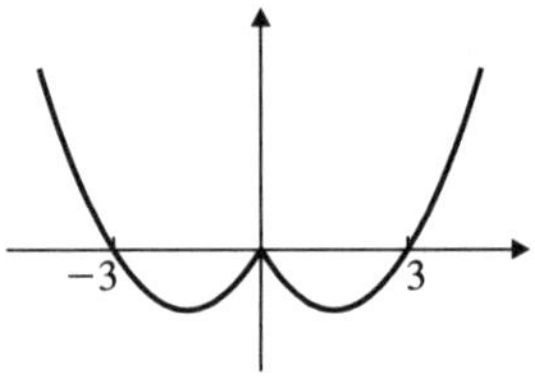

EXERCISE 8 *page 21*

1 a

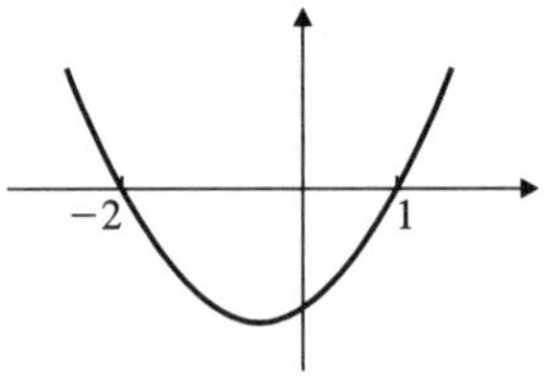

b

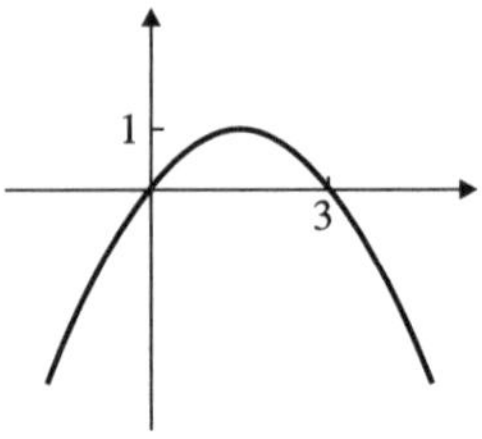

2 a

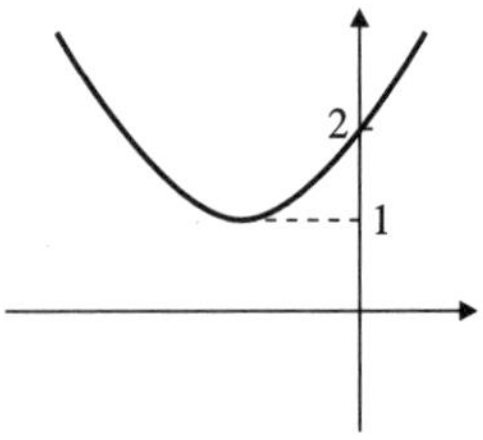

b

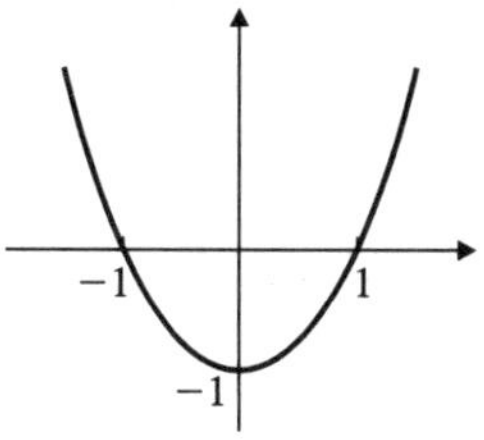

c

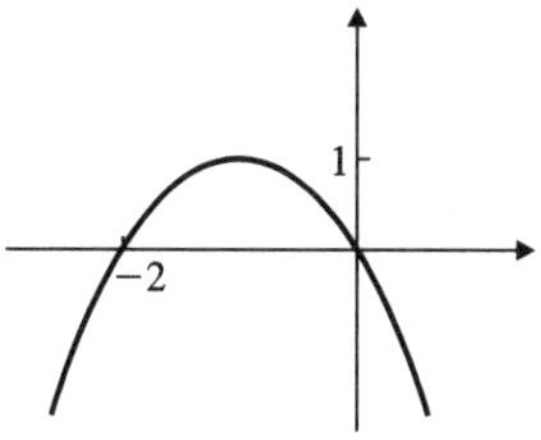

d

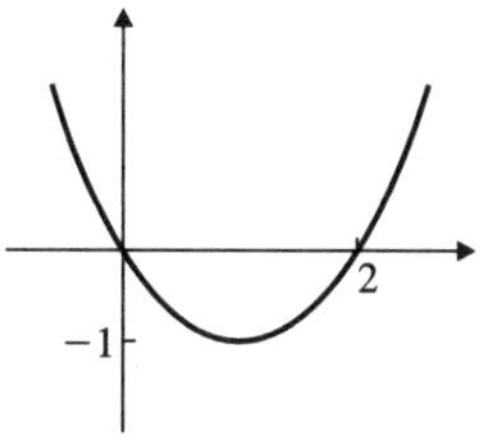

3 a

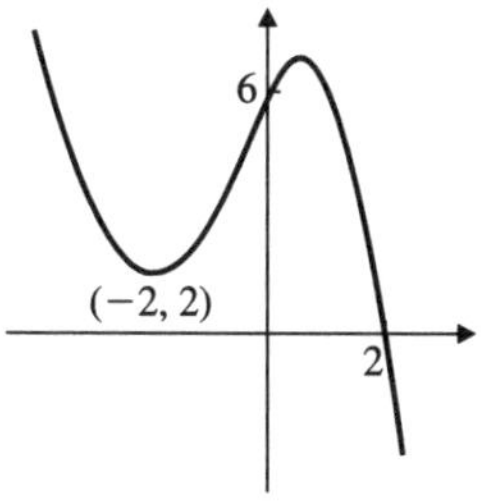

b

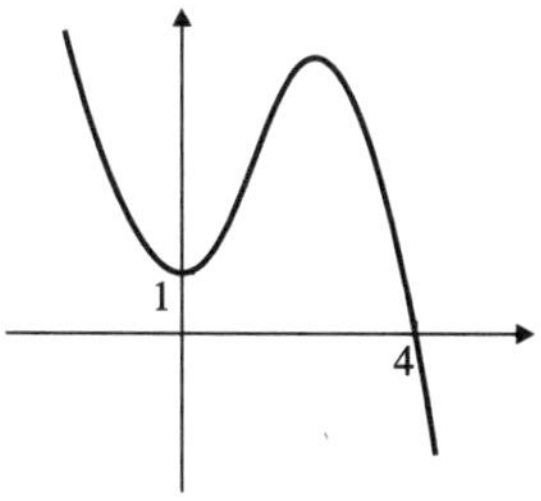

c

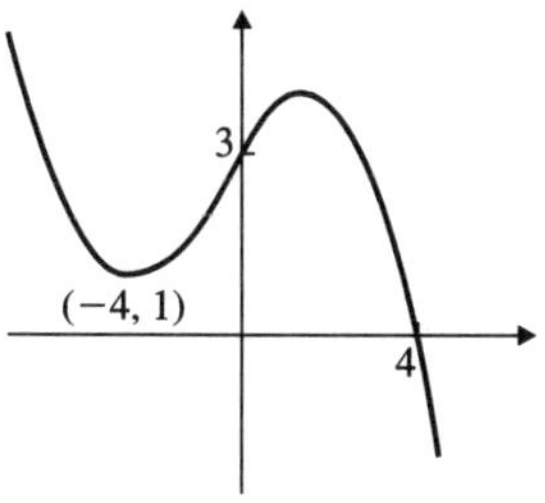

4

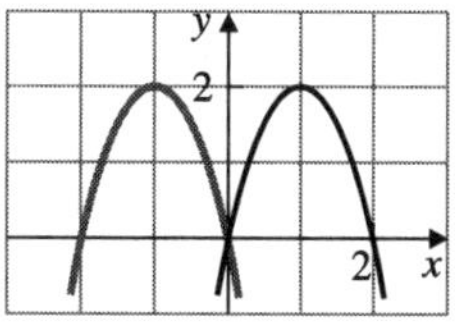

Sketch f(x + 2)

5

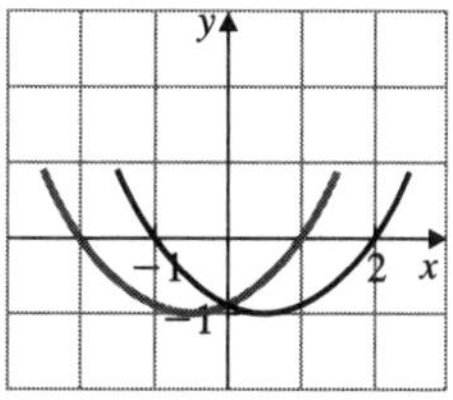

Sketch f($-x$)

6

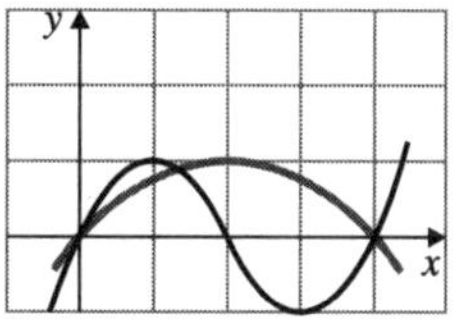

Sketch f($\frac{1}{2}x$)

7

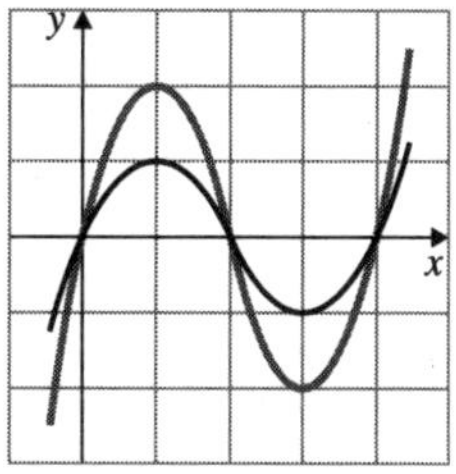

8

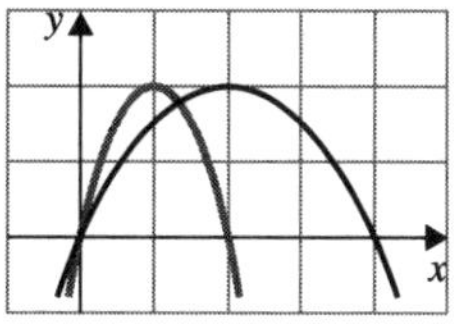

9

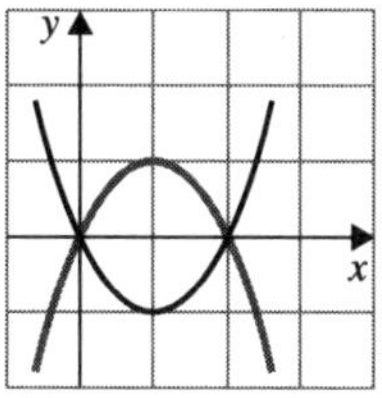

10 a

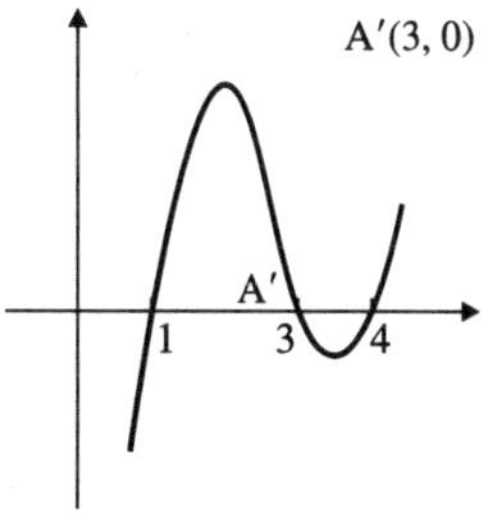

b

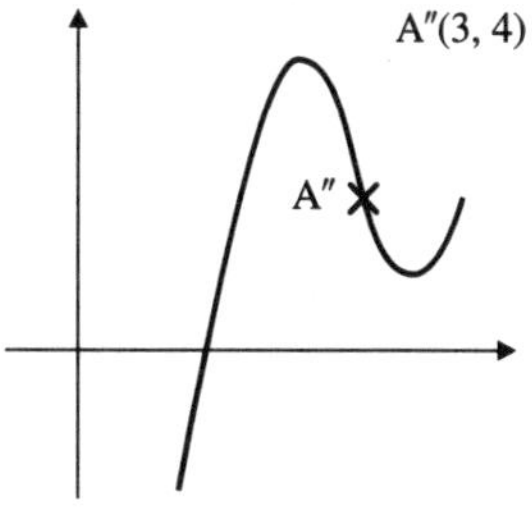

11 a

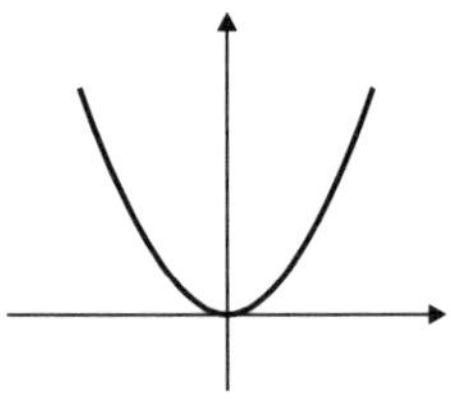

b

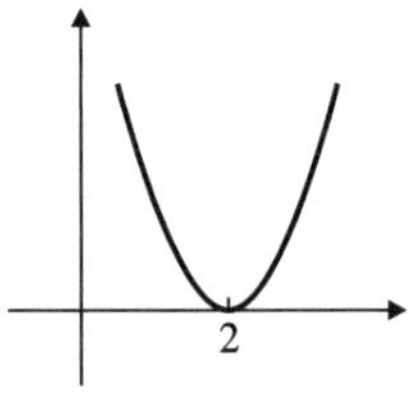

c

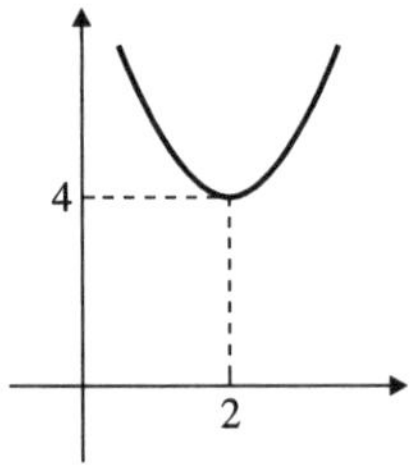

12 a

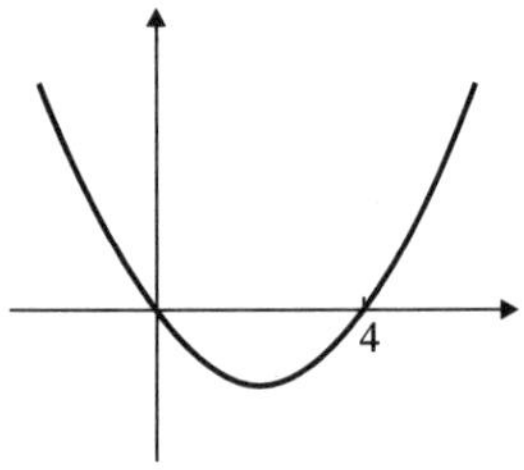

b

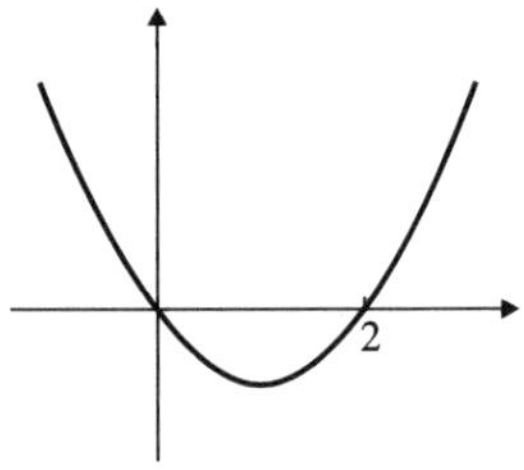

c

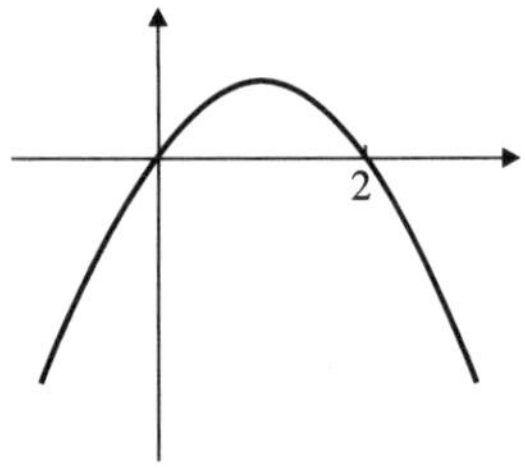

13 a

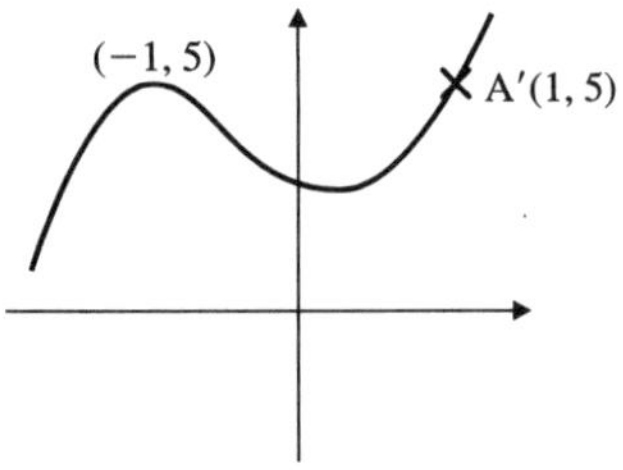

b

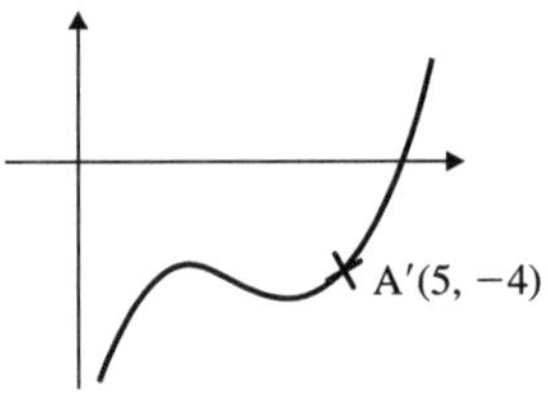

c

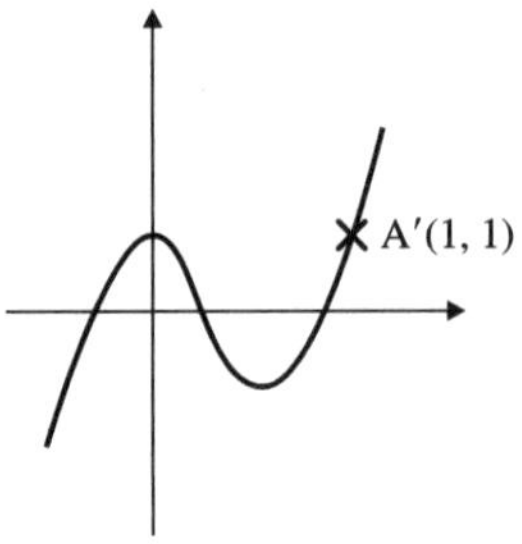

14 a $y = x^3 + 5$
b $y = (x - 2)^3$
c $y = (x - 2)^3 + 5$

15 a

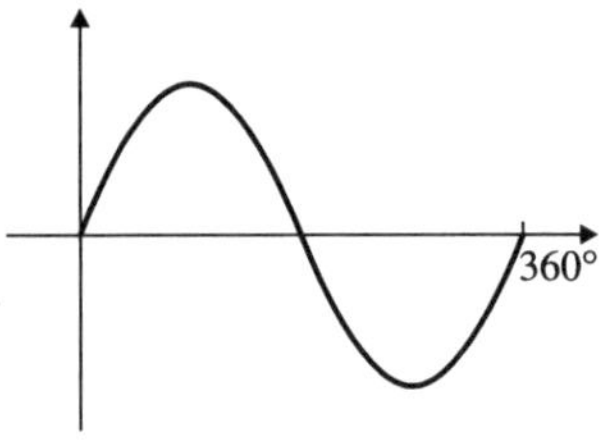

b

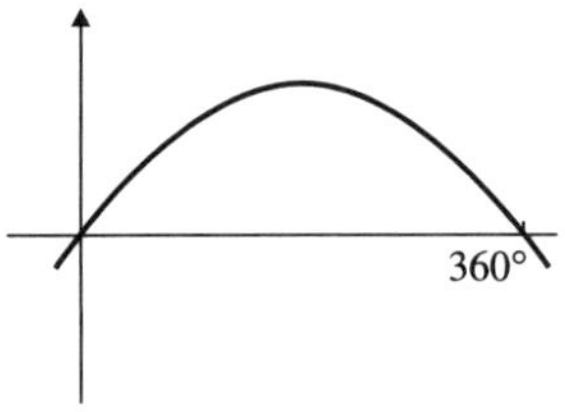

16 a

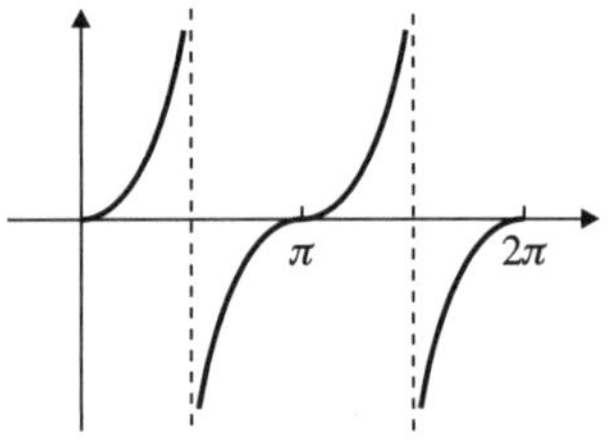

b

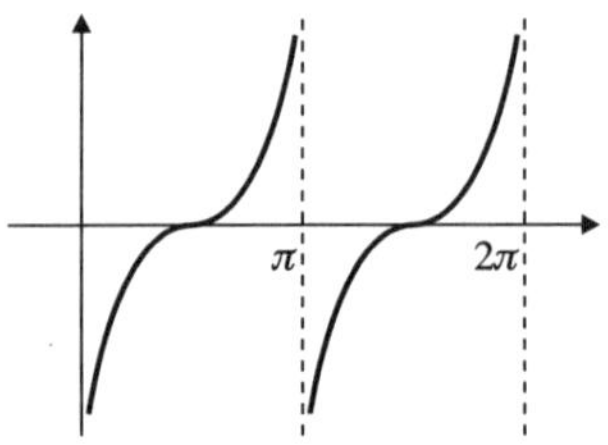

17 Stretch, scale factor 3 parallel to y-axis,
Translation $\begin{pmatrix} 0 \\ -4 \end{pmatrix}$

18 Stretch, scale factor 4 parallel to y-axis,
Translation $\begin{pmatrix} 0 \\ 9 \end{pmatrix}$

19 a $a = 2, b = 3$

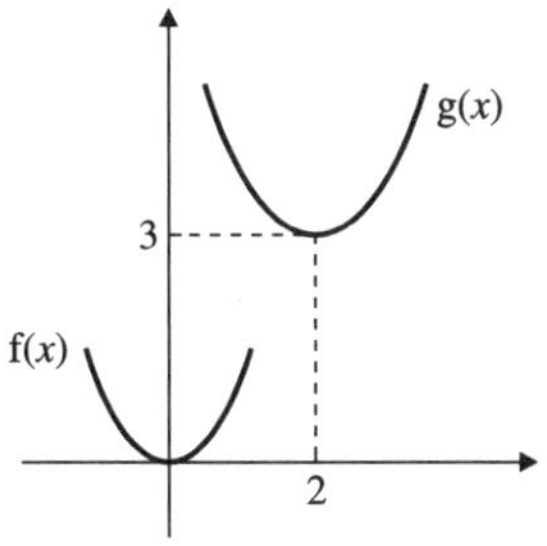

20 a $a = 4, b = 1$

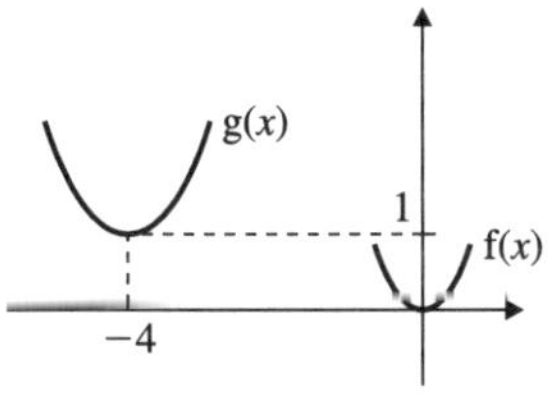

21

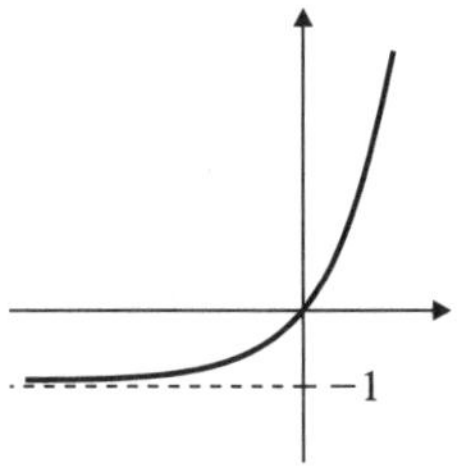

22 a

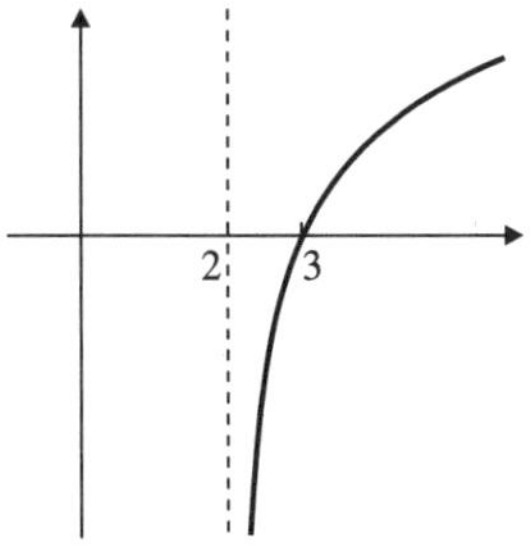

b

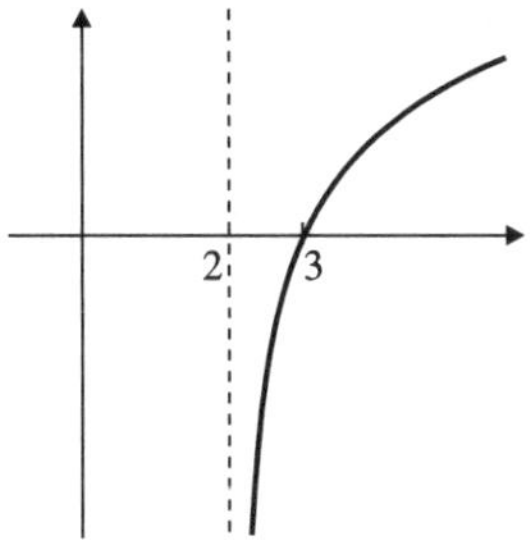

Review Exercise 1 *page 23*

1 a $\dfrac{5}{3-x}$ **b** $\dfrac{x^2}{2}$ **c** $\dfrac{3x+2}{x}$

d $1 + x$ **e** $\dfrac{1+4x}{y+y^2}$ **f** $\dfrac{3x}{2}$

2 $4(x-1)(x+1), \dfrac{(x-4)}{4(x+1)}$

3 a $\dfrac{x+2}{x}$ **b** $\dfrac{x-3}{x+1}$ **c** $\dfrac{2x-1}{x-4}$

d $\dfrac{2x^2+1}{x^2}$ **e** $3x^2 - 1$ **f** $\dfrac{4x^2-1}{x}$

4 a $\dfrac{8x+5}{6}$ **b** $\dfrac{4x+1}{x(x+1)}$

c $\dfrac{x^2+3x-2}{(x+1)(x-1)}$ **d** $\dfrac{x+4}{12}$

e $-\dfrac{(3x+5)}{10}$ **f** $\dfrac{2(x-1)}{x+3}$

5 a $(2x-1)$ is common factor

b $\dfrac{3x+1}{x(2x+1)}$

6 a $f(x) \leqslant 1$

b $x = 0, \pm\sqrt{2}$

7 a $\dfrac{5x+2}{2x+1}$ **b** $\dfrac{1}{x-2}$

8 a $\dfrac{3}{4-x}$, not defined for $x = 4$

b 1, 3

9 a $0 \leqslant f < 1$

b $\dfrac{3}{\sqrt{1-x^2}}$, range $f^{-1}(x) \geqslant 3$, domain $0 \leqslant x < 1$

10 a $g(x) \geqslant 3$ **b** $\dfrac{3}{2x^2+5}$

c $\dfrac{3+x}{2x}$ **d** $-1, \frac{3}{2}$

11 a $\dfrac{2}{2-x^2}$ **b** $\dfrac{2-x}{x}$ **c** $-2, 1$

12 a $f(x) \geqslant -9$

b domain $x \geqslant -9$, range $f^{-1}(x) \geqslant 2$

c

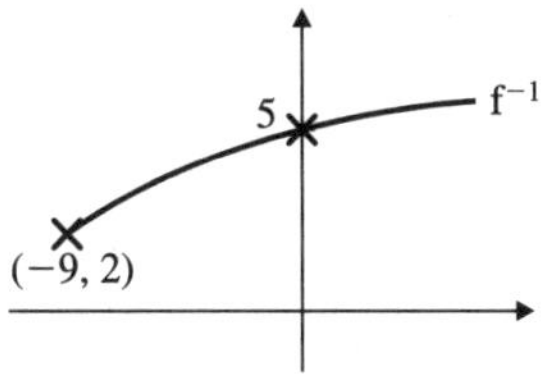

13 a

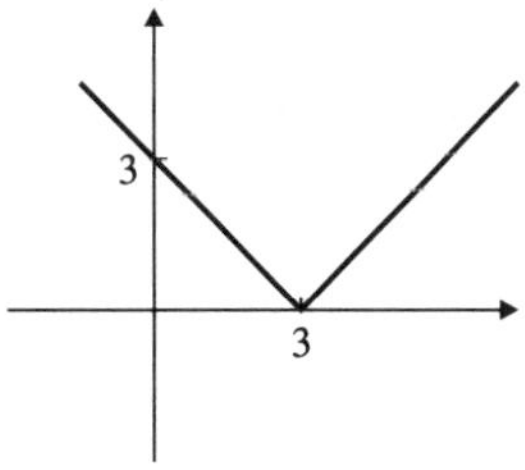

b

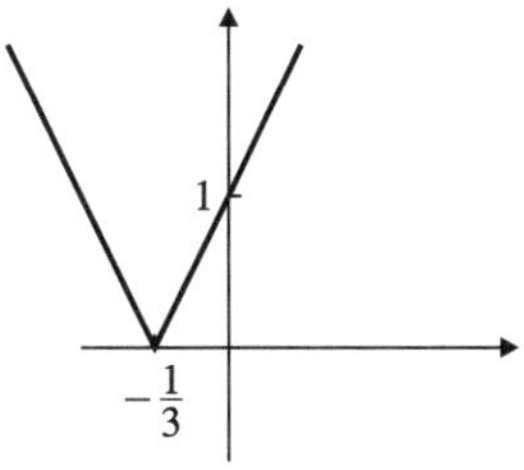

c

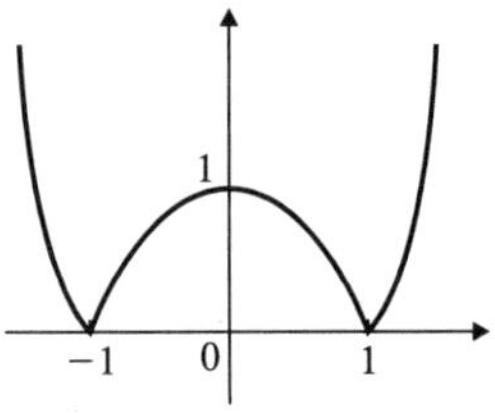

d

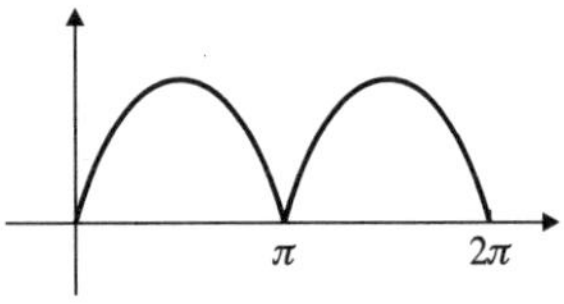

14 a

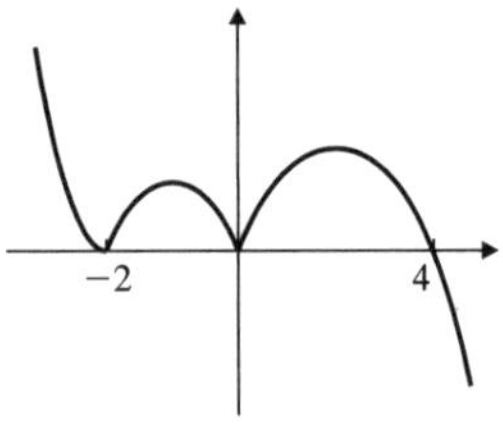

b

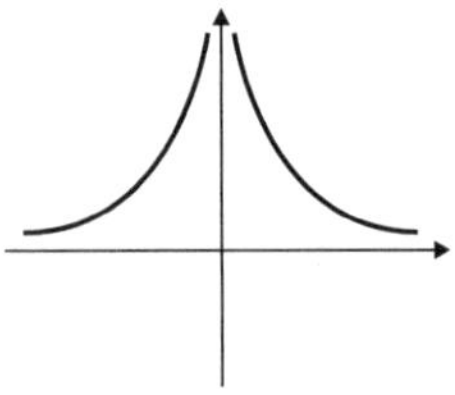

c

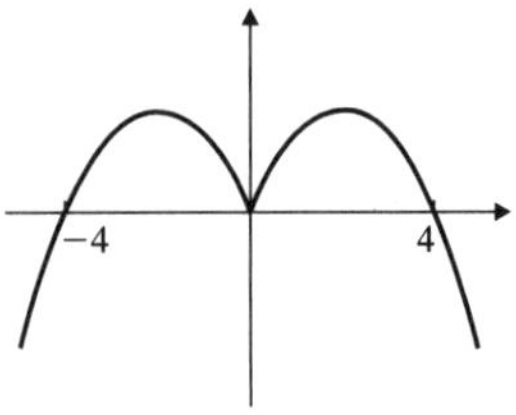

d

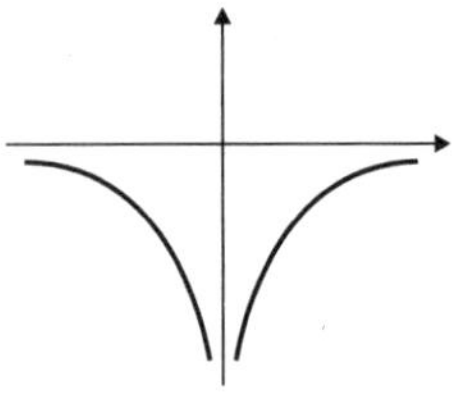

15 a $x > -\frac{1}{2}$ **b** $x < -2, x > 0$

16 a $-4 < x < 6$ **b** $x < 1, x > 2$

17 b $x > 3, x < -3$ **18** one solution

19 a

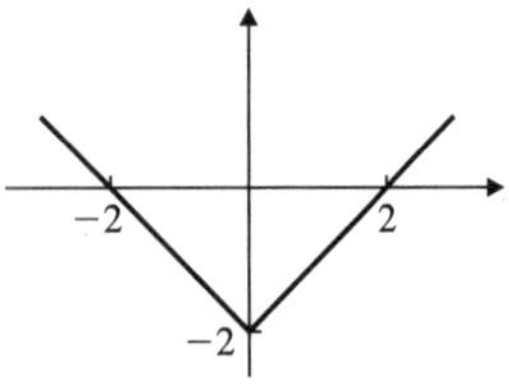

b

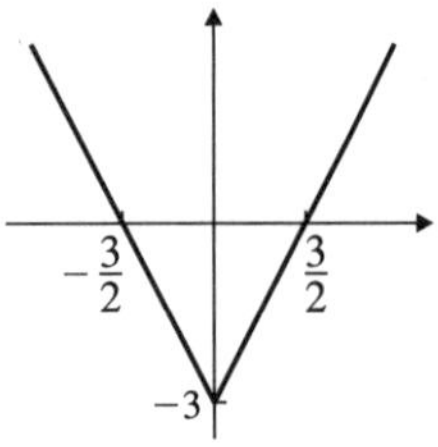

c

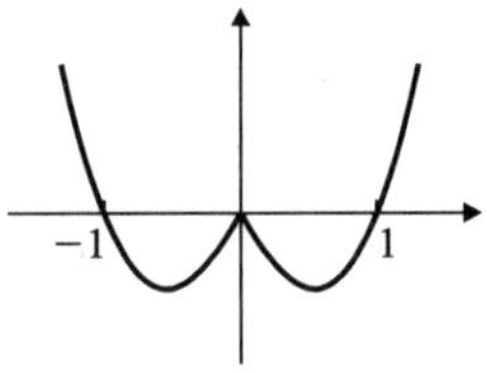

20 a

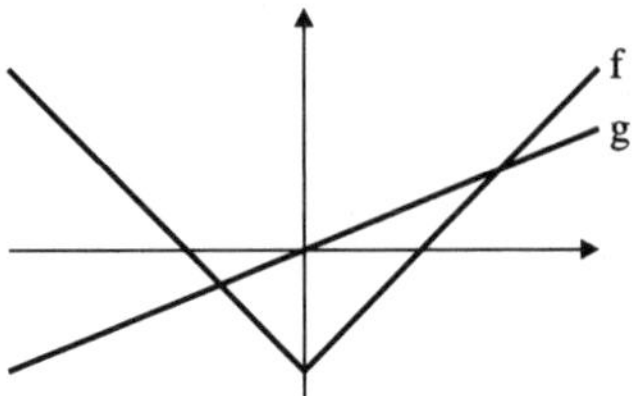

b $x > 10, x < -\frac{10}{3}$

21 a

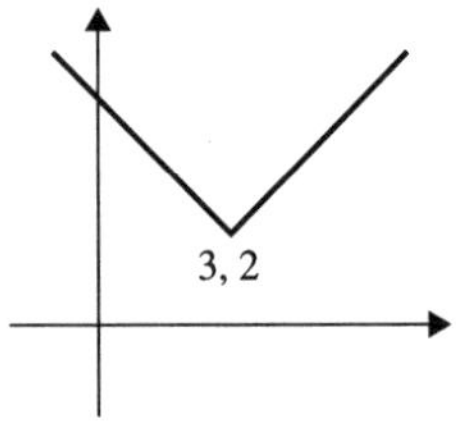

b $x = 0, 6$ **c** 3 range: $f(x) \geqslant 2$

22 a

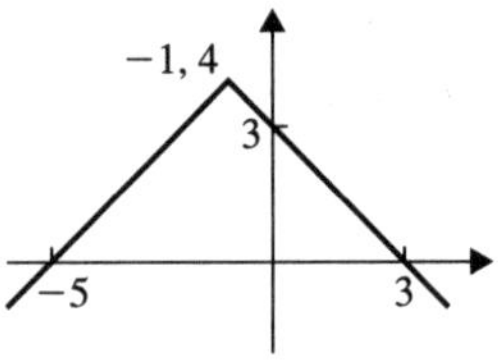

$f(x) \leqslant 4$

23 a

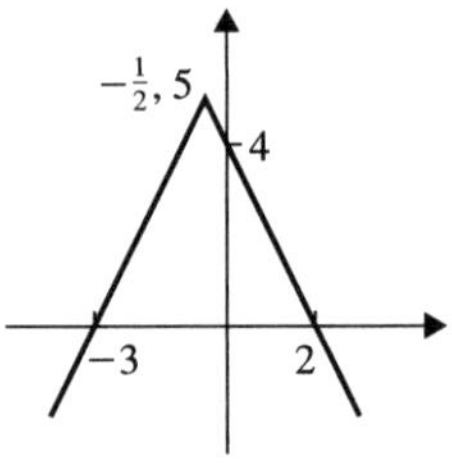

b 2 roots **c** $g(x) \leqslant 5$

24 a

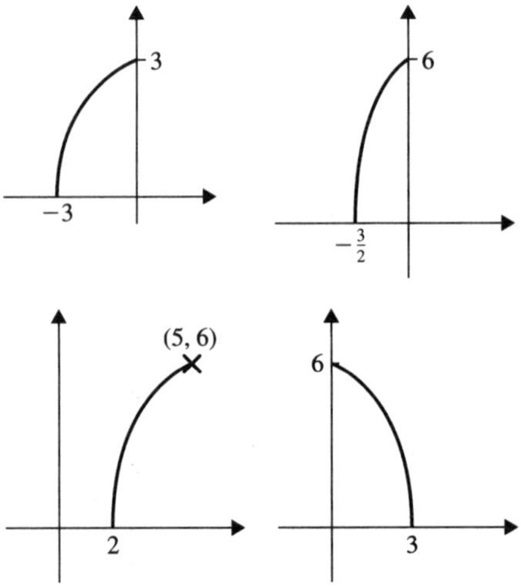

25 a i $(2, -9)$ ii $(3, -4)$ iii $(3, 9)$
iv $(1, -9)$ v $(-3, -9)$

26 a i $(-1, 5)$ ii $(2, -1)$ iii $(-2, 5)$
b $1 + 4x - x^2$

27 translation $\begin{pmatrix}3\\0\end{pmatrix}$;
stretch scale factor 2 parallel to y-axis;
translation $\begin{pmatrix}0\\-4\end{pmatrix}$

28 stretch, scale factor $\frac{1}{2}$, parallel to x-axis;
translation $\begin{pmatrix}0\\-1\end{pmatrix}$

29 stretch, scale factor 2, parallel to x-axis;
translation $\begin{pmatrix}0\\3\end{pmatrix}$

30 f : translation $\begin{pmatrix}4\\0\end{pmatrix}$, g : translation $\begin{pmatrix}0\\-3\end{pmatrix}$

EXAMINATION EXERCISE 1 *page 28*

1 $\dfrac{x}{x+3}$ 2 $\dfrac{2(x+2)}{(x+1)(x-3)}$ 3 $\dfrac{x-3}{x(x-2)}$

4 i $f(x) > 2$ ii $\dfrac{1}{(x-2)^2}$

5 a $\dfrac{2}{6-x^2}$ b ii $g(x) \leqslant 9$
c i $x = \pm 2$ ii g is not one–one
d i $\dfrac{4-3x}{x}$ ii $x = 1$

6 b $0 < f(x) < \frac{4}{3}$ c $\dfrac{4-x}{2x}$ d $f^{-1}(x) > 1$

7 b $f(x) > 0$ c $x = 6$

8 i $f(x)$ all real values, $g(x) \geqslant 0$
iii $\frac{17}{3}$ iv $\frac{13}{3}$ or $\frac{11}{3}$

9 i $fg(x) = \ln x^2$, $gf(x) = (\ln x)^2$ iv $k = 2$

10 a

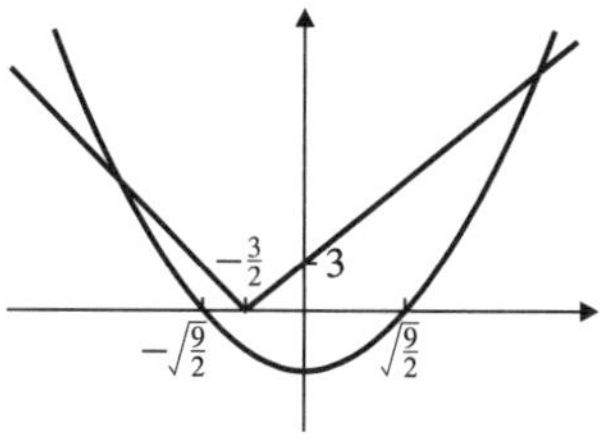

b 2 roots, $x = 3$, $-\dfrac{1-\sqrt{13}}{2}$, (-2.303)

11 i 3
ii f is one–one, $(4, 0)$ and $(0, -2)$
iii

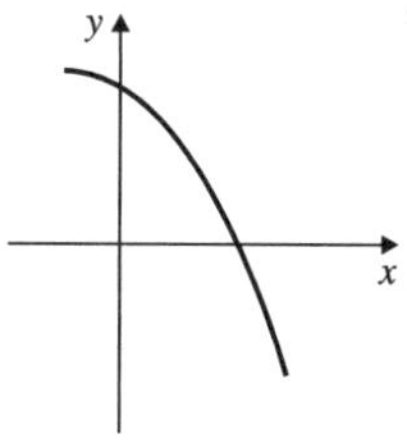

or

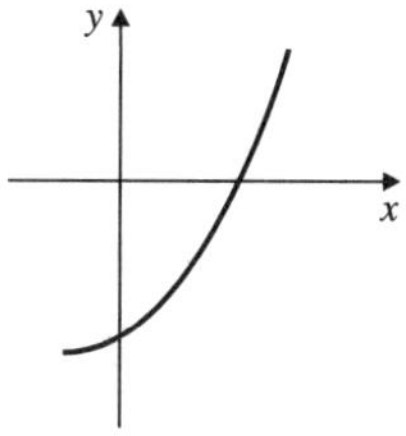

12 a $2 \leqslant f(x) \leqslant 11$
b $\lambda = 5$

13 a i $f(4) = 2$

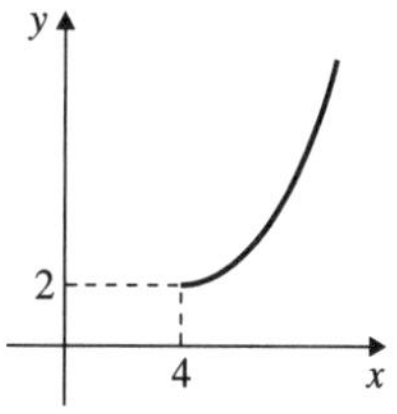

ii $f(x) \geqslant 2$
b $f(x) \geqslant 2$
c $3 + \sqrt{x-1}$

14 a

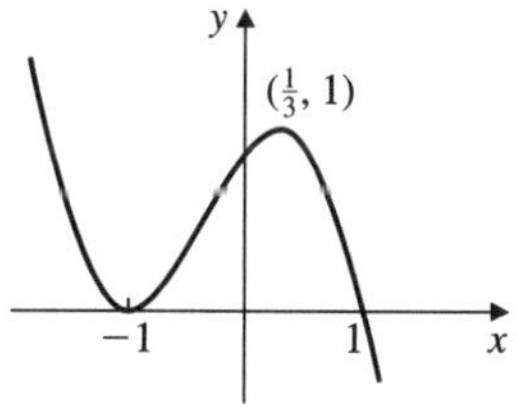

b

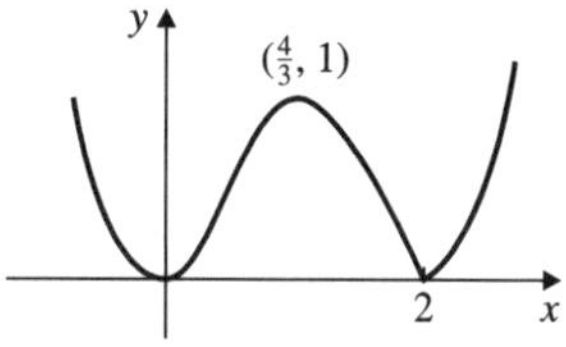

c

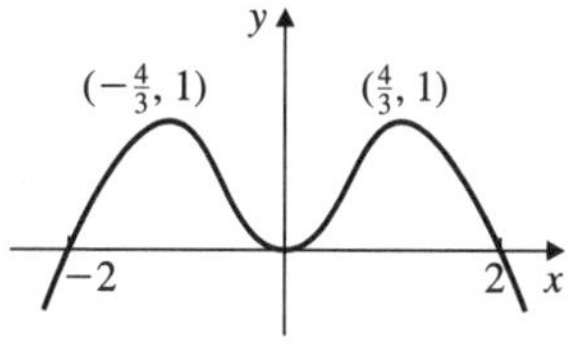

15 $(\frac{1}{2}, 0)$, $(0, 1)$

16 $-\frac{7}{2}, \frac{5}{2}$

17 $x > 2$, $x < -\frac{2}{3}$

18 i $f > -6$, $g \geqslant 0$
ii $2\ln\left(\dfrac{x+6}{5}\right)$
iii $2\ln 3$, $2\ln(\frac{1}{5})$

19 a

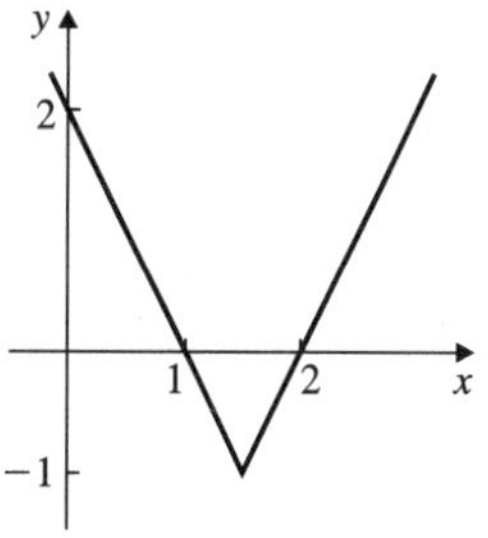

b $f(x) \geqslant -1$ c $\frac{2}{3}, 4$

20 $-\frac{3}{14}$

21 a stretch, scale factor $\frac{1}{3}$, parallel to x-axis
b translation by $\begin{pmatrix}3\\0\end{pmatrix}$
c reflection in $y = x$

22 i

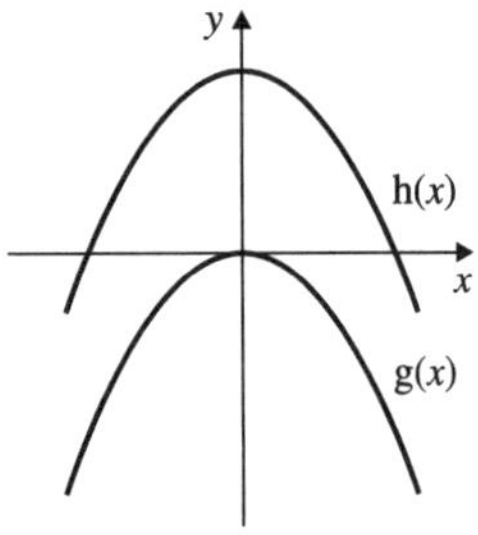

ii $g(x) = -(x^2)$, $h(x) = 2 - x^2$

23 a i reflection in $y = x$

b i stretch, scale factor 3, parallel to y-axis

ii $e^{\frac{1}{3}x}$

EXERCISE 1 *page 35*

1 a $\sqrt{3}$ b $\dfrac{2}{\sqrt{3}}$ c 1.015

d $\sqrt{2}$ e 0.325 f $\frac{4}{3}$

2 a $\sec^2 x$ b $\sec^2 2x$ c $\sin^2 x$

d $\sec x$ e $\sin\theta\cos\theta$ f $\operatorname{cosec}^2\theta$

4 a 60°, 300° b 30°, 210° c 45°, 135°

d 33.6°, 326.4° e 18.4°, 198.4° f 90°

6

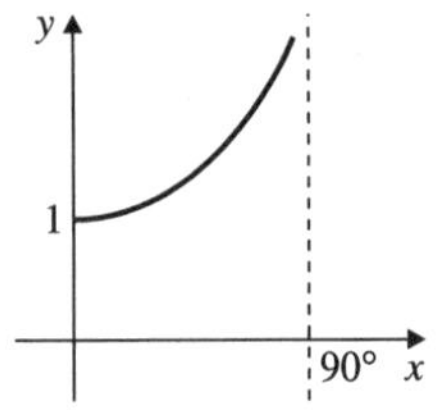

7 See page 33

8 a 45°, 225°, 63.4°, 243.4°

b −180°, 0, 180°, −135°, 45°

c 90°, 30°, 150°

10 $30° < \theta < 150°$

11 a $\cos\theta$ b $-60° < \theta < 60°$

12 a 35.3°, 144.7°, 215.3°, 324.7°

b 30°, 60°, 120°, 150°

13 $(22\frac{1}{2}, 7\frac{1}{2})$, $(67\frac{1}{2}, 142\frac{1}{2})$

EXERCISE 2 *page 39*

1 a 30° b 90° c 45°

d 120° e −30° f 60°

2 a $-\dfrac{\pi}{4}$ b $\dfrac{\pi}{3}$ c $\dfrac{\pi}{4}$

d 0 e $\dfrac{\pi}{3}$ f $-\dfrac{\pi}{2}$

3/4 See text

5 a 60°, 300° b 45°, 225° c 0, 360°

6 a $\frac{1}{2}$ b 0 c θ d θ

EXERCISE 3 *page 42*

1 b 1

2 b 0

3 a False b True

4 a $\sin 3\theta$ b $\cos 2A$ c $\sin 3x$ d 1

5 a $\dfrac{1+\sqrt{3}}{2\sqrt{2}}$ b $\dfrac{1+\sqrt{3}}{2\sqrt{2}}$ c $\dfrac{\sqrt{3}-1}{2\sqrt{2}}$

d $\dfrac{\sqrt{3}+1}{2\sqrt{2}}$ e $\sqrt{3}-2$

6 a $\dfrac{63}{65}$ b $\dfrac{56}{65}$ c $\dfrac{63}{16}$

7 73.2°, 286.8°

8 a $-\sin x$ b $-\cos x$

c $-\sin x$ d $-\cos x$

9 $\frac{1}{3}$

10 1

11 45°, 225°

12 a 1, $\theta = 50°$ b 1, $\theta = 105°$

14 $\tan(x + 45°)$

15 a $\dfrac{2\tan A}{1-\tan^2 A}$ b $\frac{1}{2}$

17 a $2+\sqrt{3}$ b $\sqrt{6}+\sqrt{2}$ c $\sqrt{6}-\sqrt{2}$

23 $\frac{1}{3}$

24 b 0.625, 2.195

EXERCISE 4 *page 45*

1 a 1 b $\dfrac{\sqrt{3}}{2}$ c $\frac{1}{2}$ d 1

2 a $\sin 20°$ b $\cos 34°$ c $\cos 70°$ d $\tan 22°$

e $\frac{1}{2}\sin 2\theta$ f $\frac{1}{2}\tan 4\theta$ g $2\sin A$ h 1

3 $\sin 2\theta = \frac{24}{25}$, $\cos 2\theta = \frac{7}{25}$

4 $\sin 2\theta = \frac{120}{169}$, $\cos 2\theta = -\frac{119}{169}$

5 a $\frac{4}{3}$ b $\frac{24}{7}$

6 $\dfrac{\pm\sqrt{7}}{4}$

7 $\pm\dfrac{4}{5}$

8 $\frac{1}{3}$ or −3

9 0, 180°, 360°, 60°, 300°

10 90°, 270°, 194.5°, 345.5°

11 60°, 300°, 75.5°, 284.5°

12 31.7°, 121.7°, 211.7°, 301.7°

13 0°, 180°, 360°, 30°, 150°

14 60°, 300°, 109.5°, 250.5°

15 0°, 60°, 180°, 240°, 300°, 360°

16 0°, 360°, 240°

17 0°, 120°, 360°

18 14.5°, 165.5°

37 $2\sin\dfrac{x}{2}\cos\dfrac{x}{2}$

38 $\dfrac{2\tan\frac{x}{2}}{1-\tan^2\frac{x}{2}}$

41 b $2+\sqrt{3}$

42 18.4°, 161.6°, 198.4°, 341.6°

43 A = 1, B = 4, C = 3

Exercise 5 *page 49*

1 a $5\cos(\theta-36.9)°$ b $13\cos(\theta-67.4)°$
c $\sqrt{5}\cos(\theta-63.4)°$
2 a $5\sin(\theta+53.1)°$ b $5\cos(\theta+36.9)°$
c $\sqrt{2}\cos(\theta-26.6)°$ d $17\sin(\theta-61.9)°$
e $10\sin(\theta+36.9)°$ f $\sqrt{2}\cos(\theta-45)°$
3 a $2\sin(\theta+60)°$
b Maximum value = 2 at $\theta=30°$
4 a 5, 53.1° b 13, 157.4°
c $\sqrt{2}$, 45° d 17, 298.1°
5 a $5\cos(\theta-36.9)°$ b 103.3°, 330.5°
6 a $3\sin(\theta+70.53)°$ b 67.7°, 331.3°
7 a 1.9°, 121.9° b 80.0°, 325.2°
c 90°, 330° d 60.4°, 193.3°
e 257.6°, 349.8° f 40.8°, 201.1°
g 78.4°, 244.8°
h 39.0°, 162.8°, 219.0°, 342.8°
8 a 2.25, 6.44 b 0.38, 1.97
9 a $\sqrt{7}\cos(x-0.714)$ b 1.42
10 a 0.36, 2.14 b 0, 4.07, 6.28
c 0, 4.71, 6.28 d 1.70, 3.29
11 a $\sqrt{10}\sin(\theta-71.6)°$ b $\dfrac{4}{\sqrt{10}}>1$
12 $-\sqrt{3}\leqslant k\leqslant\sqrt{3}$
13 a $c=12$ b $13\cos(2\theta+1.176)$
14 a $5\cos(\theta+0.93)°$ b 2
15 a $13\cos(\theta-1.176)$ b 2
16 a $5\sin(2\theta+36.9)°-3$
b 8.1°, 45.0°
17 a $5\cos(x-0.6435)$
b $\frac{1}{2}, \frac{1}{12}$
c 1.429, 3.000

Review Exercise 2 *page 51*

1 a 41.8°, 138.2° b 138.6°, 221.4°
c 145.0°, 325.0° d 210°, 330°
e $\dfrac{\pi}{3}, \dfrac{5\pi}{3}$ f $\dfrac{\pi}{4}, -\dfrac{3\pi}{4}$
g 60°, 300° h 0°, 180°, 360°
2 a 0, 132°, 228°, 360°
b 58°, 238°, 148°, 328°
3 $\dfrac{\pi}{4}, \dfrac{\pi}{3}, \dfrac{2\pi}{3}, \dfrac{3\pi}{4}$
4 a 45° b 0 c 60°
5 a $\dfrac{\pi}{6}$ b $\dfrac{\pi}{6}$ c $\dfrac{\pi}{6}$
7 a 18.4°, 198.4°, 26.6°, 206.6°
b 60°, 300°
c 90°, 270°, 199.5°, 340.5°
8 $-\frac{16}{65}$
9 a $-\frac{3}{5}$
b $\sin 2\theta=-\frac{24}{25}$, $\cos 2\theta=-\frac{7}{25}$
11 a 60°, 300°, 180°
b 0, 180°, 360°, 80.4°, 279.6°
c 25.2°, 154.8°
d 90°, 270°, 30°, 150°
e 0, 360°, 120°, 240°
f 0, 180°, 360°, 30°, 330°, 150°, 120°
12 a $-\frac{7}{25}$ b $\frac{24}{25}$ c $-\frac{24}{7}$
13 a i $\dfrac{\sqrt{6}-\sqrt{2}}{4}\left(=\dfrac{\sqrt{3}-1}{2\sqrt{2}}\right)$
ii $2+\sqrt{3}\left(=\dfrac{\sqrt{3}+1}{\sqrt{3}-1}\right)$
iii $\dfrac{\sqrt{2}-\sqrt{6}}{4}\left(=\dfrac{1-\sqrt{3}}{2\sqrt{2}}\right)$
12 a $-\frac{119}{169}$ b $\frac{2}{13}\sqrt{13}$ c $-\frac{120}{119}$
15 $\frac{1}{3}$
16 a 30°, 150°, 270°
b 0, 180°, 360°, 30°, 150°, 210°, 330°
c 90°, 120°, 240°, 270°
d 0, 120°, 240°, 360°
e 0, 60°, 300°, 360°
17 $3\sin\theta-4\sin^3\theta$
18 a $\sqrt{2}\cos\left(x+\dfrac{\pi}{4}\right)$ b $0, \dfrac{3\pi}{2}, 2\pi$ c $\sqrt{2}$
19 a $\sqrt{3}\sin(\theta+54.74)°$ b 90°, 340.5°
20 a $2\cos(\theta+60)°$ b $\frac{1}{4}$
21 a $5\sin(\theta+36.9)°$ b 120°, 347°

Examination Exercise 2 *page 53*

1 1.1, 1.3, 4.2, 4.5 (radians)
2 i $\dfrac{1}{\cos^2\theta\sin^2\theta}$ iii 0.342, 1.23
4 b $2-\sqrt{3}$
5 b $\dfrac{\pi}{12}, \dfrac{5\pi}{12}$
6 b 90°, 270°, 14.0°, 194.0°
7 $1+\dfrac{1}{\sqrt{2}}$
8 a $3-3\cos 2\theta$
b $\dfrac{\pi-3}{4}$
c $\theta=0.766, 2.376$
9 ii 29°, 209°
10 b $-2-\sqrt{3}$
11 i $\dfrac{2\tan x}{1-\tan^2 x}$ iii $\dfrac{\pi}{3}, \dfrac{2\pi}{3}, \dfrac{4\pi}{3}, \dfrac{5\pi}{3}$
12 i $\dfrac{\sqrt{3}-1}{2\sqrt{2}}$ ii $\dfrac{1+\tan x}{1-\tan x}$
13 a $2\sin(x+30°)$
14 b 114.3°, 335.7°
15 a i $2\sin\theta+4\cos\theta$ ii $\sqrt{20}\sin(\theta+1.107)$
b i $\sqrt{20}$ ii 0.46 radians
16 ii $\sqrt{58}\sin(\theta+23.2)°$ iii 17.8
17 a $\sqrt{29}\cos(\theta-1.19)$ b 5.39 when $\theta=1.19$
c 20.4°C, $t=4.55$ h d 0100 and 0830

18 i $\cos^{-1}(-1) = \pi$, $\cos^{-1}(0) = \dfrac{\pi}{2}$, $\cos^{-1}(1) = 0$

$\tan^{-1}(-1) = -\dfrac{\pi}{4}$, $\tan^{-1}(0) = 0$, $\tan^{-1}(1) = \dfrac{\pi}{4}$

ii (q, p) **iv** $p = \sqrt{\dfrac{(\sqrt{5} - 1)}{2}}$

EXERCISE 1 *page 60*

1

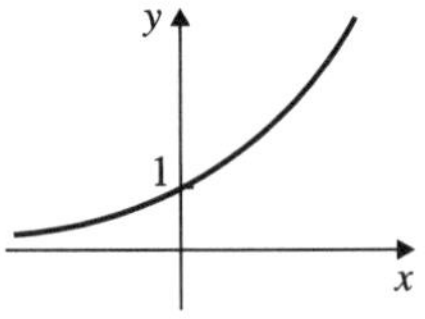

2

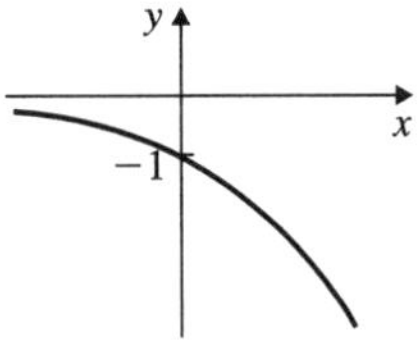

3

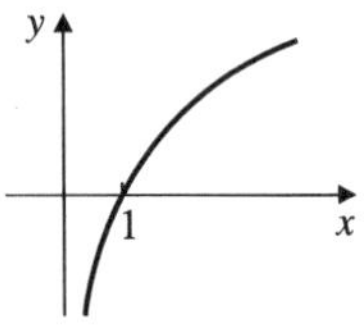

4

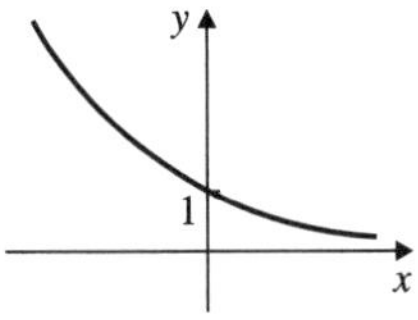

5

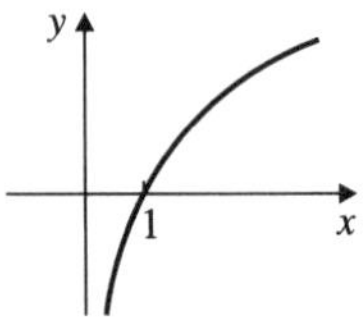

6

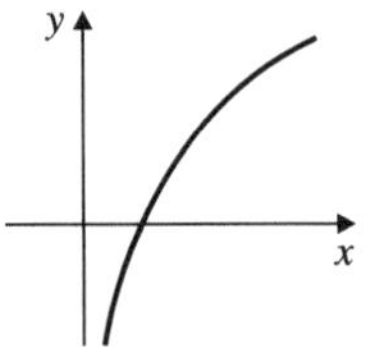

7

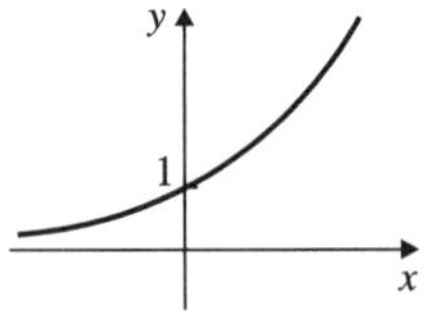

8

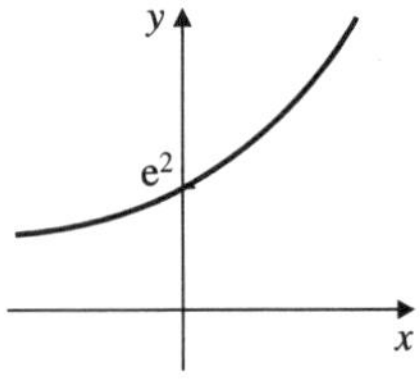

9

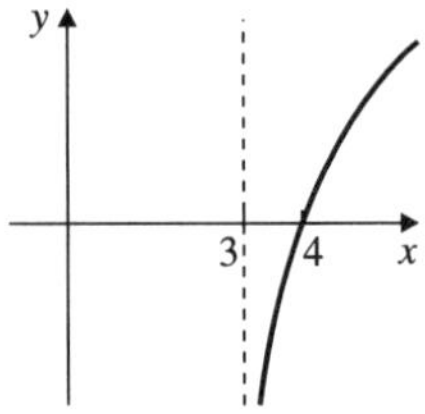

10

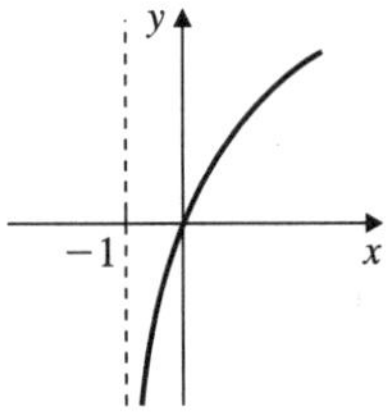

11

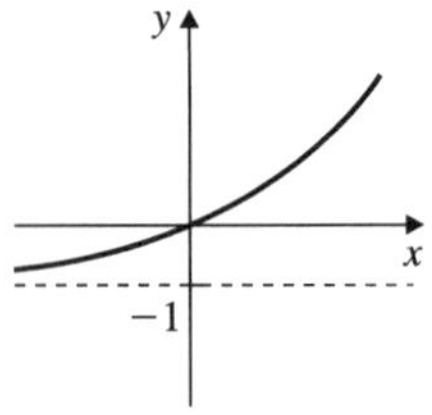

12

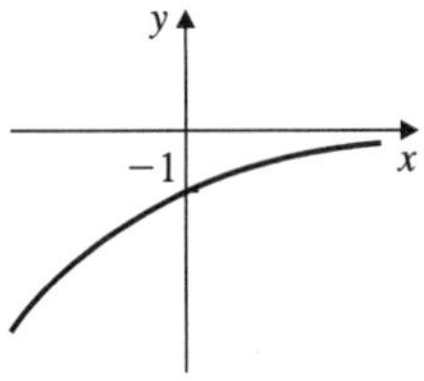

13 a

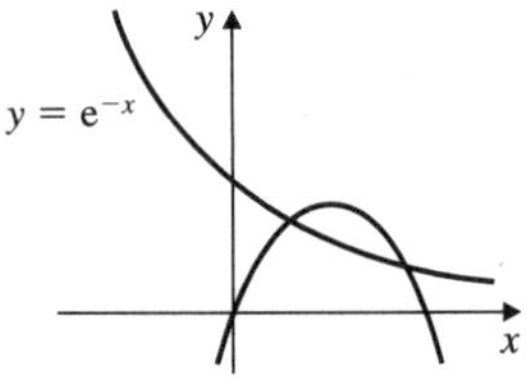

b (1, 1)
c 2

14 a

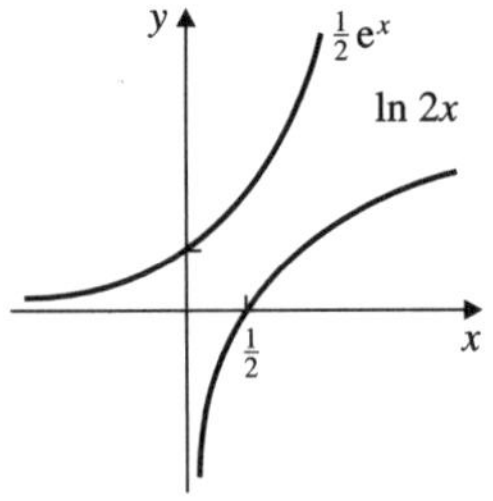

b $f^{-1}(x) = \frac{1}{2}e^x$
c reflection in $y = x$

15

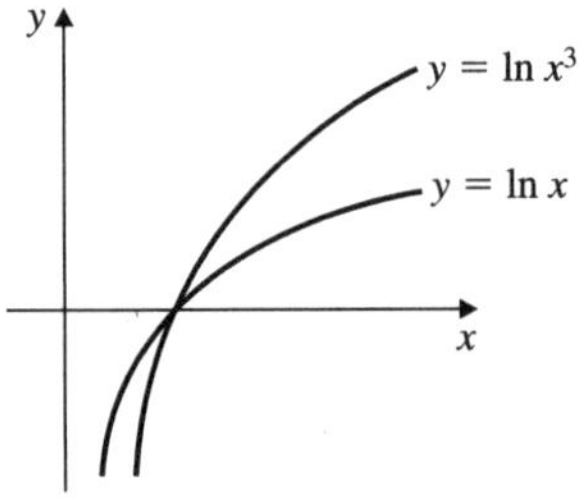

16 a $fg(x) = 2\ln x$, $gf(x) = (\ln x)^2$
b $k = 2$

17 a

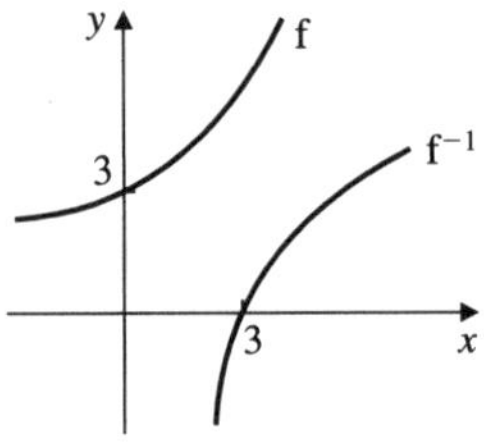

b $f^{-1}(x) = \ln\left(\dfrac{x}{3}\right)$.
c f^{-1}: domain $x > 0$, range: all real values of x.

18 a $f(x) \in \mathbb{R}$
b domain $x \in \mathbb{R}$, range $f^{-1}(x) > 0$
c $f^{-1}(x) = e^{x-3}$
e $x = e^3$

19 a 3 **b** $\dfrac{1}{2}\ln\dfrac{x}{3}$

c

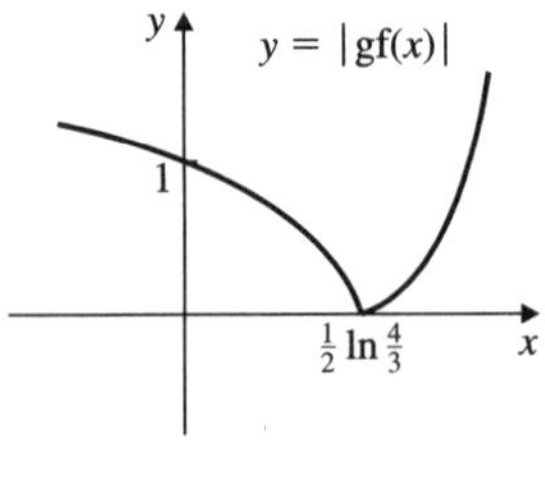

EXERCISE 2 *page 62*

1 a e^{2x} **b** e^{6x} **c** e^{3x} **d** 1
e $\ln 2x$ **f** $\ln 3$ **g** $3\ln x$ **h** 0
i 1 **j** e^{x+2} **k** $\frac{1}{2}$ **l** x
m a **n** 7 **o** e **p** 0

2 a 12.18 **b** 2.30 **c** 2.30
d 1.79 **e** 0.39 **f** 0.76

3 1.95 (= ln 7) **4** $\frac{1}{2}\ln 8$ (= 1.04)

5 $\ln\frac{9}{2}$ (= 1.50) **6** 1 + ln 11

7 0.19 **8** 1.03
9 0.52 **10** $\frac{1}{3}$
11 218.39 **12** 285.27
13 20 **14** ± 3
15 1.43
16 a $\ln\left(\dfrac{x+2}{x}\right)$ **b** $\dfrac{2}{e^4-1}$
17 a $\ln\left(\dfrac{x}{y}\right)$ **b** 0 **c** $\ln 2$ **d** x
18 a 0.65 **b** -2.39 **c** $0, \ln 2$ **d** 4
19 $x = e^3, y = e^2$
20 $x = 10$
21 a 25°C **b** 0.9 min

REVIEW EXERCISE 3 *page 64*

1 a

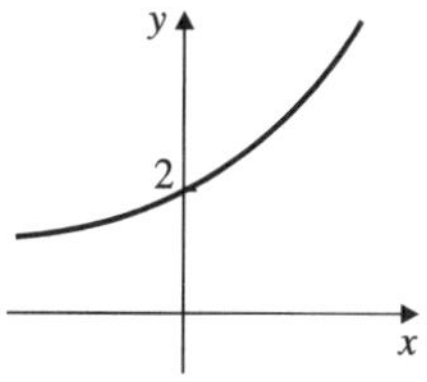

b

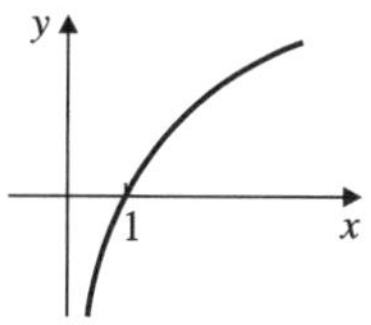

c

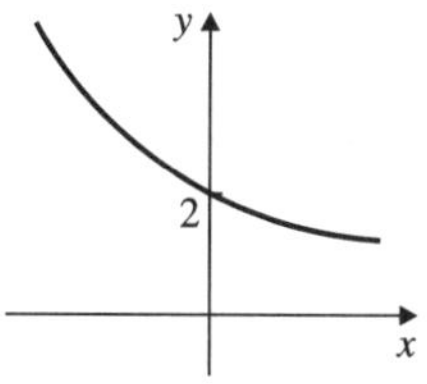

d

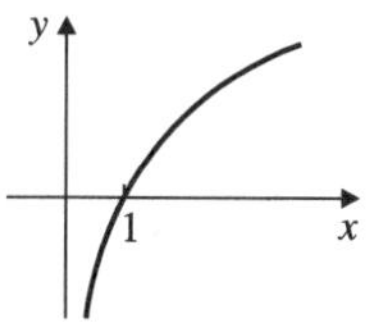

e

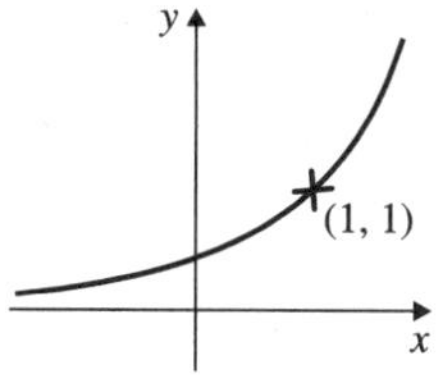

f

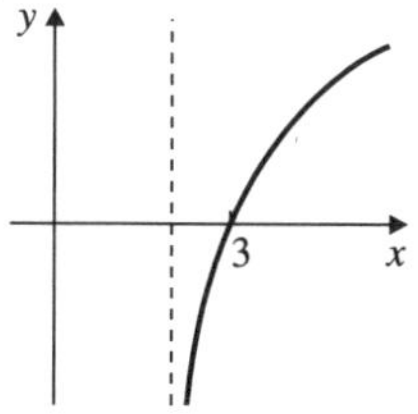

2 a $f^{-1}(x) = e^x$ **b** $x = 1$
3 $f^{-1}(x) = e^{x-4}$
4 a $g(x) > 0$ **b** $-\frac{1}{2}\ln x$
c

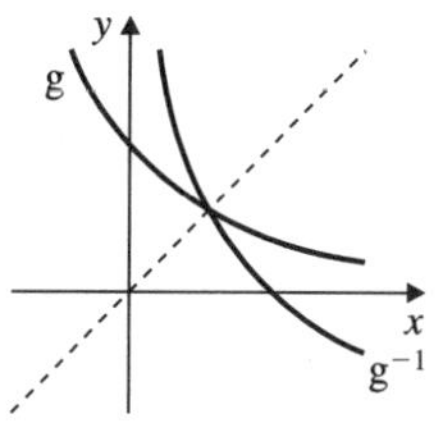

d 9.20
5 1
6

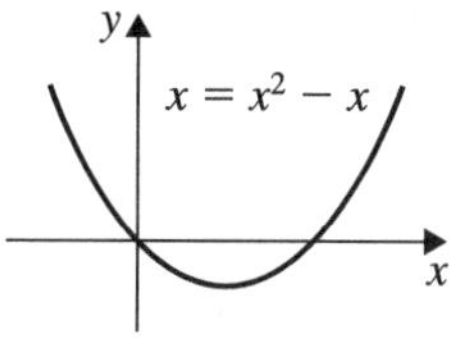

8 a $\frac{1}{3}\ln 7\ (= 0.65)$ **b** $\ln\left(\frac{2}{3}\right)$ **c** $(\ln 5) - 1$
9 a $e^2 - 1$ **b** $2e^3$ **c** $\frac{1}{3}e^{\frac{1}{2}}$ **d** e^3
10 a $a = 4$ **b** $b = 2.32$
c

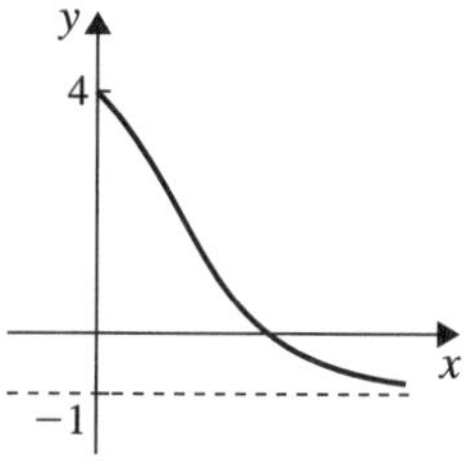

d $f(x) \to -1$
11 $\sqrt{\dfrac{3e}{4}}$
12 0.0707
13 333.5°C
14 b $x = 0, 2$

EXAMINATION EXERCISE 3 *page 65*

1 $x = 4$
2 a $\log\left(\dfrac{x}{y}\right)$ **b** 0 **c** 2
3 $-\frac{1}{6} + \log 3$
4 $x = 2$
5 a $f(x) > k$
b $2k$
c $f^{-1}: x \mapsto \ln(x-k), x > k$
d

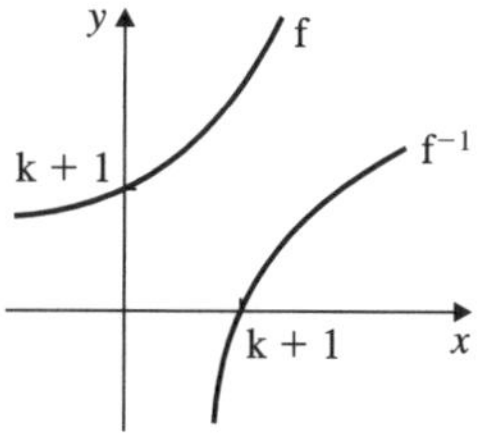

6 a $\dfrac{e^x + 2}{5}$ **b** $\mathbb{R}$ **c** 1.878
7 i a $4p + 2$ **b** $3p + 6$ **c** $3p - 1$
ii $y = \dfrac{x^3}{e^6}$
8 a reflection in y-axis; stretch, scale factor 2, parallel to y-axis
c $\ln 2 - \ln x$
d domain $x > 0$, range: $f^{-1}(x) \in \mathbb{R}$
9 a ii $f(x) > 0$ **iii** $\frac{5}{6}$
b ii $gf(x) > 10$ **iii** translation of $f(x)\begin{pmatrix}0\\10\end{pmatrix}$
10 i 2 **ii** $\dfrac{1}{3}\ln\left(\dfrac{x}{2}\right)$ **iv** $k = 8, \frac{1}{3}\ln 4$
11 a i $\dfrac{1}{x}$ **ii** $\dfrac{1}{e}$ **b** translation by $\begin{pmatrix}0\\2\end{pmatrix}$
c i $\mathbb{R}$
ii domain $x \in \mathbb{R}$, range: $f^{-1}(x) > 0$
iii e^{x-2}

Exercise 1 *page 69*

1 a $6(2x+5)^4$ **b** $8x(x^2+7)^3$

2 a $12(3x-4)^3$ **b** $40(8x+11)^4$
c $4x(x^2-3)$ **d** $27x^2(3x^3+1)^2$
e $-12(1-3x)^3$ **f** $-10(3-x)^9$

3 a $-3(1+3x)^{-2}$ **b** $-4(2x+1)^{-3}$
c $-15(5x+2)^{-4}$ **d** $-24(4x-1)^{-3}$
e $-10x(x^2+1)^{-2}$ **f** $40(1-x)^{-5}$

4 b $\frac{4}{7}\sqrt{7}$

5 a $\dfrac{5x}{\sqrt{5x^2+3}}$ **b** $(3x+1)^{-\frac{2}{3}}$
c $-x(x^2-3)^{-\frac{3}{2}}$ **d** $-\dfrac{3x^2}{(x^3+1)^2}$
e $-4(8x+7)^{-\frac{3}{2}}$ **f** $-\dfrac{1}{2\sqrt{x}(\sqrt{x}+2)^2}$

6 $y=64x-48$
7 $x+18y-1157=0$
8 $(0, 8)$
9 48
10 $2y=x+3$

11 a $3(2\sqrt{x}-3x)^2\left(\dfrac{1}{\sqrt{x}}-3\right)$ **b** $\dfrac{1}{2x^2}\left(1-\dfrac{1}{x}\right)^{-\frac{1}{2}}$
c $-3\sqrt{x}(x^{\frac{3}{2}}+2)^{-2}$ **d** $-\dfrac{1}{\sqrt{x}}(1+\sqrt{x})^{-3}$
e $\dfrac{1}{2}\left(x^2-\dfrac{1}{x}\right)^{-\frac{1}{2}}\left(2x+\dfrac{1}{x^2}\right)$ **f** $20x+(2x-3)^{-\frac{1}{2}}$

12 $(1, 7)$ minimum and $(-3, -9)$ maximum
13 $(1, 2)$

Exercise 2 *page 71*

1 $(2x+1)(6x+1)$ **2** $2x(3x-1)(6x-1)$
3 $x^2(x-1)(5x-3)$ **4** $2(x+2)(x+3)(2x+5)$
5 $2(1-x)^2(1-4x)$ **6** $2(x-1)(10x^3-6x^2+1)$
7 $5(1+2x)^3(10x+1)$ **8** $(3x+1)^3(15x-35)$
9 $\dfrac{\sqrt{x}}{2}(5x+6)$ **10** $\dfrac{(4x+1)}{2\sqrt{x}}(20x+1)$
11 $\dfrac{x}{\sqrt{2x+1}}(5x+2)$ **12** $\dfrac{x^2}{\sqrt{4x-1}}(14x-3)$

13 $y=19x-1$
14 16
15 $2, \frac{2}{5}$
16 a $-3, -\frac{1}{3}$
b maximum at $x=-3$, minimum at $x=-\frac{1}{3}$
17 a i $(-2, 0), (4, 0), (0, 32)$
ii $(0, 32)$ maximum; $(4, 0)$ minimum
18 a $(2, -4)$ **b** $(-1, -4)$
19 i $k=3$ **ii** $\frac{16}{3}$

Exercise 3 *page 75*

1 $\dfrac{2}{(x+1)^2}$ **2** $-\dfrac{15}{(2x-1)^2}$
3 $\dfrac{3x^2+2x}{(3x+1)^2}$ **4** $\dfrac{4-10x-4x^2}{(x^2+1)^2}$
5 $\dfrac{-2(x+1)}{x^3}$ **6** $\dfrac{-4x-1}{(x+1)^4}$
7 $\dfrac{-6(3x+2)(2x+3)}{(4x+1)^4}$ **8** $\dfrac{-x^3-9x^2}{(x-3)^5}$
9 $\dfrac{(3x-4)^2(6x+25)}{(2x+1)^3}$ **10** $\dfrac{-x-2}{2x^2\sqrt{x+1}}$
11 $\dfrac{2x+3}{(4x+3)^{\frac{3}{2}}}$ **12** $\dfrac{3x-4}{(2x-1)^{\frac{3}{2}}}$
13 $16y=x+1$ **14** $y=-3x-1$
15 $y=4-x$ **16** $\frac{3}{2}$
17 $(3, -\frac{1}{6})$ minimum, $(-1, \frac{1}{2})$ maximum

Exercise 4 *page 79*

1 a e^x **b** $3e^{3x}$ **c** $2e^x$
d $2x\,e^{x^2}$ **e** $2e^{2x+1}$ **f** $-e^{-x}$
g $20e^{4x}$ **h** e^x+2e^{2x} **i** $2x+e^{-x}$
j $-2e^{-2x}$ **k** $-4e^{-x}$ **l** $3x^2e^{x^3}+3x^2$

2 a xe^x+e^x **b** $x^2e^x+2xe^x$
c $2e^{2x}(2x+1)$ **d** $e^x(x^2+2x+1)$
e $4x^2e^{2x}(2x+3)$ **f** $3x^2e^{3x}+2xe^{3x}-3x^2$
g $\dfrac{e^{3x}}{x^2}(3x-1)$ **h** $\dfrac{xe^x-2e^x-2}{x^3}$
i $\dfrac{2x-x^2}{e^x}$ **j** $e^x(1+x)^2(4+x)$
k $e^{-x}(3x^2-x^3-1)$ **l** $\dfrac{2e^{2x}-e^{3x}}{(1-e^x)^2}$

3 a $3e^2$ **b** 3
c 1 **d** $-\dfrac{4e^2}{(e^2-1)^2}$

4 $y=e^2(x-1)$
5 $y=x$
6 b $y+e^2x=\dfrac{1}{e^2}$ **c** $\left(\dfrac{1}{e^4}, 0\right)$
7 $\left(-1, -\dfrac{1}{e}\right)$ minimum
8 $(1, e)$ minimum
9 a $(0, 2)$
10 a $(0, 0)$ minimum, $(-2, 4e^{-2})$ maximum
16 $(\frac{1}{3}\ln 2, 8)$ minimum

Exercise 5 *page 82*

1 a $\dfrac{1}{x}$ **b** $\dfrac{3}{x}$ **c** $\dfrac{6}{x}$
d $\dfrac{3}{3x-1}$ **e** $\dfrac{-2}{1-2x}$ **f** $\dfrac{3x^2+1}{x^3+x}$
g $\dfrac{1}{x+1}$ **h** $\dfrac{2}{x}$ **i** $\dfrac{3}{x+2}$
j $-\dfrac{1}{x}$ **k** $\dfrac{1}{2x}$ **l** $\dfrac{2x+1}{x^2+x-2}$

2 a $\ln(x+4) - \ln(x-2)$

b $\dfrac{1}{x+4} - \dfrac{1}{x-2}$

3 a $\dfrac{1}{x} - \dfrac{1}{x+1}$ **b** $\dfrac{2}{2x+3} - \dfrac{4}{4x-1}$

4 a $1 + \ln x$ **b** $x + 2x\ln x$

c $\dfrac{x}{1+x} + \ln(1+x)$ **d** $\dfrac{1 - \ln x}{x^2}$

e $\dfrac{(x+1) - 2x\ln x}{x(x+1)^3}$ **f** $\dfrac{\ln x - 1}{(\ln x)^2}$

g $2x + \dfrac{1}{x}$ **h** $\dfrac{3}{x} - \dfrac{1}{x^2}$

i $\dfrac{1}{2x} + \dfrac{1}{2\sqrt{x}}$ **j** $-\dfrac{\ln x}{x^2}$

k $\dfrac{2x^3}{1+x^2} + 2x\ln(1+x^2)$ **l** $\dfrac{1}{x+1} - \dfrac{1}{x+2}$

5 $x + 2x\ln x,\ 3 + 2\ln x$

7 $\dfrac{e^x}{1+e^x}$

8 b $3y = x + 3\ln 3 - 2$

9 $y = -x + \ln 2 + 1$

10 a $y = x - 2$

b 2 square units

11 $A(1, 0)$ $B\left(e, \dfrac{1}{e}\right)$

12 $\left(\dfrac{1}{e}, -\dfrac{1}{e}\right)$

13 $\dfrac{1}{2e}$

14 $(3.67, -4.57)$

15 a $\dfrac{f'(x)}{f(x)}$ **b** $\ln(x^3+1)$

16 $y = e(x-1)$

17 $[1, \ln(e+1)]$

Exercise 6 *page 87*

1 a $\cos x$ **b** $-3\sin 3x$

c $4\sec^2 x$ **d** $6\cos 6x$

e $-\frac{3}{2}\sin\frac{3}{2}x$ **f** $10\cos 2x$

g $\frac{1}{2}\sec^2\frac{1}{2}x$ **h** $-10\sin x + \cos x$

i $\sec^2 x + \sin x$ **j** $\cos(x+1)$

k $2\sec^2(2x-1)$ **l** $\frac{1}{2}\sin\frac{1}{2}x$

2 a $2\sin x\cos x$ **b** $-6\cos x\sin x$

c $3\sin^2 x\cos x$ **d** $-6\cos^2 x\sin x$

e $-\frac{1}{2}(\cos x)^{-\frac{1}{2}}\sin x$ **f** $-8\cos 4x\sin 4x$

g $18\tan 9x\sec^2 9x$ **h** $1 + \dfrac{\cos x}{2\sqrt{\sin x}}$

i $3(\sin 2x)^{\frac{1}{2}}\cos 2x$

3 a $\sin x + x\cos x$ **b** $\cos 2x - 2x\sin 2x$

c $x^2\cos x + 2x\sin x$ **d** $\cos^2 x - \sin^2 x$

e $-\left(\dfrac{x\sin x + \cos x}{x^2}\right)$ **f** $\dfrac{2x\cos 2x - 2\sin 2x}{x^3}$

g $\dfrac{\sin x - x\cos x}{\sin^2 x}$ **h** $\dfrac{3x^2\cos x + x^3\sin x}{\cos^2 x}$

i $-\dfrac{1}{\sin^2 x} = -\operatorname{cosec}^2 x$

6 a -6 **b** 4 **c** -2 **d** $\dfrac{\pi}{2} + 1$

9 $y = 2x + 1 - \dfrac{\pi}{2}$

10 a $\pi\cos\pi x$ **b** $2\pi\sec^2 2\pi x$

c $-\sin(x - \pi)$ **d** $-\dfrac{\pi}{2}\sin\dfrac{\pi}{2}x$

e $2x\cos x^2$ **f** $3x^2\sec^2 x^3$

g $2\cos(2x - \pi)$ **h** $\dfrac{1}{\pi}\cos\dfrac{x}{\pi}$

i $4x\sin(x^2)\cos(x^2)$

11 $0, \sqrt{\dfrac{\pi}{2}}$ **13** $2y + x = \dfrac{3\pi}{2}$

15 $\dfrac{\pi}{3} + \dfrac{\sqrt{3}}{2}$ **16** $-3\sqrt{3}$

18 b -4 **19** $\dfrac{\pi\sqrt{3} + 3}{6}$

20 -4

21 $\left(\dfrac{\pi}{4}, \sqrt{2}\right)$ maximum $\left(\dfrac{5\pi}{4}, -\sqrt{2}\right)$ minimum

22 b 0.62 1.57 2.53

23 a $2y + 3 = -x + \dfrac{4\pi}{3}$

b $\left(0, \dfrac{4\pi - 3\sqrt{3}}{6}\right)$

24 a $-\pi$ **b** $y\pi = x - 1$

Exercise 7 *page 89*

1 a $-\operatorname{cosec} x\cot x$ **b** $-\operatorname{cosec}^2 x$

c $2\sec 2x\tan 2x$ **d** $-3\operatorname{cosec}^2 3x$

e $-4\operatorname{cosec} 4x\cot 4x$ **f** $-2\operatorname{cosec}^2 x$

g $2\sec(2x+1)\tan(2x+1)$

h $2x + \cos x + \sec x\tan x$

2 a $-2\cot x\operatorname{cosec}^2 x$ **b** $2\sec^2 x\tan x$

c $\sec x\tan^2 x + \sec^3 x$ **d** $6\sec^2 3x\tan 3x$

e $\sec x(x\tan x + 1)$

f $2x\cot 2x - 2x^2\operatorname{cosec}^2 2x$

g $\dfrac{\sec x(x\tan x - 1)}{x^2}$

h $\dfrac{2x\cot x + x^2\operatorname{cosec}^2 x}{\cot^2 x}$

i $2\sec x\tan x(1 + \sec x)$

6 $-2\cot x - \operatorname{cosec} x + c$

Exercise 8 *page 92*

1 a $\dfrac{1}{x}$ **b** $\dfrac{1}{x}$ **c** $\dfrac{2}{x}$

d $1 + \dfrac{1}{x}$ **e** $5e^x + \dfrac{1}{x}$ **f** $2e^x + \dfrac{5}{x}$

2 a $(\frac{1}{3}, \ln\frac{1}{3})$
3 a $(1, \ln 4 - 3)$
4 b $4x + 12y = 5$
5 $y = 6x - 2$
6 a $5e^x$ **b** $7e^x$ **c** $20e^{4x}$
d $\dfrac{1}{x+1}$ **e** $\dfrac{-4}{e^x}$ **f** $6e^{3x}(e^{3x}+1)$
g $2e^x + \dfrac{1}{x}$ **h** $5e^x$ **i** $\dfrac{-2e^x}{(e^x-1)^2}$
7 e^x
8 a $\dfrac{2}{x}$ **b** $\dfrac{3}{x}$ **c** $\dfrac{1}{2x}$
d $-\dfrac{1}{x}$ **e** $\dfrac{1}{2x}$ **f** $-\dfrac{1}{3x}$
9 a $2(x+1)$ **b** $14(2x+1)^6$
c $24(3x-5)^7$ **d** $45(3x-7)^2$
e $-12(4x+3)^{-4}$ **f** $-20(5x-2)^{-5}$
g $2(4x+11)^{-\frac{1}{2}}$ **h** $5(15x-17)^{-\frac{2}{3}}$
i $na(ax+b)^{n-1}$
11 $(1, 1)$
12 $(0, -1)$
13 a $1 + \ln x$ **b** $x^2e^x + 2xe^x$
c $\cos^2 x - \sin^2 x$ **d** $e^x\left(\sqrt{x} + \dfrac{1}{2\sqrt{x}}\right)$
e $e^x(\sec^2 x + \tan x)$ **f** $\cot x - x\,\text{cosec}^2 x$
g $e^x\,\text{cosec}\,x(1 - \cot x)$ **h** $x^2\sec x(x\tan x + 3)$
i $x^2(3\cos x - x\sin x)$
15 a $\dfrac{1-\ln x}{x^2}$ **b** $\dfrac{e^x}{x^3}(x-2)$
c $\dfrac{\sec x}{e^x}(\tan x - 1)$ **d** $\sec^2 x$
e $-\text{cosec}^2 x$ **f** $\dfrac{x}{e^x}(2-x)$
g $\dfrac{x-1}{2x\sqrt{x}}$ **h** $\dfrac{x^2(3\cos x + x\sin x)}{\cos^2 x}$
i $-\dfrac{(x\,\text{cosec}^2 x + 2\cot x)}{x^3}$
17 $a = 1, b = -1, c = -1$
18 a $2\cos 2x$ **b** $5e^{5x}$
c $\cos x\, e^{\sin x}$ **d** $-2x\sin(x^2)$
e $\sec x\tan x\, e^{\sec x}$
f $-3\,\text{cosec}\left(3x + \dfrac{\pi}{3}\right)\cot\left(3x + \dfrac{\pi}{3}\right)$
g $-3\,\text{cosec}^2 3x$ **h** $\dfrac{3x^2}{x^3+3}$
i $2\cot 2x$ **j** $-\tan x$
k $3x^2e^x$ **l** $\tan x$
m $6x(x^2+1)^2$ **n** $3\sin^2 x\cos x$
o $6\tan^2(2x)\,.\,\sec^2 2x$
19 $-\dfrac{1}{\sqrt{1-x^2}}$ **20** $\dfrac{1}{2\sqrt{1-x^2}}$
21 a $\dfrac{1}{1+x^2}$ **b** $\dfrac{1}{3(1+x^2)}$
c $\dfrac{1}{2x}$ **d** $-\dfrac{1}{x}$
e $\dfrac{e^x}{5}$ **f** $\dfrac{e^{\frac{x}{2}}}{2}$
g $\dfrac{1}{2\sqrt{1-x^2}}$
22 a $\dfrac{1}{2\sqrt{x}}$ **b** $\dfrac{1}{2\sqrt{x}}$
23 a $nf(x)^{n-1}f'(x)$ **b** $f'(x)e^{f(x)}$
c $f'(x)\cos[f(x)]$
24 a $x = a^p, a = e^q, x = e^r$
c $\log_a x = \dfrac{\log_e x}{\log_e a}$

Review Exercise 4 *page 94*

1 a $12(3x-1)^3$ **b** $45(3x-4)^2$
c $-8(4x+3)^{-3}$ **d** $-15(5x-1)^{-4}$
e $12(2-3x)^{-5}$ **f** $\frac{3}{2}(3x+1)^{-\frac{1}{2}}$
2 a $8x(x^2+3)^3$ **b** $-15x^2(2-x^3)^4$
c $30x(3x^2+1)^4$ **d** $4(2x+1)(x^2+x)^3$
e $-\dfrac{3}{x^2}\left(1+\dfrac{1}{x}\right)^2$ **f** $-3\left(x^2+\dfrac{1}{x}\right)^{-4}\left(2x-\dfrac{1}{x^2}\right)$
3 a $(x+1)^2(4x+1)$ **b** $5(2x-3)^3(2x+1)$
c $(x+1)^2(x-1)(5x-1)$ **d** $4x(3x-2)(3x-1)$
4

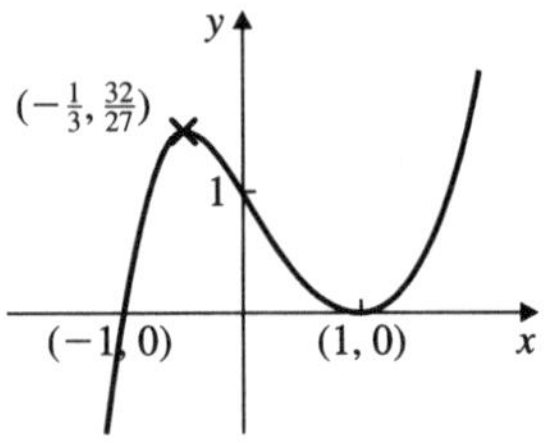

5 a $(-3, 0)$ maximum, $(\frac{1}{3}, -\frac{500}{27})$ minimum
6 a $\dfrac{x^2+2x}{(x+1)^2}$ **b** $\dfrac{-(x+2)}{2x^2\sqrt{x+1}}$
c $-\dfrac{1}{(x+1)^2}$ **d** $\dfrac{e^{3x}}{x^3}(3x-2)$
e $\dfrac{2x-6}{x^3}$ **f** $\dfrac{x\cos x - \sin x}{x^2}$
8 a e^x **b** $-3e^x$
c $\frac{1}{2}e^x$ **d** $6x^2 - 4e^x$
e $3e^x - \frac{3}{2}x^{-\frac{1}{2}}$ **f** $\frac{1}{6}x^{-\frac{2}{3}} - \frac{1}{4}e^x$
9 a $\dfrac{1}{x}$ **b** $\dfrac{4}{x}$ **c** $\dfrac{1}{x}$ **d** $\dfrac{2}{x}$
e $\dfrac{1}{2x}$ **f** $\dfrac{1}{x}$ **g** $-\dfrac{1}{x}$
10 a $e^2 - 1$ **b** 1 **c** 0 **d** 9
11 a $y = x + 1$ **b** $y = x - 1$
c $y = e^2 + 1 - e^2x$
12 a $ey = 1 + e^2 - x$ **b** $y = \frac{9}{2} + \ln 9 - \frac{3}{2}x$
13 a $(0, 1)$ minimum
b $(0, 2)$ maximum
c $(1, -1)$ maximum
d $(\frac{1}{2}, \frac{1}{2} + \ln 2)$ minimum
e $(\frac{1}{8}, \ln\frac{1}{8})$ maximum

14 A(1, 0) $B\left(e, \frac{1}{e}\right)$

16 a $\cos x$ b $-2\sin x$
c $\cos x - \sin x$ d $2\cos x + 3\sin x$
e $\sec^2 x + 2x$ f $3\cos x - \sec^2 x$

17 a $\frac{1}{2}$ b -1 c 0 d $\frac{1}{2}$ e 1

18 $y = 2x - \frac{\pi}{2}$ 19 $y = x + 2 - \frac{\pi}{2}$

20 a $3\cos 3x$ b $-4\sin 4x$
c $5\sec^2 5x$ d $2\sin x\cos x$
e $4\tan^3 x\sec^2 x$ f $2\cos\left(2x + \frac{\pi}{4}\right)$
g $\frac{1}{2}\frac{\cos x}{\sqrt{\sin x}}$ h $6\sin 3x\cos 3x$
i $-6\cos^2 2x\sin 2x$

21 a $2e^{2x}$ b $-3e^{-3x}$
c $2x\,e^{x^2}$ d $-e^{-x}$
e $\cos x\, e^{\sin x}$ f $\frac{2x}{x^2+1}$
g $\frac{3x^2}{x^3-2}$ h $\frac{3x^2+2}{x^3+2x-1}$
i $\frac{2x}{x^2+1} - \frac{1}{x-1}$ j $-\left(\frac{2x}{x^2+3}\right)$
k 1 l $\frac{1}{2\sqrt{x}}e^{\sqrt{x}}$

23 $\frac{e^x\cos x}{(1+\sin x)} + e^x\ln(1+\sin x)$

24 a $e^x(x+1)$ b $1 + \ln x$
c $e^x\left(\frac{1}{x} + \ln x\right)$ d $x\cos x + \sin x$
e $-x^2\sin x + 2x\cos x$ f $e^x\sec^2 x + e^x\tan x$

25 $\frac{2\pi\ell^3\sqrt{3}}{27}$ at $\theta = \tan^{-1}\sqrt{2}$

26 $\left(\frac{\pi}{6}, \frac{3\sqrt{3}}{16}\right)$ maximum $\left(\frac{\pi}{2}, 0\right)$ shoulder
$\left(\frac{5\pi}{6}, -\frac{3\sqrt{3}}{16}\right)$ minimum

27 a $\sin x(2\cos^2 x - \sin^2 x)$
b $\frac{2}{2x+5}$
c $2\cos 2x\cos 3x - 3\sin 2x\sin 3x$
d $2x\,e^{x^2+2}$
e $\frac{2}{x}\ln x$
f $2x\cos x^2$

28 a $\sec x\tan x$ b $-\operatorname{cosec} x\cot x$
c $-\operatorname{cosec}^2 x$

29 $6\sqrt{3}$

30 $(2, 4 - \ln 16)$ minimum

31 a $\frac{1}{\sqrt{1-x^2}}$ b $\frac{1}{3x}$
c e^x d $\frac{1}{2}e^x$

32 a $y^2 + 3y^2\ln 2y$ b $\frac{1}{(4e^2 + 3e^2\ln 2)}$

33 -1 and 3

EXAMINATION EXERCISE 4 *page 98*

1 i $-10(3-2x)^4$ ii $\frac{4}{4x+7}$

2 a $3e^{2x}\cos 3x + 2e^{2x}\sin 3x$
b $20x(2x^2+1)^4$

3 $y = 240x - 448$

4 a $x(1+x^2)^{-\frac{1}{2}}$ b $\frac{2}{(1-x)^2}$

5 a i $x^{-\frac{1}{2}}$ ii $\frac{1}{x+1}$ 6 $(\frac{1}{2}, \frac{1}{2}e^{-1})$

7 i $(-1, 0)$
iii $\left(-\frac{2}{3}, \frac{-2}{3\sqrt{3}}\right)$, gradient is infinite at P

8 $y = 2x + 4$

10 $y = x + 2$

11 i $A(-\sqrt{3}, 0)$, $C(0, -3)$, $D(\sqrt{3}, 0)$
iii $-2e$, $6e^{-3}$

12 a $\frac{4}{x} + \frac{1}{x^2}$
b $x > 0$ so both terms are (+)ve.
c f^{-1} does exist because $f(x)$ is one–one function.

13 i $\frac{3\sqrt{2}}{2}$
iii $\frac{3}{2}$, 3, at $(a, 0)$ gradient is infinite.

14 a $-3\sin 3x(x^2+5x+4) + \cos 3x(2x+5)$
b $y = 5x + 4$

15 a $\frac{2\sin x - 2x\cos x}{\sin^2 x}$ b ii $y = -\frac{1}{2}x + \frac{5\pi}{4}$

16 a $\frac{1}{x} - 4x$ c -8 d maximum

17 a i $2x - 3 + \frac{1}{x}$
b ii $\frac{1}{2}$, 1 iii $2 - \frac{1}{x^2}$ iv -2, 1

18 b $\frac{1}{2} + \frac{1}{2}\ln 2$
c 4
d minimum $\left(\frac{d^2y}{dx^2} > 0\right)$

19 a i $-2e^{-2x} - \frac{3}{x^2}$
ii $e^{-2x} > 0, x^2 > 0 \Rightarrow f^{-1}(x) < 0$
b $f(x) > 3$

20 b $3e^{3x} - 24$, $9e^{3x}$ c $\ln 2$, minimum point

21 b $\frac{1}{x} - \frac{1}{2}e^x$
c i $y - (10 - \ln 3 - \frac{1}{2}e) = (1 - \frac{1}{2}e)(x - 1)$
ii $9 + \ln 3$

EXERCISE 1 *page 104*

1 b 3.4
2 b 4.0
3 a $-1, -2$ b -1.2 c 1.4

4 c 1.8 **d** $(-3, -2)$
5 b $a = 9$ **c** 0.9
6 a 5.85, -173, -8030
b $(1, 2)$ **c** 1.52
7 b 0.6, 8.4

EXERCISE 2 *page 107*

1 a 2.571 **b** -2.714 **c** 0.143
d 8.602 **e** 2.714 **f** 2.554
2 a $x^3 - 7x + 1 = 0$ **b** $x^3 - 7x + 1 = 0$
c $x^3 - 7x + 1 = 0$ **d** $x^2 - 74 = 0$
e $x^3 - 20 = 0$ **f** $x - \sin x - 2 = 0$
3 b $y = 1 - 2x$
d -2.414
e $x_1 = 0, x_2 = 1, x_3$ does not exist
f 0.414
4 a i 0.562 **ii** -7.140 **iii** 0.372
b i $x^2 + 3x - 2 = 0$
ii $x^2 + 7x - 1 = 0$
iii $x^2 + 5x - 2 = 0$
c i $\frac{\sqrt{17} - 3}{2}$ **ii** $\frac{-7 - \sqrt{53}}{2}$ **iii** $\frac{-5 + \sqrt{33}}{2}$
5 a $f(-2) = 2, f(-3) = -14$ **c** -2.196
6 c 1.7
7 c 5.82
8 $p = 7.2\%$

REVIEW EXERCISE 5 *page 108*

2 b 0.629 **3** 1.78 **4 c** 2.21
5 -1.67 **6** 0.697 **7** 1.70
8 1.31 **9** 0.821 **10** -1.84
11 1.87

EXAMINATION EXERCISE 5 *page 110*

2 b one root
c ii closer to 4 (try $x = 4.5$)
3 ii 4.062
4 i 1.94 **ii** $x^3 + 5x - 17 = 0$
5 a 2.431, 2.418, 2.424, 2.422, 2.422…
Answer = 2.422
6 ii draw $y = x$ **iii** -2.02
7 iii 0.92
8 a

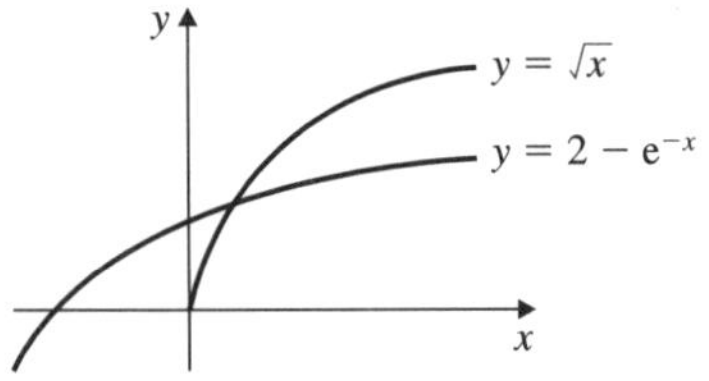

d $x_1 = 3.92707\ldots$
$x_2 = 3.92158\ldots$
$x_3 = 3.92115\ldots$
$x_4 = 3.92111$
Answer 3.921 (to 3 d.p.)

9 a -2.000 **b** 0.504
10 a $p = \frac{1}{3}$ **d** 0.304
11 ii 2.1
12 ii $a_2 = 2.539, a_3 = 2.533, a_4 = 2.534$;
$a = 2.53$ (2 d.p.)
13 a $y = 0.5x + 0.5$ **b** $x_3 = 2.1533$

EXERCISE 1 *page 117*

1 a A = 1, B = 2 **b** A = 3, B = 2
c A = 2, B = 3 **d** A = 5, B = 2
e A = 7, B = −2 **f** A = 4, B = −3
2 a $\frac{1}{x-3} + \frac{8}{x-2}$ **b** $\frac{1}{x-1} + \frac{2}{x-2}$
c $\frac{3}{x+1} + \frac{5}{x+2}$ **d** $\frac{1}{x+2} + \frac{6}{x+3}$
e $\frac{1}{x-1} + \frac{2}{x+2}$ **f** $\frac{2}{x-1} - \frac{1}{x+3}$
g $-\frac{4}{x} + \frac{5}{x-2}$ **h** $\frac{2}{x} + \frac{3}{x+1}$
3 a A = 3, B = 1 **b** A = 4, B = −3
c A = 1, B = 3 **d** A = 2, B = 1
e A = −2, B = 5 **f** A = 7, B = −1
4 a $\frac{3}{2x-3} + \frac{4}{x-3}$ **b** $\frac{3}{2x+1} + \frac{1}{x-1}$
c $\frac{5}{3x-2} + \frac{2}{3x+4}$ **d** $\frac{1}{2x-1} + \frac{1}{5x-1}$
e $\frac{4}{x+3} + \frac{3}{2x-1}$ **f** $\frac{2}{3x+5} + \frac{3}{2x-3}$
g $\frac{7}{2x-1} - \frac{1}{2x+1}$ **h** $\frac{2}{5x-2} + \frac{9}{5x+2}$

EXERCISE 2 *page 119*

1 a A = 1, B = 3, C = 2
b A = 1, B = −1, C = 2
c A = 2, B = 3, C = 6
d A = 4, B = −1, C = 5
e A = 3, B = 2, C = 1
2 a $\frac{1}{x-2} + \frac{3}{2x+1} + \frac{1}{x+2}$
b $\frac{5}{x-3} + \frac{1}{x+1} - \frac{1}{2x+3}$
c $\frac{2}{x-1} - \frac{4}{x} + \frac{1}{3x-4}$
d $\frac{3}{5x-2} + \frac{1}{4x-1} + \frac{2}{3x-2}$
e $\frac{5}{x-2} + \frac{5}{x-1} - \frac{3}{x+1}$
f $\frac{1}{5x+1} + \frac{2}{4x+3} + \frac{3}{4x-1}$
3 a $\frac{2}{x+3} + \frac{5}{x-2} + \frac{1}{(x-2)^2}$
b $\frac{2}{x-1} - \frac{1}{x+2} - \frac{3}{(x+2)^2}$

c $\dfrac{3}{2x-1}-\dfrac{7}{(3x+1)^2}-\dfrac{1}{(3x+1)}$

d $\dfrac{2}{3x-2}+\dfrac{4}{2x-1}+\dfrac{1}{(2x-1)^2}$

4 a $\dfrac{1}{x-2}+\dfrac{2}{(x-2)^2}$ b $\dfrac{1}{2x-1}+\dfrac{2}{(2x-1)^2}$

c $\dfrac{5}{x+1}+\dfrac{4}{(x-1)^2}$ d $\dfrac{3}{x-1}-\dfrac{1}{(x+3)^2}$

e $\dfrac{5}{(x+5)^2}+\dfrac{4}{3x-2}$ f $\dfrac{5}{x}+\dfrac{3}{x+1}+\dfrac{7}{x-1}$

g $\dfrac{7}{2x-1}-\dfrac{3}{x}-\dfrac{1}{2x+1}$

h $\dfrac{2}{x+4}+\dfrac{1}{(x-1)^2}-\dfrac{1}{(x-1)}$

EXERCISE 3 *page 123*

1 a $1+\dfrac{1}{x+3}+\dfrac{1}{x-2}$

b $3-\dfrac{1}{5x+1}+\dfrac{2}{2x-1}$

c $x+\dfrac{1}{x-3}+\dfrac{1}{x-1}$

d $2x+\dfrac{2}{x-2}+\dfrac{3}{x-1}$

e $x+1+\dfrac{3}{x+3}+\dfrac{4}{x+1}$

f $2x+3-\dfrac{2}{x-3}+\dfrac{5}{x+4}$

g $3x+2+\dfrac{2}{x+1}-\dfrac{1}{2x-1}$

h $2x-1+\dfrac{4}{2x-3}-\dfrac{3}{3x-5}$

2 a $3+\dfrac{1}{x+1}+\dfrac{2}{(x-1)^2}$

b $2-\dfrac{3}{2x-1}+\dfrac{4}{3x-2}-\dfrac{5}{(3x-2)^2}$

c $x+\dfrac{1}{x-2}+\dfrac{1}{x-1}+\dfrac{2}{(x-1)^2}$

d $x+\dfrac{2}{x+1}+\dfrac{3}{x-2}+\dfrac{4}{(x-2)^2}$

3 a $\dfrac{4}{5(x+3)}+\dfrac{6}{5(x-2)}$

b $f'(x)=\dfrac{-4}{5(x+3)^2}-\dfrac{6}{5(x-2)^2}$ etc.

4 a $\dfrac{1}{2}\ln\left(\dfrac{x-2}{x}\right)+c$

b $\dfrac{1}{6}\ln\left(\dfrac{x-5}{x+1}\right)+c$

5 a $\dfrac{1}{1-2x}-\dfrac{1}{1-x}+\dfrac{1}{(1-x)^2}$

b $1+3x+6x^2+11x^3$

EXAMINATION EXERCISE 6 *page 123*

1 a $\dfrac{1}{x+2}+\dfrac{2}{x-1}$

2 a $-\dfrac{1}{2-x}+\dfrac{3}{(2-x)^2}$

3 a $\dfrac{4}{2x-3}-\dfrac{3}{x+1}$ b $y=\dfrac{108(2x-3)^2}{(x+1)^3}$

4 a A = 1, B = −1, C = 2 b $r=\dfrac{35}{216}$

5 a $\dfrac{2}{1+2x}+\dfrac{1}{4-x}$

b ii $1-2x+16x^2$

iii $\dfrac{9}{4}-\dfrac{63}{16}x+\dfrac{513}{64}x^2$

iv $-\dfrac{1}{2}<x<\dfrac{1}{2}$

c i $\ln(1+2x)-\ln(4-x)$ ii 0.011

6 a A = −1, B = 0.5, C = 2.5

b $-2+3\ln 3$

7 a $\dfrac{2}{2-x}-\dfrac{1}{1+x}$ b $|x|<1$

c $y=\dfrac{4}{(2-x)^2(1+x)}$

EXERCISE 1 *page 127*

1 a $y=(x-1)^2$ b $y=x^{\frac{3}{2}}$

c $y=\dfrac{9}{x}$ d $y=x^3-4x$

e $y=x^2+2x+2$ f $y\sqrt{x}=1$

g $y=\dfrac{18}{x^2}+1$ h $y=\dfrac{3x^4}{16}+4$

i $y=\dfrac{x}{2x-1}$ j $y=3x^2-10x+9$

2 a $y=t^2+1$ b $y=\sin t$

c $y=\dfrac{2}{t}$ d $y=9t^3-t$

3 $y=2x-3$

4 $4x^2+y^2=4$

5 a $x^2+y^2=1$

b circle, centre (0, 0) radius 1

6 $x+y=1$

7 $x=2y^2-1$

8 a $y=1-2x^2$

9 $(x+y)(x-y)^2=8$

EXERCISE 2 *page 129*

1 a $\dfrac{t}{2}$ b $-\dfrac{3}{2}\tan t$

2 a $\dfrac{1}{t}$ b $-\dfrac{3}{5t^2}$ c $-\dfrac{4}{7}\cot t$

d $\dfrac{1}{2\sin t}$ e $\dfrac{e^t+e^{-t}}{e^t}$

3 b $3y = x + 1$

4 a $t = -1$ b $\frac{2}{3}$ c $2y + 3x = 10$

5 b $y = x$ c $16y = -9x + \frac{9}{4}$

6 a $\left(-\frac{20}{9}, -\frac{40}{27}\right), (-3, -9)$ b $(-4, -8)$

7 a $\frac{t^2 - 1}{t}$ b $(8, -4)$ and $(8, 4)$

8 $(1, 0)$ and $(-1, -4)$

9 $y + x = \sqrt{2}$ 11 $4y = 4x + a$

12 $\frac{2y}{\sin\theta} - \frac{x}{\cos\theta} = 3$

13 a $-\frac{\cos 2t}{\sin t}$ b $\frac{\pi}{6}, \frac{5\pi}{6}, \frac{3\pi}{2}$

14 $(0, 0), (\pi, 2), (2\pi, 0)$

15 a $(x - 2)^2 + (y - 2)^2 = 9$; circle, centre $(2, 2)$, radius 3

b $x + y = 4 + 3\sqrt{2}$

Exercise 3 *page 132*

1 a $\frac{26}{3}$ b 818.4 c $25 \ln 3$

2 a $\frac{28}{3}$ b 9 c 8

3 a $9 \ln 4$ b 174 c $7\frac{1}{2}$ d 261.2

4 $A = 3\pi$

Exercise 4 *page 133*

1 $\frac{243\pi}{5}$ 2 3π 3 24π 4 1580

5 $\frac{3\pi}{2}$ 6 a $7u^2$ b $55\pi u^3$

Review Exercise 7 *page 134*

1 a $y = (x - 1)^2$ b $y = \frac{3}{x + 1}$ c $xy = 8$

d $y = 1 - x$ e $\frac{x^2}{25} + y^2 = 1$ f $y = 4x^2 - 2$

3 a $\frac{28}{3}$ b $9\frac{3}{4}$ c 261.2

4 a $\frac{8}{3}$ b $\frac{32\pi}{5}$ 5 $\frac{15\pi}{2}$

6 a $\frac{1}{2t}$ b $-\tan t$ c $-\frac{1}{t^2}$ d $\frac{2}{3t}$

7 $(\frac{1}{2}, -\frac{3}{4})$ 8 $(1, -2), (-1, 2)$

9 a $y = -2x - 1$ b $y + x = \sqrt{2}$

c $x + y = 2$ d $9y = 24\sqrt{3} - 4\sqrt{3}x$

e $2y = 9 - 3x$ f $y = (\sqrt{2} - 1)x + 5 - 3\sqrt{2}$

Examination Exercise 7 *page 135*

1 i $t = 2$ ii 3

2 a $-\frac{1}{t}$ b $y = 3x + 30$

3 $y = 4x - 27$

4 i $\frac{2t + 3t^2}{1 + 2t}$ ii $16x - 5y - 36 = 0$

5 b $y = -\frac{1}{3}x + \sqrt{2}$ 6 a $y = \frac{2}{9}x^2 - 1$

7 ii $(-12, 1)$ iii $t = \frac{1}{2}$

8 ii a $\tan\alpha = 0.75$

b $\theta = -0.232, 2.09$ radians

9 b $\sqrt{2} + \pi + 2$

10 b i $-t\cos t + \sin t + c$ ii $2\pi + \pi^5$

11 a $y^2 = 4x^2(9 - x^2)$ b $A = 27$

c 18 d 36 cm^2

Exercise 1 *page 140*

1 a $1 - 2x + 3x^2$ b $1 - 3x + 6x^2$

c $1 - 6x + 21x^2$ d $1 - 10x + 55x^2$

e $1 + \frac{1}{2}x - \frac{1}{8}x^2$ f $1 + \frac{3}{2}x + \frac{3}{8}x^2$

g $1 - \frac{1}{2}x + \frac{3}{8}x^2$ h $1 - \frac{3}{4}x + \frac{21}{32}x^2$

i $1 - \frac{2}{3}x - \frac{1}{9}x^2$ j $1 + \frac{2}{5}x + \frac{7}{25}x^2$

k $1 - \frac{1}{2}x - \frac{1}{8}x^2$ l $1 + \frac{1}{3}x + \frac{2}{9}x^2$

2 0.980 3 1.010

4 a $-\frac{1}{2} < x < \frac{1}{2}$ b $-1 < x < 1$

c $-\frac{1}{5} < x < \frac{1}{5}$ d $-\frac{1}{2} < x < \frac{1}{2}$

e $-\frac{1}{2} < x < \frac{1}{2}$ f $-\frac{1}{4} < x < \frac{1}{4}$

g $-\frac{1}{3} < x < \frac{1}{3}$ h $-\frac{1}{2} < x < \frac{1}{2}$

i $-3 < x < 3$ j $-2 < x < 2$

5 a $1 - 6x + 24x^2 - 80x^3$

b $1 + 2x + 3x^2 + 4x^3$

c $1 - 20x + 250x^2 - 2500x^3$

d $1 - 14x + 112x^2 - 672x^3$

e $1 + x - \frac{1}{2}x^2 + \frac{1}{2}x^3$

f $1 + 10x + 30x^2 + 20x^3$

g $1 + \frac{3}{2}x + \frac{27}{8}x^2 + \frac{135}{16}x^3$

h $1 - \frac{1}{2}x - \frac{3}{8}x^2 - \frac{7}{16}x^3$

i $1 - \frac{x}{6} - \frac{x^2}{72} - \frac{x^3}{432}$

j $1 + \frac{x}{6} + \frac{x^2}{18} + \frac{7x^3}{324}$

6 $-\frac{77}{128}x^4$ 7 $\frac{4389}{256}x^5$ 8 $-\frac{1}{125}x^3$

9 a $x^3 + 6x^2 + 12x + 8$ b $15\sqrt{3} + 26$

10 b $264\sqrt{2}$

11 $1 + \frac{1}{2}x - \frac{1}{8}x^2 + \frac{1}{16}x^3$, 1.0392 12 -20

13 $1 - \frac{1}{2x} - \frac{1}{8x^2} - \frac{1}{16x^3}$, 9.94987

14 a $1 - \frac{x}{3} - \frac{x^2}{9}$ b 3.332222

Exercise 2 *page 142*

1 a $\frac{1}{2} - \frac{x}{4} + \frac{x^2}{8}, |x| < 2$

b $\frac{1}{64}\left(1 + \frac{3x}{4} + \frac{3}{8}x^2\right), |x| < 4$

c $\frac{1}{9}\left(1 - \frac{2}{3}x + \frac{x^2}{3}\right), |x| < 3$

d $\frac{1}{625}\left(1 - \frac{4x}{5} + \frac{2x^2}{5}\right), |x| < 5$

e $2\left(1 - \frac{1}{8}x - \frac{1}{128}x^2\right), |x| < 4$

f $\frac{1}{3}\left(1 + \frac{1}{18}x + \frac{1}{216}x^2\right), |x| < 9$

g $2\left(1 - \frac{1}{24}x - \frac{1}{576}x^2\right), |x| < 8$

h $\frac{1}{4}\left(1 - \frac{1}{12}x + \frac{5}{576}x^2\right), |x| < 8$

i $5^5\left(1 + \frac{x}{10} + \frac{3x^2}{1000}\right), |x| < 25$

2 a $2\left(1 + \frac{3x}{8} - \frac{9x^2}{128}\right), |x| < \frac{4}{3}$

b $\frac{1}{3}\left(1 + \frac{x}{9} + \frac{x^2}{54}\right), |x| < \frac{9}{2}$

c $2(1 - \frac{5x}{24} - \frac{25x^2}{576}), |x| < \frac{8}{5}$

d $125\left(1 - \frac{21x}{50} + \frac{147}{5000}x^2\right), |x| < \frac{25}{7}$

3 a $a = \frac{1}{2}, b = -\frac{3}{16}$ **b** $a = 2, b = \frac{1}{4}$

c $a = \frac{1}{\sqrt{2}}, b = \frac{7}{4\sqrt{2}}$ **d** $a = \frac{1}{3}, b = -\frac{2}{243}$

4 $3 + \frac{2x}{3} - \frac{2x^2}{27} + \frac{4x^3}{243}, |x| < \frac{9}{4}$

5 a $1 - x - \frac{x^2}{2}, |x| < \frac{1}{2}$ **b** 0.990

6 a $1 - \frac{3x}{5} - \frac{18}{25}x^2$ **b** 1.961

Exercise 3 *page 143*

1 a $-\left(\frac{13}{3} + \frac{19}{9}x + \frac{28}{27}x^2\right), |x| < 2$

b $-\left(2 + \frac{3x}{2} + \frac{5x^2}{4}\right), |x| < 1$

c $\frac{11}{2} - \frac{17x}{4} + \frac{29x^2}{8}, |x| < 1$

d $\frac{5}{2} - \frac{11x}{12} + \frac{25x^2}{72}, |x| < 2$

e $-\frac{3}{2}x - \frac{3}{4}x^2, |x| < 1$

f $-\frac{7}{3} - \frac{17x}{9} - \frac{55x^2}{27}, |x| < 1$

2 a $1 + \frac{1}{x-2} + \frac{1}{x-1}$

b $-\left(\frac{1}{2} + \frac{5x}{4} + \frac{9x^2}{8} + \frac{17x^3}{16}\right)$ **c** $|x| < 1$

3 a $1 - x - 4x^2 - 6x^3$ **b** $|x| < \frac{1}{4}$

4 $\lambda = 2, n = -3$ **5** Both equal -2.

6 $1 + \frac{3x}{2} + \frac{7x^2}{8}, |x| < 1$

7 a $1 + x + \frac{1}{2}x^2, |x| < 1$

8 a $a = \frac{3}{2}, b = \frac{1}{2}$ **b** $\frac{1970}{1393}$ or $\frac{1393}{985}$

9 a $q = 3$ **b** $-x^3$ **c** $-1 < x < \frac{1}{3}$

10 a $\frac{[1 - (-x)^n]}{1 + x}$ **b** $-1 < x < 1$ **c** $\frac{1}{1 + x}$

11 b $1 + 2x + 3x^2 + 4x^3$

c $1 + 3x + 6x^2 + 10x^3$

d $1 + 4x + 10x^2$

Review Exercise 8 *page 145*

1 a $|x| < \frac{2}{3}$ **b** $|x| < \frac{5}{3}$ **c** $|x| < \frac{2}{7}$

d $|x| < \frac{9}{2}$ **e** $|x| < \frac{3}{2}$ **f** $|x| < \frac{5}{3}$

2 $2 - \frac{x}{4} - \frac{x^2}{32}, |x| < \frac{3}{8}$

3 a $a = 2, b = -\frac{1}{12}$

b $a = \frac{1}{8}, b = \frac{9}{64}$

4 a $\frac{1}{3(x+1)} + \frac{2}{3(1-2x)}$

b $1 + x + 3x^2$

c $|x| < \frac{1}{2}$

5 a $\frac{1}{2x+1} + \frac{3}{x-1}$

b $-2 - 5x + x^2$

c $|x| < \frac{1}{2}$

6 a $\frac{4}{3(x-1)} + \frac{8}{x+2}$

c $-108\frac{1}{2}$

d $|x| < \frac{1}{3}$

Examination Exercise 8 *page 146*

1 i $1 - x - \frac{1}{2}x^2 - \frac{1}{2}x^2$ **ii** $|x| < \frac{1}{2}$

2 a $1 + \frac{1}{2}x - \frac{1}{8}x^2$ **b i** $2 + \frac{x}{2} - x^2$ **ii** $-2 < x < 2$

3 i $1 + 2x - 2x^2$

ii $|x| < \frac{1}{4}$

iii $k = 5$, coef. of $x^2 = 8$

4 a $1 - 6x + 27x^2 - 108x^3$

b $4 - 23x + 102x^2$

5 a $A = 2, B = 16$

b $10 + 10x^2 + 15x^3$

7 b $1 - \frac{5}{2}x + \frac{75}{8}x^2 - \frac{625}{16}x^3$

c 0.863426

d 0.002594

8 a $1 - 3x + 9x^2 - 27x^3$

b 0.98058

9 i $2 + \frac{1}{12}y - \frac{1}{288}y^2$

ii $|y| < 8$

iii $2 - \frac{1}{6}k + \frac{5}{72}k^2$

10 a $a = 3, n = -2$

b -108

c $|x| < \frac{1}{3}$

11 a $\frac{1}{1-x} + \frac{2}{2+x}$

b ii $1 + x + x^2$

c $2 + \frac{x}{2} + \frac{5}{4}x^2$

12 a $A = 1, B = 2$

b $3 - x + 11x^2$

c Not valid, require $-\frac{1}{3} < x < \frac{1}{3}$

13 a $1 + 2x^2 + 6x^4, |x| < \frac{1}{2}$

b $1 + 2x + 2x^2 + 4x^3 + 6x^4 + 12x^5$

Exercise 1 *page 151*

1 a $-\frac{x}{y}$ **b** $-\frac{y}{x}$ **c** $-\left(\frac{2x+y}{x+2y}\right)$

d $\frac{3-x}{y+4}$ **e** $-\frac{2x}{(6y-4)}$ **f** $\frac{2x}{3y^2}$

g $-\left(\frac{9x^2+4y^2}{8xy+3y^2}\right)$

2 a -3 **b** -1 **c** $-\frac{3}{2}$ **d** 3

3 $x+y=4$ **4** $y=x$ **5** $x+y=4$

6 $y=x$ **7** $4y+3x=20$ **8** $y=x$

10 a -1 **b** 6 **c** -4 **d** -8

11 b $2, -2$ **c** $(2, 4), (-2, -4)$

12 $(-4, 2), (4, -2)$

13 a $\frac{2-x-y}{3y-x}$ **c** $(5, 3)$

14 $(-7, 7)$

15 a $-\frac{17}{8}$ **b** $\frac{1}{7}$ **c** $-\frac{4}{3}$

d $-\frac{2}{5}$ **e** $-\frac{1}{5}$

16 a i $\frac{-1}{\sin y}$ **ii** $-\frac{1}{\sqrt{1-x^2}}$ **b** $-\frac{1}{\sqrt{1-x^2}}$

17 a i $\cos^2 y$ **ii** $\frac{1}{1+x^2}$ **b** $\frac{1}{1+x^2}$

18 $x^{x^2}(x+2x\ln x)$

19 a $\frac{1}{x}$ or $\frac{1}{e^y}$ **b** $-\frac{y}{x}$ **c** $-\frac{y^2}{x^2}$

d $\tan x \tan y$ **e** $\frac{1}{\cos y - \sin y}$

Exercise 2 *page 156*

1 a 1.73 **b** 4.25 **c** 3.86

d 2.94 **e** 0.928 **f** -1.30

g 1.95 **h** $\frac{1}{3}$ **i** $e^3(=20.1)$

j $\frac{e^5}{2}(=74.2)$ **k** $e^{-\frac{1}{2}}(=0.607)$ **l** $2e^4(=109)$

2 26.8 **3** 7330 **4** 1390

5 a $3^x \ln 3$ **b** $5^{x-1}\ln 5$ **c** $3(4^x)\ln 4$

d $2x(\ln 2)2^{x^2}$ **e** $-\frac{2x}{3}e^{-\frac{x^2}{3}}$

6 a $7^t(\ln 7)$ **b** $4^t \ln 4$ **c** $-3^{-t}\ln 3$

7 a $-15e^{-\frac{t}{10}}$ **b** $2e^{-2t}$ **c** $6^t(10\ln 6)$

8 a 1000 **b** 45.8

9 a -3.94 **b** 1.82 **c** 120

10 a -0.0693 **b** 5°C

11 372°C/min

12 a 1000 g **b** 1660 years

13 a 2000 **c** 217/218

14 a 0.0563 **b** 8

15 a 220 **b** 1.36 g/year

16 a $\frac{1}{3}\ln\left(\frac{m}{a}\right)$ **b** $\ln\left(\frac{b}{10v}\right)$ **c** $\sqrt{\ln\left(\frac{P}{150}\right)}$

17 a 2000 **b** 12.8 min

c $\frac{180}{e}$ (≈66.2°C/min)

18 a 190°C **b** 22.0 min **c** 11.2°C/min

Exercise 3 *page 159*

1 $0.6\ \text{cm s}^{-1}$ **2** $10\pi\ \text{cm}^2\,\text{s}^{-1}$

3 $0.2\ \text{cm s}^{-1}$ **4** $1.92\ \text{cm}^3\,\text{s}^{-1}, 0.96\ \text{cm}^2\,\text{s}^{-1}$

5 $300\pi\ \text{cm}^3\,\text{s}^{-1}$ **6** $0.0637\ \text{cm s}^{-1}$

7 $800\pi\ \text{cm}^3\,\text{s}^{-1}$ **8** $225\ \text{cm}^3\,\text{s}^{-1}$

9 $2\ \text{cm s}^{-1}$ **10** $5\ \text{cm s}^{-1}$

11 $\frac{4}{\pi}\ \text{cm s}^{-1}$ **12 b** $0.15\ \text{m}^3\,\text{s}^{-1}$

13 $\frac{3}{2}$ units s^{-1} **14** $\frac{1}{3\pi}\ \text{cm s}^{-1}$ (≈0.106)

15 $\frac{2}{\pi}\ \text{cm s}^{-1}$

Review Exercise 9 *page 161*

1 a $\frac{1}{2y+1}$ **b** $-\left(\frac{y}{1+x+2y}\right)$ **c** $-\frac{x}{y}$

2 a $-\frac{x}{3y}$ **b** $\frac{2+8x}{3y^2-3}$ **c** $\frac{3x+2y}{5y-2x}$

d $\frac{1-y^3-3x^2y}{1+3xy^2+x^3}$ **e** $\frac{\cos x}{2\sin y}$ **f** $\frac{2xy^2}{1-2x^2y}$

3 $\frac{3}{4}$

4 a $-\left(\frac{2x+7y}{7x+6y}\right)$ **b** $-\frac{16}{19}$

5 $7x+11y-32=0$

6 $16x-10y-33=0$

7 a $\frac{2-x}{4y-4}$

b At (2, 2) gradient = 0. At (0, 1) gradient of curve is infinite.

10 b 3

11 a $-\frac{3}{2}$ **b** $-\frac{4}{3}$

12 a $8^x \ln 8$ **b** $(2\ln 3)3^x$ **c** $(2x\ln 4)4^{x^2}$

d $-12e^{-3x}$ **e** $x^x(1+\ln x)$

14 a 300 **b** 22.0

15 0.121°C/min

16 a $\frac{1}{2}\ln\frac{3}{5}$ (= -0.2554)

b 21.6

17 a $\frac{1}{2}\ln\frac{m}{A}$ **b** $\ln\left(\frac{a}{4s}\right)$ **c** $\sqrt{\ln\left(\frac{v}{20}\right)}$

18 a 22.3 years

b 9.05 g/year

19 i $50\ \text{m s}^{-1}$ **ii** 4.58 s **iii** $4\ \text{m s}^{-2}$

20 a 1000 **c** 4000

21 $100\pi\ \text{cm}^3\,\text{s}^{-1}$

22 $216\ \text{cm}^3\,\text{s}^{-1}$

23 $64\pi\ \text{cm}^3\,\text{s}^{-1}$

24 $3\ \text{cm s}^{-1}$

25 $0.25\ \text{cm s}^{-1}$

26 b $\frac{45}{2\pi}\ \text{cm s}^{-1}$

27 a $3.08\times10^{-7}\ \text{cm s}^{-1}$

b $7.4\times10^{-6}\ \text{cm}^2\,\text{s}^{-1}$

28 $2\ \text{cm s}^{-1}$

29 $12\ \text{cm s}^{-1}$

EXAMINATION EXERCISE 9 *page 164*

1 $-\frac{4}{9}$

2 i $-\left(\frac{4x+y}{x+2y}\right)$ ii $(1, -4), (-1, 4)$

3 a ± 3.73 b ± 1.5

4 b $(-2, 1), (2, -1)$

5 a $\frac{7}{5}$ b $y = -\frac{5}{7}x + \frac{19}{7}$

6 a $(3, \frac{7}{3}), (3, -\frac{1}{3})$ b $1, -1$

8 a $P = 200, q = 60$ b i 198 ii 1.6

9 i 120 g ii -0.0021 iii -0.66

10 b 11.9 g

11 i 1787 ii 139

12 a 1000

c i $t \approx 24.14 \ln\left(\frac{N}{1000}\right)$ ii 167 mins

13 0.027 cm s^{-1}

14 a $2\pi r$ b $942 \text{ cm}^2/\text{min}$

15 a $15e^{3x} - \left(\frac{5}{2-5x}\right)$ b $\frac{1}{21}$ cm/min

16 a i $10 - 2x$ b $1 \text{ m}^2 \text{s}^{-1}$

17 i a 32.2 b 0.5 ii a 35 b $\frac{15}{14}$

18 a 32 m^3

b $-1.25 \text{ m}^3/\text{h}$, decreasing

c $e^{3.2}$ (or 24.5…)

19 a i $x = a^y$

c $y - 1 = \frac{x - 10}{10 \ln 10}$

d $10 - 10 \ln 10$

20 ii $e^{y \ln 10}(\ln 10)$

21 iii $(\ln \frac{4}{3}, \ln \frac{2}{3})$

EXERCISE 1 *page 170*

Integration constants omitted.

1 a $\frac{(1+x)^4}{4}$ b $\frac{1}{7}(x+4)^7$

c $\frac{1}{10}(2x+1)^5$ d $-\frac{1}{5}(1-x)^5$

e $-\frac{1}{(x+3)}$ f $\frac{2}{9}(3x-1)^3$

g $-2(x+1)^{-2}$ h $-\frac{1}{5}(5x+3)^{-1}$

i $7e^x$ j $\frac{x^2}{2} + 5e^x$

k $\frac{1}{5}e^{5x}$ l $\frac{e^{6x}}{6} - x$

m $\frac{3}{2}e^{2x}$ n $-9e^{-x}$

o $-\frac{4}{e^x}$ p $-\frac{1}{e^x} + x^2$

2 a $-\cos x$ b $\frac{1}{4}\sin 4x$

c $-\frac{1}{10}\cos 10x$ d $\frac{2}{5}\sin 5x$

e $-\frac{1}{12}\cos 6x$ f $\frac{1}{2}\sin 2x + x^2$

g $-\frac{1}{4}\cos(4x-1)$ h $\frac{x^2}{2} - \sin(x+1)$

i $x^2 + \ln x$ j $\sin x + 2\ln x$

k $\frac{1}{4}\ln x$ l $\ln(x+3)$

m $6\ln(1+x)$ n $\frac{1}{3}\ln(3x+2)$

o $\frac{1}{2}e^{2x} + 2\ln x$

3 a $6(4x+1)^{\frac{1}{2}}$ b $\frac{1}{6}(4x+1)^{\frac{3}{2}}$

4 a $\frac{5}{2}(5x-2)^{-\frac{1}{2}}$ b $\frac{2}{5}(5x-2)^{\frac{1}{2}}$

5 a $\frac{1}{8}(6x+1)^{\frac{4}{3}}$ b $2(x+2)^{\frac{1}{2}}$

c $\frac{1}{5}(1+2x)^{\frac{5}{2}}$

6 $e^2 - 1$

7 a $\frac{1}{3}e^{3x+1}$ b $2e^{2x-3}$

c $-\frac{5}{e^x}$ d $\frac{1}{8}(2x+1)^4$

e $-\frac{1}{4}(4x-3)^{-1}$ f $e^x - e^{-x}$

g $\frac{1}{3}(2x+1)^{\frac{3}{2}}$ h $\frac{1}{2}(4x+5)^{\frac{1}{2}}$

i $\frac{1}{2}e^{2x} + 2e^x + x$

8 a $2\ln(2x+7)$ b $2x + e^x - e^{-x}$

c $-\frac{2}{3}(3x-2)^{-1}$ d $\frac{1}{6}(4x+3)^{\frac{3}{2}}$

e $\frac{1}{4}(8x-1)^{\frac{1}{2}}$ f $\frac{1}{8}(6x+5)^{\frac{4}{3}}$

9 $x^2 + 5\ln x - \frac{1}{x} + c$

10 a $x - \frac{4}{x} - 4\ln x$ b $x + 5\ln x - \frac{3}{x}$

c $\frac{1}{2}(e^x + x)$ d $x - 5\ln x - \frac{6}{x}$

e $\frac{x^2}{2} - 25\ln x$ f $e^{x+2} - \frac{1}{2}e^{-2x}$

11 a $\frac{\pi}{4}$ b $\frac{1}{2}u^2$

12 $\frac{1}{2}$ 13 $\frac{1}{3}(e^3 - 1)$ 14 $\ln\frac{5}{2}$

15 $-\frac{1}{12}$ 16 $\frac{1}{4}$ 17 $\frac{1}{2}e^3 - \frac{1}{2}e^{-1}$

18 $10\ln 2$ 19 $\ln\frac{3}{2}$ 20 $12\frac{2}{3}$

21 $0.0339 = \frac{1}{2}(\sin 2^c - \sin 1^c)$ 22 $8 + \frac{1}{2}\ln 3$

23 1 24 $3 - \sqrt{3}$ 25 $4\frac{5}{6}$

26 $\frac{1}{2}\ln 5$ 27 $\frac{1}{2} + \frac{3\pi}{4}$ 28 $\frac{7}{12}$

30 a 2, 3 b $6\ln\frac{3}{2} - \frac{5}{2}$

31 $4 - 3\ln 3$

32 a 0, 5 b $\frac{1}{6}u^2$

34 a $\ln 2, \ln 6$ b $12\ln 3 - 16$

EXERCISE 2 *page 174*

1 $\ln(x+3)$ 2 $\ln(2x+1)$ 3 $\ln(x^2+5)$

4 $\ln(x^3+2)$ 5 $\frac{1}{4}\ln(4x+1)$ 6 $\frac{1}{7}\ln(7x-1)$

7 $\ln(e^x+3)$ 8 $\frac{1}{2}\ln(x^2+3)$ 9 $\ln(\sin x)$

10 $\ln(x+1)(x-2)$

11 $\ln(2x+1)(5x+2)$

12 $3\ln(2x+1) - \ln(x+1)$

13 $\frac{1}{3}\ln\frac{5}{2}$

14 a $2\ln 4$ **b** $\ln 2$ **c** $6 + \ln 2$
d $12 - \ln 2$ **e** $\ln 4$ **f** $2\ln 3$
g $\ln\frac{27}{10}$ **h** $-\ln 2$ **i** $10 + \ln 3$
j $\ln 30$ **k** $10 + 3\ln 2$ **l** $4 - 4\ln 3$

15 $\ln(\frac{3}{2})$

16 a -1.10 **b** -0.24 **c** -0.32

17 a $-\ln\cos x$ **b** $\frac{1}{2}\ln\sin 2x$ **c** $\ln(\tan x)$
d $-\ln(\cos x + 3)$ **e** $-\frac{1}{2}\ln\cos 2x$ **f** $\ln(\ln x)$

EXERCISE 3 *page 177*

1 a $\dfrac{(x+2)^3(3x-2)}{12}$ **b** $\dfrac{(x-3)^3(x+1)}{4}$
c $\frac{2}{5}(x+4)^4(x-1)$ **d** $\dfrac{(x-1)^3}{30}(6x^2+3x+1)$

2 a $\dfrac{(x+1)^4(4x-1)}{20}$ **b** $\dfrac{(x-1)^6(6x+1)}{42}$
c $\dfrac{2(x+3)^{\frac{3}{2}}}{5}(x-2)$
d $\dfrac{2(x-2)^{\frac{1}{2}}}{15}(3x^2+8x+32)$
e $\dfrac{(2x+1)^4}{80}(8x-1)$ **f** $\dfrac{(4x-1)^{\frac{3}{2}}}{60}(6x+1)$
g $\frac{2}{75}(5x+1)^{\frac{1}{2}}(5x-2)$ **h** $\dfrac{(3x-2)^5}{54}(3x+4)$

3 a $\frac{3}{2}(x^2+1)^4$ **b** $2(x^3-3)^3$
c $\frac{1}{4}(e^x-1)^4$ **d** $2(e^x+2)^{\frac{1}{2}}$
e $\frac{1}{3}\sin^3 x$ **f** $-\frac{1}{4}\cos^4 x$
g $\frac{1}{4}\tan^4 x$ **h** $\sin^{-1} x$

4 a $\frac{49}{20}$ **b** $\frac{23}{30}$ **c** $21\frac{11}{15}$ **d** $3\frac{1}{3}$

5 a $e - \dfrac{1}{e}$ **b** $2\sqrt{3} - \frac{2}{3}\sqrt{5}$
c $\frac{1}{2}\ln 2$ **d** $\frac{2}{3}$

7 0.18

EXERCISE 4 *page 180*

1 a $\dfrac{x}{3}(1+x)^3 - \dfrac{(1+x)^4}{12}$
b $x\sin x + \cos x$

2 a $xe^x - e^x$ **b** $\dfrac{x}{4}(1+x)^4 - \dfrac{(1+x)^5}{20}$
c $-xe^{-x} - e^{-x}$ **d** $\dfrac{x}{3}e^{3x} - \frac{1}{9}e^{3x}$
e $-x\cos x + \sin x$ **f** $\dfrac{3x}{2}\sin 2x + \frac{3}{4}\cos 2x$
g $\dfrac{x}{2}(x-1)^4 - \frac{1}{10}(x-1)^5$
h $\dfrac{2x}{3}(x+1)^{\frac{3}{2}} - \frac{4}{15}(x+1)^{\frac{5}{2}}$
i $\dfrac{x^2}{2}\ln 2x - \dfrac{x^2}{4}$ **j** $\dfrac{x^3}{3}\ln x - \dfrac{x^3}{9}$
k $-\dfrac{1}{2x^2}\ln x - \dfrac{1}{4x^2}$ **l** $e^x(x^2-2x+2)$

3 $1 - \dfrac{2}{e}(= 0.264)$

4 a $\frac{1}{9} - \dfrac{4}{9e^3}$ **b** $\frac{1}{4}$ **c** $\frac{17}{6}$
d $71\frac{11}{15}$ **e** $\dfrac{2e^3+1}{9}$ **f** $\pi - 2$

5 a $7\frac{1}{10}$

7 $e - 2$

8 a $2x\,e^{x^2}$ **b** $\dfrac{e^{x^2}}{2}(x^2-1) + c$

9 b $-e^x\cos x + \int e^x\cos x\,dx$
c $\dfrac{e^x}{2}(\sin x + \cos x)$

10 a $\dfrac{1}{34}e^{5x}[-3\cos 3x + 5\sin 3x]$
b $\dfrac{e^{ax}}{a^2+b^2}[a\sin bx - b\cos bx]$

EXERCISE 5 *page 183*

1 a $\frac{1}{2}x - \frac{1}{4}\sin 2x$ **b** $\frac{1}{2}x + \frac{1}{4}\sin 2x$
c $\tan x - x$ **d** $\frac{1}{2}x - \frac{1}{12}\sin 6x$
e $\frac{1}{2}x + \frac{1}{8}\sin 4x$ **f** $3x + 4\cos x - \sin 2x$
g $\frac{5}{2}x + \frac{1}{4}\sin 2x + \tan x$
h $\frac{1}{2}\tan 2x - x$ **i** $\tan x - 2\ln\cos x$

2 a $\dfrac{\pi}{8} - \frac{1}{4}$ **b** $\frac{1}{2} + \dfrac{\pi}{4}$ **c** $\sqrt{3} - \dfrac{\pi}{3}$
d $\dfrac{3\pi}{4} + 2$ **e** $\dfrac{\pi}{12}$

EXERCISE 6 *page 184*

1 $\dfrac{2\pi}{3}$ **2** $\pi\ln 4$ **3** $\dfrac{32\pi}{3}$
4 $\dfrac{124\pi}{3}$ **5** $\dfrac{14\pi}{3}$ **6** $\dfrac{512\pi}{15}$
7 168π **8** $\dfrac{\pi}{4}(e^4-1)$ **9** 18π
10 $\dfrac{\pi}{4}e^2(3e^2-1)$ **11** $\dfrac{\pi}{3}\ln 3$ **12** $\dfrac{\pi^2}{4}$
13 $\dfrac{\pi^3}{16} + \dfrac{\pi}{4}$ **14** $\dfrac{\pi^4}{3} + 2\pi^2 + 8\pi$
15 $\pi\ln 4$ **16** $\dfrac{9\pi}{2}$ **17** $\dfrac{32\pi}{5}$

EXERCISE 7 *page 186*

1 a $\dfrac{1}{1+x} - \dfrac{1}{1+2x}$
b $\ln(1+x) - \frac{1}{2}\ln(1+2x)$

2 a $\ln(x-1) - \ln(x+1)$
b $2\ln(x+4) - \ln(x+5)$
c $5\ln(x-7) - 4\ln(x-2)$

4 0.235

5 a $\dfrac{4}{2x+1} + \dfrac{2}{x+1} + \dfrac{1}{(x+1)^2}$

9 a $(x+1)(x+2)(x+3)$

b $\frac{1}{x+1} - \frac{2}{x+2} + \frac{1}{x+3}$

c $\ln\frac{32}{27}$

EXERCISE 8 *page 189*

Section A

1 $\frac{1}{4}(x+2)^4$ **2** $\frac{1}{3}\sin 3x$
3 $4e^x$ **4** $\ln x$
5 $-\frac{1}{x}$ **6** $\ln\left(\frac{x-1}{x+1}\right)$
7 $x\sin x + \cos x$ **8** $\ln(5x-1)$
9 $\frac{1}{2}\tan 2x$ **10** $\ln(x^2+a)$
11 $\frac{(x+4)^3}{12}(3x \quad 4)$ **12** $\frac{x}{5}e^{5r} \quad \frac{1}{25}e^{5r}$
13 $\frac{1}{3}\ln\sin 3x$ **14** $\frac{1}{3}e^{3x+2}$
15 $\frac{1}{6}(x^4-1)^{\frac{3}{2}}$ **16** $5\ln(x-7)$
17 $2\ln(x+1) + 3\ln(x+2)$
18 $\frac{1}{2}x + \frac{1}{4}\sin 2x$ **19** $\frac{1}{2}x^2\ln x^{-\frac{1}{4}x^2}$
20 $x + \ln x$ **21** $\ln(1+x)$
22 $e^x(x^2-2x+2)$ **23** $\frac{1}{4}\cos(3-4x)$
24 $4\tan x$ **25** $\frac{1}{2}e^{2x} - 2x - \frac{1}{2}e^{-2x}$
26 $\frac{x^2}{2} + 3x + \ln x$ **27** $-\ln\cos x$
28 $\tan x - x$ **29** $\frac{1}{8}(x^2+1)^4$
30 $-\frac{1}{2}x\cos 2x + \frac{1}{4}\sin 2x$
31 a e^{x^3} **b** $e^{\sin x}$ **d** $\frac{34}{3}$
32 a $e-1$ **b** $\frac{1}{2}(e^2-1)$
c $\ln 3$ **d** $\frac{1}{2}\ln 4$
33 a $2+\ln 3$ **b** $3+4\ln 2$
c $6\ln 2 + 27\frac{1}{6}$ **d** $2e^2+4e-2$
34 $y = \frac{1}{2}\ln x + 3$
35 $y = \frac{1}{3}(5\ln x + 1)$
36 $y = \frac{1}{6}(x^2+5+2\ln x)$
37 $y = e^x + x - 1$
38 a $\cos 2x = 1 - 2\sin^2 x$ **b** $\frac{\pi}{8} + \frac{1}{4}$
39 b $\sqrt{3} - 1 - \frac{\pi}{12}$
40 $2x\sin x + (2-x^2)\cos x$

Section B

1 a 2 **b** 0 **c** $\frac{1}{3} - \frac{\sqrt{2}}{6}$
d $\sqrt{3}$ **e** $\frac{e}{2}(e^2-1)$ **f** $\frac{1}{3}$
g $44\frac{1}{3}$ **h** $\frac{98}{3}$ **i** 4
2 $\frac{3}{2}\ln(2x+5)$ **3** e^{x^2}
4 $\ln(4+\sin x)$ **5** $2\ln x + \ln(x-3)$
6 $\frac{3}{2}x^2\ln x - \frac{3}{4}x^2$ **7** $\frac{(1+x)^{11}}{132}(11x-1)$
8 $x\ln x - x$ **9** $-e^{-3x}$
10 $-2\cos\left(\frac{x}{2}\right)$ **11** $x - 3\ln(x+3)$
12 $\sin x + \ln\cos x$ **13** $\frac{1}{2}x - \frac{1}{8}\sin 4x$
14 $\frac{2}{15}(5x+1)^{\frac{3}{2}}$ **15** $2(x-4)^{\frac{1}{2}}$
16 $2x^3 - \frac{3x^2}{2}$
17 $e - \frac{1}{e}$ **18** $\frac{1}{2}e^4 + \frac{3}{2}e^2 - 2e + 1$
19 $2 + \frac{1}{e} - \frac{1}{e^3}$ **20** 38
21 $64\frac{3}{4}$ **22** 11005
23 $\frac{8}{9}$ **24** $\frac{2}{9}$
25 $\frac{4}{5}\ln 4$ **26** $\frac{2e^3+1}{9}$
27 -2 **28** $2\ln 2 - \frac{3}{4}$
29 $29.6 = \frac{148}{5}$ **30** $2\ln 4$
31 $\ln 2$
32 b $\frac{8}{3}$
33 $\frac{1}{2}$
34 b $1 - \frac{\sqrt{3}}{3} - \frac{\pi}{12}$
35 a $-\frac{1}{3}(1-x^2)^{\frac{3}{2}}$
b $-\frac{x^2}{3}(1-x^2)^{\frac{3}{2}} - \frac{2}{15}(1-x^2)^{\frac{5}{2}}$
38 3 **39** $\frac{1}{2}\ln(\sec 2x + \tan 2x)$

EXERCISE 9 *page 192*

1 a $y = x + \ln x + c$ **b** $y = x^3 - x + c$
c $y = e^x - x + c$ **d** $y = \sin x + c$
e $y = x - 3e^{-x} + c$ **f** $y = \ln(\sin x) + c$
2 a $y = \frac{1}{6}(2x+1)^3 + \frac{5}{6}$ **b** $y = \frac{2}{3}(x-1)^{\frac{3}{2}} + \frac{4}{3}$
c $y = \frac{3}{2} - \frac{1}{2}\cos 2x$ **d** $y = \frac{1}{2}\ln(2x-1) + 1$

EXERCISE 10 *page 195*

1 a $\ln y = \frac{x^3}{3} + c$ **b** $y^2 = x^2 + c$
c $y = kx$ **d** $y = x + \ln x + c$
e $y^2 = x^3 + c$ **f** $y^2 = \frac{1}{2}e^{4x} + c$
g $y = -\frac{1}{(e^x + c)}$ **h** $y^2 = 2\sin x + c$
i $y = \ln(1+x) + c$ **j** $y = e^{2x+c}$
k $y = e^{x^3+c} - 3$ **l** $\sin y = \ln x + c$
m $y = \ln(2x^2 + c)$ **n** $y = e^{x^2+c}$
o $y = -\frac{1}{(x+c)}$
2 $y = 4x^2$
3 a $x = e^{3t+10}$ **b** $y = 1000e^{-10t}$
c $x = 15e^t + 5$ **d** $y = 18e^{-t} - 8$
4 a $x = 100e^{5t}$ **b** $y = 1000\,e^{-2t}$
c $x = x_0\,e^{3t}$ **d** $y = y_0\,e^{-4t}$
e $x = x_0\,e^{kt}$ **f** $y = y_0\,e^{-kt}$
5 a $-e^{-y} = \frac{x^2}{2} - 3$ or $y = \ln\left(\frac{2}{6-x^2}\right)$
b $y = e^{-\cos x}$

c $y = \dfrac{3e^{x^2} - 5}{2}$

d $y = \sqrt{(1 - \ln 2 - 2\ln\cos x)}$

e $\tan y = 2 - \cos x$

f $y = 2e^{\left(\frac{x^2-1}{2}\right)} - 1$

6 a $x = 20 - 15e^{-\frac{t}{10}}$ b 4.05

7 $y = \dfrac{2}{1 + e^{-2x}}$

8 a $\dfrac{1}{1+x} - \dfrac{1}{2+x}$ b $y = \dfrac{4}{3}\left(\dfrac{1+x}{2+x}\right)$

Exercise 11 *page 198*

2

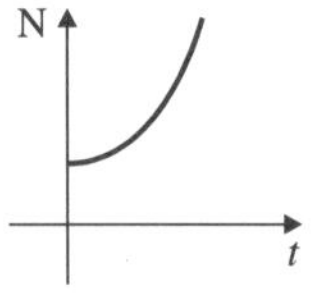

4 $P = 50\,000\,000\,e^{-0.00912t}$

5 b 46.1 min c $100e^3 \approx 2008$

6 b 12 min c 45°C

7 b £7746 c 9 years

8 10 hours

9 20 minutes ($c = 2, k = \frac{1}{10}$)

10 $V = 300\,e^{-\frac{t}{10}} + 200$ c 200 cm^3

Exercise 12 *page 201*

1 $21\frac{1}{2}\,u^2$, exact value = 21

2 5.83, 4 ln 4, 0.29

3 a 58.0 b 8.2%

4 a 0.977 b 0.994

c 2.35% with 3 trapeziums, 0.58% with 6

5 a 0.076 b 0.720 c 0.785

6 a 3.92 b 2.78 c 3.39

7 0.937

Exercise 13 *page 203*

1 a 18.00 b 72.67

2 33.3

3 a 53.86

b Simpson's rule gives more accurate result

4 a 2.38 b 0.591 c 2 970 000

5 2.501

Review Exercise 10 *page 204*

1 a $-\cos x$ b $\frac{1}{2}\sin 2x$

c $-4\cos\dfrac{x}{4}$ d $\tan x$

e $\frac{1}{7}e^{7x}$ f $-\frac{1}{3}e^{-3x}$

g $\frac{1}{5}\sin(5x+4)$ h $\frac{1}{8}(2x+3)^4$

i $-\frac{1}{4}(1+4x)^{-1}$ j $\ln x$

k $5\ln x$ l $\frac{3}{2}\ln(2x+9)$

m $5\ln(x-7)$ n $\sin x - \cos x$

o $\frac{1}{9}(6x-1)^{\frac{3}{2}}$

3 a $\dfrac{1}{2} - \dfrac{\sqrt{3}}{4}$ b $\ln 2$ c 14

4 a $2\ln 2 - 2$ b maximum

5 a $2\ln(x+1) + 3\ln(x+2)$

b $4\ln(x+1) + 6\ln(x+2)$

6 a $\ln(x^3+5)$ b $\frac{1}{4}\ln(4x^2-7)$

c $\frac{1}{2}\ln(1+e^{2x})$ d $-\ln\cos x$

e $\ln f(x)$ f $\frac{1}{2}\ln\sin 2x$

7 a $\frac{71}{10}$ b $\frac{41}{288}$

c $\frac{19}{108}$ d $-\frac{2048}{45}$

e 18 f $-e$

g $e^8 - 1$ h $\frac{3}{16}$

8 $\frac{7}{9}$

9 $\frac{46}{15}$

11 a $\dfrac{x}{2}e^{2x} - \frac{1}{4}e^{2x}$ b $\sin x - \cos x$

c $\frac{3}{2}x^2\ln x - \frac{3}{4}x^2$ d $\dfrac{x^3}{9}(3\ln x - 1)$

12 $e^2 + 1$ 13 $\dfrac{e^4 - 1}{2}$

15 c $\dfrac{\pi}{4}$

16 a $\frac{1}{2}x - \frac{1}{4}\sin 2x$ b $\tan x - x$

17 a $\frac{14}{3}$ b $\dfrac{15\pi}{2}$

18 a $\frac{1}{3}\pi r^2 h$ b cone

19 $\dfrac{\pi^2}{2} + \pi$

20 a (1, 2), (2, 1) b $\frac{3}{2} - 2\ln 2$ c $\dfrac{\pi}{3}u^3$

21 b $\pi\left(\dfrac{\pi}{2} - 1\right)$

22 a $y = [\frac{3}{2}(x+14)]^{\frac{2}{3}}$ b $y = \sqrt{40x + 4}$

c $y = e^{x^2-4}$ d $y = 2e^{\frac{1}{2}(x^2-1)} - 1$

e $y = x + \ln x + 2$

f $y = \sqrt{5e^{2\tan x} - 1}$ or $\ln\left(\dfrac{y^2+1}{5}\right) = 2\tan x$

24 $y = e^{2x^2-2}$

25 c $\frac{1}{10}\ln(\frac{3}{2})$ d 17.1 hours

26 $\frac{75}{2} - 50\ln 2$

28 a $\sec x$ b $\frac{1}{3}\cot 3x$

c $x\ln x - x$

d $x^2\sin x + 2x\cos x - 2\sin x$

e $\ln(x+2) + \dfrac{3}{x+2}$

29 a 4.5×10^6 b 5.3 c 0.95

Examination Exercise 10 *page 208*

2 $\frac{1}{2}e^2 + \frac{1}{6}$

5 i $x = 1$

ii $-\dfrac{1}{(x-1)^2}$

iii $1 + \ln 2$

iv symmetrical about $y = x$

6 $x \tan x + \ln \cos x + c$

7 $\frac{1}{2}\left[-\frac{1}{(1+2x)} + \frac{1}{2(1+2x)^2}\right] + c$

8 a i $\frac{1}{2}x^{-\frac{1}{2}}$ ii $\frac{1}{4}$

b i $\frac{2}{3}x^{\frac{3}{2}} + 2x + c$

d i $y = x$ ii $\frac{32}{3}u^2$

9 $\frac{1}{2}\ln\left[\frac{\sqrt{3}+1}{\sqrt{3}-1}\right]$

12 a $\ln 4$ b $\ln 16$

13 ii $\frac{58}{15}$

14 $\frac{1}{4}$

15 a i $-\frac{1}{2}\cos 2x + c$

16 $a = \frac{5}{2}, b = 9$

17 i $2 + \frac{1}{u-1} - \frac{1}{u+1}$

18 ii 0.0219

19 $\frac{5\pi}{6}$

20 i $\frac{1}{2}(\sqrt{7} - \sqrt{3})$ ii $\frac{\pi}{4}(\ln 7 - \ln 3)$

21 ii 1 iv $\frac{\pi^4}{24} + \pi^2 - 4\pi$

22 b $\frac{\pi^2}{4} - \frac{\pi}{2}\ln 2$ (≈ 1.38)

23 a $\frac{1}{4}$ b $\frac{\pi}{16}(1 - 5e^{-4})$

24 $y^3 = x^3 + \frac{3}{x} + 4$

25 a $y = \sqrt{3x + c}$ b $y = \sqrt[3]{3x - 4}$

26 i $(2 - x^2)\cos x + 2x \sin x + c$

ii $\sec^2 x - 1$

iii $\tan y - y = (2 - x^2)\cos x + 2x \sin x + c$

27 a $x e^x - e^x$

28 b $V = 225 + 775 e^{-\frac{2t}{15}}$

c 225

29 a i $\frac{dh}{dt} = \pm k\sqrt{h}$ iii 0.293

b 6 hours 50 mins

30 a $\frac{dN}{dt} = -kN$ c 0.3937 d 1.29×10^{16}

31 ii $x = -\frac{1}{(kt + c)}$ iii 2

32 a $p = 1.357, q = 1.382$

b 2.59

33 13.6

34 a $\frac{1}{44}(4x + 3)^{11} + c$

b 12.3

35 3.14

EXERCISE 1 *page 217*

1 a $2\mathbf{a} + \mathbf{b}$ b $2\mathbf{a} + 2\mathbf{b}$ c $-\mathbf{a} - \mathbf{b}$

d $4\mathbf{a} + 2\mathbf{b}$ e $2\mathbf{a} - 2\mathbf{b}$ f $2\mathbf{a} + \mathbf{b}$

2 a $\overrightarrow{CO}$ b $\overrightarrow{TN}$ c $\overrightarrow{FT}$ d $\overrightarrow{KC}$

3 a) $-\mathbf{a}$ b) $\mathbf{a} + \mathbf{b}$ c) $2\mathbf{a} - \mathbf{b}$ d) $-\mathbf{a} + \mathbf{b}$

4 a) $\mathbf{a} + \mathbf{b}$ b) $\mathbf{a} - 2\mathbf{b}$ c) $-\mathbf{a} + \mathbf{b}$ d) $-\mathbf{a} - \mathbf{b}$

5 a) $-\mathbf{a} - \mathbf{b}$ b) $3\mathbf{a} - \mathbf{b}$ c) $2\mathbf{a} - \mathbf{b}$ d) $-2\mathbf{a} + \mathbf{b}$

6 a) $\mathbf{a} - 2\mathbf{b}$ b) $\mathbf{a} - \mathbf{b}$ c) $2\mathbf{a}$ d) $-2\mathbf{a} + 3\mathbf{b}$

7 a) and d

8 a) AD, BE, CF

EXERCISE 2 *page 219*

1 a) $\mathbf{a}$ b) $-\mathbf{a} + \mathbf{b}$

c) $2\mathbf{b}$ d) $-2\mathbf{a}$

e) $-2\mathbf{a} + 2\mathbf{b}$ f) $-\mathbf{a} + \mathbf{b}$

g) $\mathbf{a} + \mathbf{b}$ h) $\mathbf{b}$

i) $-\mathbf{b} + 2\mathbf{a}$ j) $-2\mathbf{b} + \mathbf{a}$

2 a) $\mathbf{a}$ b) $-\mathbf{a} + \mathbf{b}$

c) $3\mathbf{b}$ d) $-2\mathbf{a}$

e) $-2\mathbf{a} + 3\mathbf{b}$ f) $-\mathbf{a} + \frac{3}{2}\mathbf{b}$

g) $\mathbf{a} + \frac{3}{2}\mathbf{b}$ h) $\frac{3}{2}\mathbf{b}$

i) $-\mathbf{b} + 2\mathbf{a}$ j) $-3\mathbf{b} + \mathbf{a}$

3 $\frac{1}{2}\mathbf{s} - \frac{1}{2}\mathbf{t}$

4 $\frac{1}{3}\mathbf{a} - \frac{2}{3}\mathbf{b}$

5 $\mathbf{a} + \mathbf{c} - \mathbf{b}$

6 $2\mathbf{m} + 2\mathbf{n}$

7 a) $\mathbf{b} - \mathbf{a}$ b) $\mathbf{b} - \mathbf{a}$

c) $2\mathbf{b} - 2\mathbf{a}$ d) $\mathbf{b} - 2\mathbf{a}$

e) $\mathbf{b} - 2\mathbf{a}$ f) $2\mathbf{b} - 3\mathbf{a}$

8 a) $\mathbf{y} - \mathbf{z}$ b) $\frac{1}{2}\mathbf{y} - \frac{1}{2}\mathbf{z}$

c) $\frac{1}{2}\mathbf{y} + \frac{1}{2}\mathbf{z}$ d) $-\mathbf{x} + \frac{1}{2}\mathbf{y} + \frac{1}{2}\mathbf{z}$

e) $-\frac{2}{3}\mathbf{x} + \frac{1}{3}\mathbf{y} + \frac{1}{3}\mathbf{z}$ f) $\frac{1}{3}\mathbf{x} + \frac{1}{3}\mathbf{y} + \frac{1}{3}\mathbf{z}$

EXERCISE 3 *page 222*

1 a 5 b $5\sqrt{10}$ c $\sqrt{61}$ d $\sqrt{61}$

2 a $5, \frac{1}{5}(4\mathbf{i} + 3\mathbf{j})$

b $13, \frac{1}{13}(-5\mathbf{i} + 12\mathbf{j})$

c $25, \frac{1}{25}(7\mathbf{i} - 24\mathbf{j})$

3 a 7 b 9 c $3\sqrt{5}$ d $3\sqrt{3}$

4 a $\frac{1}{3}(2\mathbf{i} + 2\mathbf{j} + \mathbf{k})$

b $\frac{1}{6}(4\mathbf{i} - 2\mathbf{j} + 4\mathbf{k})$

c $\frac{3}{7}\mathbf{i} + \frac{6}{7}\mathbf{j} - \frac{2}{7}\mathbf{k}$

d $\frac{2}{3}\mathbf{i} - \frac{1}{3}\mathbf{j} + \frac{2}{3}\mathbf{k}$

5 $a = 5$

6 $12\mathbf{i} + 16\mathbf{j}$

7 $20\mathbf{i} - 48\mathbf{j}$

8 3

9 4

10 a $\begin{pmatrix} 2 \\ 3 \\ 1 \end{pmatrix}$ b $\begin{pmatrix} 1 \\ -1 \\ 4 \end{pmatrix}$ c $\begin{pmatrix} 0 \\ 2 \\ 1 \end{pmatrix}$

d $\begin{pmatrix} -1 \\ 1 \\ 4 \end{pmatrix}$ e $\begin{pmatrix} 0 \\ 2 \\ 0 \end{pmatrix}$ f $\begin{pmatrix} 2 \\ 0 \\ 3 \end{pmatrix}$

11 -1

12 a $-2\mathbf{i} - 8\mathbf{j}$ b $2\mathbf{i} + 8\mathbf{j}$

13 a $\begin{pmatrix}2\\6\\0\end{pmatrix}$ **b** $\begin{pmatrix}1\\3\\0\end{pmatrix}$

14 $\begin{pmatrix}7\\3\end{pmatrix}$ **15** $\begin{pmatrix}3\\1\\4\end{pmatrix}$

16 a $\begin{pmatrix}3\\3\\5\end{pmatrix}$ **b** $\sqrt{43}$ **c** $\begin{pmatrix}-5\\1\\1\end{pmatrix}$

17 a $\begin{pmatrix}-6\\-9\\9\end{pmatrix}$ **b** $\begin{pmatrix}-12\\-13\\-9\end{pmatrix}$ **c** $\begin{pmatrix}-5\\-4\\18\end{pmatrix}$

d 13 **e** $\begin{pmatrix}-0.8\\0\\0.6\end{pmatrix}$

18 a $\begin{pmatrix}3\\4\\4\end{pmatrix}$ **b** 13 **c** (7, 6, 11)

d (31, 12, 19) **e** 2 : 1

EXERCISE 4 *page 228*

1 a $\overrightarrow{OA} = \begin{pmatrix}-2\\1\end{pmatrix}, \overrightarrow{OB}\begin{pmatrix}4\\4\end{pmatrix}$ **b** $\begin{pmatrix}6\\3\end{pmatrix}$

c $\mathbf{r} = \begin{pmatrix}-2\\1\end{pmatrix} + \lambda\begin{pmatrix}6\\3\end{pmatrix}$

2 a $\begin{pmatrix}2\\4\end{pmatrix}; \mathbf{r} = \begin{pmatrix}3\\2\end{pmatrix} + \lambda\begin{pmatrix}2\\4\end{pmatrix}$

b $\begin{pmatrix}5\\1\end{pmatrix} : \mathbf{r} = \begin{pmatrix}-3\\4\end{pmatrix} + \lambda\begin{pmatrix}5\\1\end{pmatrix}$

c $\begin{pmatrix}-3\\5\end{pmatrix} \mathbf{r} = \begin{pmatrix}-1\\-2\end{pmatrix} + \lambda\begin{pmatrix}-3\\5\end{pmatrix}$

d $\begin{pmatrix}-6\\-5\end{pmatrix} : \mathbf{r} = \begin{pmatrix}2\\3\end{pmatrix} + \lambda\begin{pmatrix}-6\\-5\end{pmatrix}$

3 $\mathbf{r} = \begin{pmatrix}1\\-3\end{pmatrix} + \lambda\begin{pmatrix}-2\\5\end{pmatrix}$

4 $\mathbf{r} = \begin{pmatrix}0\\2\end{pmatrix} + \lambda\begin{pmatrix}3\\-4\end{pmatrix}$

5 ℓ_1 only

6 a $\lambda = 2$ **b** $\lambda = 3$ **c** $\lambda = 3$ **d** $\mu = 2$

7 a $\lambda = 2, \mu = 2$ **b** $\begin{pmatrix}2\\4\end{pmatrix}$

8 a $\begin{pmatrix}3\\3\end{pmatrix}$ **b** $\begin{pmatrix}7\\6\end{pmatrix}$ **c** $\begin{pmatrix}1\\4\end{pmatrix}$ **d** $\begin{pmatrix}1\\4\end{pmatrix}$

9 a $\begin{pmatrix}-1\\5\\1\end{pmatrix}$ **b** $\mathbf{r} = \begin{pmatrix}2\\-2\\1\end{pmatrix} + \lambda\begin{pmatrix}-1\\5\\1\end{pmatrix}$

10 a $\mathbf{r} = \begin{pmatrix}3\\2\\1\end{pmatrix} + \lambda\begin{pmatrix}-2\\2\\1\end{pmatrix}$ **b** $\mathbf{r} = \begin{pmatrix}1\\4\\0\end{pmatrix} + \lambda\begin{pmatrix}2\\-2\\7\end{pmatrix}$

c $\mathbf{r} = \begin{pmatrix}3\\-2\\1\end{pmatrix} + \lambda\begin{pmatrix}-5\\2\\3\end{pmatrix}$ **d** $\mathbf{r} = \begin{pmatrix}1\\-2\\1\end{pmatrix} + \lambda\begin{pmatrix}-4\\1\\1\end{pmatrix}$

11 $\mathbf{r} = \begin{pmatrix}2\\-1\\0\end{pmatrix} + \lambda\begin{pmatrix}-3\\1\\2\end{pmatrix}$

12 ℓ_1 and ℓ_3

13 a $(k = 2)$ **b** $(\lambda = 1)$ **c** $(\lambda = 2)$

14 $\lambda = 1, \mu = 2; 5\mathbf{i} + 2\mathbf{j} + 4\mathbf{k}$

15 a $\begin{pmatrix}6\\5\\8\end{pmatrix}$ **b** $\begin{pmatrix}3\\7\\1\end{pmatrix}$ **c** $\begin{pmatrix}3\\4\\2\end{pmatrix}$

16 a $\mathbf{r} = \begin{pmatrix}1\\5\\2\end{pmatrix} + \lambda\begin{pmatrix}-2\\1\\1\end{pmatrix}$

b $\mathbf{r} = \lambda\begin{pmatrix}7\\9\\-5\end{pmatrix}$

c $\mathbf{r} = \begin{pmatrix}-3\\7\\4\end{pmatrix} + \lambda\begin{pmatrix}7\\0\\-\frac{11}{2}\end{pmatrix}$

17 a $\mathbf{r} = \begin{pmatrix}1\\4\\7\end{pmatrix} + \lambda\begin{pmatrix}0\\0\\1\end{pmatrix}$

b $\mathbf{r} = \begin{pmatrix}3\\5\\-4\end{pmatrix} + \lambda\begin{pmatrix}2\\1\\3\end{pmatrix}$

c $\mathbf{r} = \begin{pmatrix}2\\-7\\1\end{pmatrix} + \lambda\begin{pmatrix}2\\4\\1\end{pmatrix}$

d $\mathbf{r} = \begin{pmatrix}3\\2\\-5\end{pmatrix} + \lambda\begin{pmatrix}2\\-4\\5\end{pmatrix}$

18 (7, 10, 0)

19 (0, −2, −4)

20 a $\mathbf{r} = \begin{pmatrix}4\\5\\0\end{pmatrix} + \lambda\begin{pmatrix}2\\-5\\-2\end{pmatrix}$ **b** (6, 0, −2)

21 a skew

b intersect at (7, 2, −6)

c parallel

EXERCISE 5 *page 234*

1 a 11 **b** 22 **c** 0 **d** −2

2 a 5 **b** 13 **c** 25 **d** 5

3 a $\frac{63}{65}$ **b** $-\frac{117}{125}$

4 a 59.5° **b** 90°
c 36.9° **d** 132.3° (or 47.7°)
5 a 45°, 30° **b** 15°
6 a 75° **b** 60.3° **c** 98.1°
8 $\overrightarrow{AB} = \begin{pmatrix}1\\2\end{pmatrix}, \overrightarrow{CD} = \begin{pmatrix}6\\-3\end{pmatrix}; \overrightarrow{AB}.\overrightarrow{CD} = 0$
9 $\frac{1}{13}\begin{pmatrix}12\\5\end{pmatrix}$
10 −5
11 a 26 **b** 7, 9 **c** 65.6°
12 a 14.1° **b** 79.0° **c** 75.0° **d** 90° **e** 88.1°
14 a 1 **b** 4
15 a $\mathbf{i} + 0\mathbf{j} + 0\mathbf{k}$ **b** 48.2°, 64.6°, 152.7°
16 a 65.6° **b** 85.9° **c** 66.0°

Exercise 6 *page 236*

1 $\hat{A} = 122.86°, \hat{B} = 20.44°, \hat{C} = 36.70°$
2 $z = 1.33$ or -7.47
3 a −1, −8 **b** 71°
4 a 11 **b** 49.3°
c $\mathbf{r} = \begin{pmatrix}-2\\1\\2\end{pmatrix} + \lambda\begin{pmatrix}4\\2\\3\end{pmatrix}$ **d** (−3, −6, 0)
5 a $\overrightarrow{BC} = \begin{pmatrix}1\\1\\4\end{pmatrix}, \overrightarrow{BA} = \begin{pmatrix}3\\4\\5\end{pmatrix}$
c $\frac{3}{2}\sqrt{19}$
6 a $\begin{pmatrix}7\\2\\-6\end{pmatrix}$ **b** $\cos^{-1}\frac{14}{15}$
7 a $\mathbf{r} = \begin{pmatrix}7\\1\\7\end{pmatrix} + \lambda\begin{pmatrix}1\\4\\1\end{pmatrix}$ **b** $\begin{pmatrix}6\\-3\\6\end{pmatrix}$
8 a $\begin{pmatrix}3\\4\end{pmatrix}$ **b** 34.7° **c** 5 **d** 2.85
9 a 3.13 **b** 4.24 **c** 2.85
10 a $\begin{pmatrix}2\\1\\2\end{pmatrix}$ **b** 3 **c** 63.5°, 2.69
11 a $\begin{pmatrix}1\\0\\-2\end{pmatrix}$ **b** 6.95
12 a 2.69 **b** 6.35 **c** 10.4
13 a −3 **b** 35 **c** 0
d b and c **e** 32 **f** a.b + a.c
14 a $\begin{pmatrix}4\\-5\\-3\end{pmatrix}$ **b** $\begin{pmatrix}10\\15\\5\end{pmatrix}$ **c** $\begin{pmatrix}7\\5\\4\end{pmatrix}$
15 $\begin{pmatrix}1\\1\\-1\end{pmatrix}$ **16** $\begin{pmatrix}2\\1\\-1\end{pmatrix}$

17 $\mathbf{r} = \begin{pmatrix}3\\4\\6\end{pmatrix} + \lambda\begin{pmatrix}3\\4\\1\end{pmatrix}$
18 a $\begin{pmatrix}1+3p\\3+3p\\-1+2p\end{pmatrix}$ **c** $\begin{pmatrix}7\\9\\3\end{pmatrix}$
19 b (14, −1, 10)

Review Exercise 11 *page 240*

1 a −c **b** c + d **c** 2c − d **d** d − c
2 a c − 2d **b** c − d **c** 2c **d** 3d − 2c
3 a i −2a + 2b **ii** −2b + 2c **iii** b
iv c − a **v** c − a
b parallel and equal
c parallelogram
4 a i −b + a **ii** $-\frac{1}{3}\mathbf{b} + \frac{1}{3}\mathbf{a}$
iii $-\frac{1}{6}\mathbf{a} + \frac{2}{3}\mathbf{b}$ **iv** $-\frac{1}{2}\mathbf{a} + 2\mathbf{b}$
b $\overrightarrow{CE} = 3\overrightarrow{CD}$
5 a 13 **b** $\sqrt{3}$ **c** 3 **d** 3 **e** $\sqrt{29}$
6 a $\sqrt{14}$ **b** $\sqrt{40}$
7 a $\frac{1}{3}(\mathbf{i} + 2\mathbf{j} + 2\mathbf{k})$ **b** $\frac{1}{7}(6\mathbf{i} - 2\mathbf{j} + 3\mathbf{k})$
8 a i − 6j **b** −i + 6j
9 a $\begin{pmatrix}3\\3\\6\end{pmatrix}$ **b** $\sqrt{54}$ **c** $\begin{pmatrix}-7\\1\\0\end{pmatrix}$
10 a $\mathbf{r} = \begin{pmatrix}2\\3\\5\end{pmatrix} + \lambda\begin{pmatrix}3\\-1\\7\end{pmatrix}$ **b** $\mathbf{r} = \begin{pmatrix}2\\1\\-3\end{pmatrix} + \lambda\begin{pmatrix}3\\-1\\5\end{pmatrix}$
c $\mathbf{r} = \begin{pmatrix}7\\5\\2\end{pmatrix} + \lambda\begin{pmatrix}6\\6\\-2\end{pmatrix}$ **d** $\mathbf{r} = \begin{pmatrix}2\\0\\-1\end{pmatrix} + \lambda\begin{pmatrix}-4\\1\\5\end{pmatrix}$
11 a intersect at $\begin{pmatrix}9\\14\\13\end{pmatrix}$ **b** skew
c intersect at $\begin{pmatrix}6\\-5\\9\end{pmatrix}$
12 a 11 **b** 0 **c** 26
13 $t = 10$
14 a i $\sqrt{14}$ **ii** 14
15 a 53.1° **b** 22.6° **c** 8.8°
d 53.1° **e** 41.0° **f** 79.0°
16 a $\begin{pmatrix}13\\9\\23\end{pmatrix}$ **b** $\begin{pmatrix}2\\15\\4\end{pmatrix}$ **c** $\sqrt{83}$
d −2 **e** 71° **f** 93°
17 a $\mathbf{r} = \begin{pmatrix}3\\4\end{pmatrix} + \lambda\begin{pmatrix}4\\-8\end{pmatrix}$ **b** $\mathbf{r} = \begin{pmatrix}14\\7\end{pmatrix} + \mu\begin{pmatrix}8\\4\end{pmatrix}$
c $\begin{pmatrix}4\\2\end{pmatrix}$ **d** $\sqrt{125}$

18 a $\mathbf{r} = \begin{pmatrix} 3 \\ 5 \\ 1 \end{pmatrix} + \lambda \begin{pmatrix} 9 \\ 12 \\ 0 \end{pmatrix}$ **b** $\mathbf{r} = \begin{pmatrix} 3 \\ 5 \\ 1 \end{pmatrix} + \mu \begin{pmatrix} 4 \\ 4 \\ 2 \end{pmatrix}$

c $21°$ **d** 15, 6

e $16\,u^2$ **f** 2.15

19 a i (1, 3, 7)

ii (4, 7, 19)

iii (7, 9, 10)

b $48.2°$

c $43.6\,u^2$

EXAMINATION EXERCISE 11 *page 244*

1 ii $\mathbf{b} - \mathbf{a}$

2 i $\mathbf{r} = \begin{pmatrix} 3 \\ 6 \\ 1 \end{pmatrix} + t \begin{pmatrix} 2 \\ 3 \\ -1 \end{pmatrix}, \mathbf{r} = \begin{pmatrix} 3 \\ -1 \\ 4 \end{pmatrix} + t \begin{pmatrix} 1 \\ -2 \\ 1 \end{pmatrix}$

ii (1, 3, 2)

iii $56.9°$

3 $71.6°$

4 a $\mathbf{r} = \begin{pmatrix} 9 \\ -2 \\ 1 \end{pmatrix} + \lambda \begin{pmatrix} 3 \\ -4 \\ -5 \end{pmatrix}$

b $p = 6, q = 11$ **c** $39.8°$

d $\frac{1}{5}\begin{pmatrix} 36 \\ 2 \\ 20 \end{pmatrix}$

5 i $\begin{pmatrix} 2 \\ 1 \\ -2 \end{pmatrix}$ **ii** 3 **iv** -3

6 i $a = 4, b = -3$ **iii** $\begin{pmatrix} 12 \\ -2 \\ -1 \end{pmatrix}$ or $\begin{pmatrix} 0 \\ 14 \\ -1 \end{pmatrix}$

7 a $-\frac{4}{9}$ **b** $\frac{1}{2}\sqrt{65}$ **d** 2 : 1

8 a (2, 5, 4) **c** $\frac{3}{2}\sqrt{14}$

9 b i $t = -1$

10 a $\mathbf{r} = \begin{pmatrix} 1 \\ 2 \\ -3 \end{pmatrix} + \lambda \begin{pmatrix} 4 \\ -5 \\ 3 \end{pmatrix}$

c $19.5°$ **d** 1 unit

INDEX